METHODS IN MOLECULAR BIOLOGY

Series Editor
John M. Walker
School of Life Sciences
University of Hertfordshire
Hatfield, Hertfordshire, AL10 9AB, UK

For further volumes:
http://www.springer.com/series/7651

Epistasis

Methods and Protocols

Edited by

Jason H. Moore

Department of Genetics, Institute for Quantitative Biomedical Sciences, Geisel School of Medicine, Hanover, NH, USA

Scott M. Williams

Department of Genetics, Institute of Quantitative Biomedical Sciences, Geisel School of Medicine, Hanover, NH, USA

 Humana Press

Editors
Jason H. Moore
Department of Genetics
Institute for Quantitative Biomedical Sciences
Geisel School of Medicine
Hanover, NH, USA

Scott M. Williams
Department of Genetics
Institute of Quantitative Biomedical Sciences
Geisel School of Medicine
Hanover, NH, USA

ISSN 1064-3745   ISSN 1940-6029 (electronic)
ISBN 978-1-4939-2154-6   ISBN 978-1-4939-2155-3 (eBook)
DOI 10.1007/978-1-4939-2155-3
Springer New York Heidelberg Dordrecht London

Library of Congress Control Number: 2014953909

Humana Press is a brand of Springer
Springer is part of Springer Science+Business Media (www.springer.com)

Preface

Modern genetic analyses, with phenomenal technological advances, now permit deeper interrogation of genomes with the intent of constructing more accurate and comprehensive genotype to phenotype maps. However, as recognized by the authors of the chapters in this volume, a key to defining this map requires inclusion of factors not always explicitly incorporated into genetic analyses—namely, epistasis or interactions. Not doing this has led, at least in part, to less than perfect descriptions of genotype to phenotype maps and has motivated the term, missing heritability, or the amount of genetic variance of a trait that is left unexplained. The key is that even with enormous quantities of data, fully explanatory genetic models evade description, if inappropriate simplifying assumptions are made. The focus of this book is to explore how we can avoid making these assumptions and do so in ways that are practical.

One key to unraveling the role of epistasis in genotype to phenotype maps is to minimize the extraordinary number of possible interactions that can be assessed in genome-wide data sets, hence predefining the set of possible models of epistasis that are to be included in analyses. Such filtering can serve as a precursor to statistical or data mining analyses, both of which are covered in this book. With respect to appropriate statistical analyses for the detection of epistasis, it is important to precisely define the meaning of epistasis to be included in analyses, as historically more than one definition has existed, and they can create ambiguities in terms of how epistasis is tested. Therefore, several of our authors take substantial space to define epistasis with respect to how to appropriately analyze it. Lastly, genomic data can be mined using a variety of computational tools that make no a priori assumptions about the underlying genetic models. These are promising but often make interpretation difficult.

As any genotype to phenotype map is determined by the history of the genome in question, it is important to define how evolutionary processes may have shaped a trait's genetic architecture. This is addressed in Chapter 1. Methods that reduce the multiple testing burden are described in Chapters 2 and 3. An alternative approach in model systems is to perturb the "natural" genetic system by generating de novo mutations and assessing their roles via quantitative trait locus mapping in multiple backgrounds (Chapter 4). In systems not amenable to such manipulation (e.g., humans) epistasis analyses may depend on well-chosen candidates, an approach shown to work for neuropsychiatric diseases where epistasis and pleiotropy appear to overlap (Chapter 5).

In Chapter 6 the authors discuss the decomposition of genetic variance into its individual components, how this underpins our understanding of epistasis, and how this may affect the outcome of selection. Measuring epistasis is a key topic of Chapter 7, where it is argued that how epistasis is measured can appear to minimize its effects in an evolutionary context. The role of measurement of epistasis is also taken up in Chapter 8, where it is shown that the arbitrariness of epistasis or interaction can be eliminated by applying measurement theoretic constraints. Extending the allelic average excess and average effect to two or more loci is proposed as a novel analytical approach in Chapter 9. By explicitly defining capacitating epistasis in Chapter 10, the authors develop means to examine its effects.

Distinct from most other chapters, the authors of Chapter 11 take an explicitly epidemiological view of what they define as "compositional" epistasis, and how to best detect it. Chapter 12 examines Boolean function interactions in Age-related Macular Degeneration data and finds relevant gene–gene and gene–environment interactions. Using information theory to detect and characterize epistasis is the focus of Chapter 13. Chapters 14 and 15 examine the application of network building to better elucidating epistasis. Agnostic data mining methods are the core of Chapters 16 and 17 where two methods, multifactor dimensionality reduction and ReliefF, are described. Lastly, artificial intelligence methods are introduced in Chapter 18 as a means to detect epistasis in association studies.

Overall, we think that the chapters provide a comprehensive set of ideas that can help us elucidate epistasis in the context of modern data availability, and thereby help us to better understand the genetic bases of complex phenotypes and their evolutionary histories.

Hanover, NH, USA   *Jason H. Moore*
                                                                                         Scott M. Williams

Contents

Contributors

JOSÉ M. ÁLVAREZ-CASTRO • *Department of Genetics, University of Santiago de Compostela, Lugo, Galiza, Spain; Instituto Gulbenkian de Ciência, Oeiras, Portugal*

PETER C. ANDREWS • *Department of Genetics, Geisel School of Medicine, DHMC, Lebanon, NH, USA*

JON ATTIA • *School of Medicine and Public Health, The University of Newcastle, Callaghan, NSW, Australia; CREDITSS – Clinical Research Design, Information Technology and Statistical Support Unit, Hunter Medical Research Institute, New Lambton Heights, NSW, Australia*

ÖRJAN CARLBORG • *Division of Computational Genetics, Department of Clinical Sciences, Swedish University of Agricultural Sciences, Uppsala, Sweden*

ANDREW G. CLARK • *Department of Biological Statistics and Computational Biology, Cornell University, Ithaca, NY, USA; Department of Molecular Biology and Genetics, Cornell University, Ithaca, NY, USA*

CHRISTIAN DARABOS • *Department of Genetics, Geisel School of Medicine, DHMC, Lebanon, NH, USA*

CHARLES GOODNIGHT • *Department of Biology, University of Vermont, Burlington, VT, USA*

THOMAS F. HANSEN • *Department of Biology, Centre for Ecological and Evolutionary Synthesis, University of Oslo, Blindern, Norway*

DOUG P. HILL • *Department of Genetics, Geisel School of Medicine, DHMC, Lebanon, NH, USA*

ELIZABETH G. HOLLIDAY • *School of Medicine and Public Health, The University of Newcastle, Callaghan, NSW, Australia; CREDITSS – Clinical Research Design, Information Technology and Statistical Support Unit, Hunter Medical Research Institute, New Lambton Heights, NSW, Australia*

TING HU • *Department of Genetics, Geisel School of Medicine, DHMC, Lebanon, NH, USA*

ALON KEINAN • *Department of Biological Statistics and Computational Biology, Cornell Center for Comparative and Population Genomics, Cornell University, Ithaca, NY, USA*

CALEB A. LAREAU • *Department of Mathematics, University of Tulsa, Tulsa, OK, USA*

LI MA • *Department of Animal and Avian Sciences, University of Maryland, College Park, MD, USA*

TRUDY F.C. MACKAY • *Department of Biological Sciences, North Carolina State University, Raleigh, NC, USA*

BRETT. A. MCKINNEY • *Department of Mathematics, Tandy School of Computer Science, Laureate Institute for Brain Research, University of Tulsa, Tulsa, OK, USA*

PAUL MITCHELL • *Department of Ophthalmology, Westmead Centre for Vision Research, Millennium Institute, University of Sydney, Sydney, Australia*

JASON H. MOORE • *Department of Genetics, Institute for Quantitative Biomedical Sciences, Geisel School of Medicine, Hanover, NH, USA; Department of Community and Family Medicine, Geisel School of Medicine, DHMC, Lebanon, NH, USA*

PABLO A. MOSCATO • *CIBM – Centre for Bioinformatics, Biomarker Discovery, and Information-Based Medicine, Hunter Medical Research Institute, New Lambton Heights, NSW, Australia; School of Electrical Engineering and Computer Science, The University of Newcastle, Callaghan, NSW, Australia*

CHRISTOPHER OLDMEADOW • *School of Medicine and Public Health, The University of Newcastle, Callaghan, NSW, Australia; CREDITSS – Clinical Research Design, Information Technology and Statistical Support Unit, Hunter Medical Research Institute, New Lambton Heights, NSW, Australia*

MATS E. PETTERSSON • *Division of Computational Genetics, Department of Clinical Sciences, Swedish University of Agricultural Sciences, Uppsala, Sweden*

MARYLYN D. RITCHIE • *Department of Biochemistry and Molecular Biology, Center for Systems Genomics, University Park, PA, USA*

CARLOS RIVEROS • *CIBM – Centre for Bioinformatics, Biomarker Discovery, and Information-Based Medicine, Hunter Medical Research Institute, New Lambton Heights, NSW, Australia; School of Electrical Engineering and Computer Science, The University of Newcastle, Callaghan, NSW, Australia*

ARNAUD LE ROUZIC • *Laboratoire Évolution, Génomes, Spéciation; CNRS UPR 9034, CNRS, Gyf-sur-Yvette, France*

RODNEY J. SCOTT • *CIBM – Centre for Bioinformatics, Biomarker Discovery, and Information-Based Medicine, Hunter Medical Research Institute, New Lambton Heights, NSW, Australia; School of Medicine and Public Health, The University of Newcastle, Callaghan, NSW, Australia*

ETSUJI SUZUKI • *Department of Epidemiology, Graduate School of Medicine, Dentistry and Pharmaceutical Sciences, Okayama University, Kita-ku, Okayama, Japan*

TYLER J. VANDERWEELE • *Departments of Epidemiology and Biostatistics, Harvard School of Public Health, Boston, MA, USA*

RENATO VIMIEIRO • *CIBM – Centre for Bioinformatics, Biomarker Discovery, and Information-Based Medicine, Hunter Medical Research Institute, New Lambton Heights, NSW, Australia; School of Electrical Engineering and Computer Science, The University of Newcastle, Callaghan, NSW, Australia*

GÜNTER P. WAGNER • *Department of Ecology and Evolutionary Biology, and Systems Biology Institute, Yale University, West Haven, CT, USA*

JIE JIN WANG • *Centre for Vision Research, Department of Ophthalmology and Westmead Millennium Institute, University of Sydney, Sydney, Australia; Department of Ophthalmology, Centre for Eye Research Australia, University of Melbourne, Melbourne, Australia*

SCOTT M. WILLIAMS • *Department of Genetics, Institute of Quantitative Biomedical Sciences, Geisel School of Medicine, Dartmouth College, Hanover, NH, USA*

RONG-CAI YANG • *Department of Agricultural, Food and Nutritional Science, University of Alberta, Edmonton, AB, Canada*

Chapter 1

Long-Term Selection Experiments: Epistasis and the Response to Selection

Charles Goodnight

Abstract

Two common features of long-term selection experiments are that, first, there is typically no evidence for selection limits due to exhaustion of genetic variation, and second, selection plateaus are frequently observed that last multiple generations before a response to selection is resumed. These features are usually attributed to the high mutation rates of quantitative traits, and the effects of linkage disequilibrium. Using previously published theoretical results and a simple deterministic model I explore the potential role of gene interaction in generating these patterns seen in the response to long-term selection experiments. I show that epistasis provides a pool of variation that can contribute to an extended response to selection, but that this extended response will, at least under some circumstances, result in intermediate selection plateaus.

Key words Epistasis, Long-term selection experiments, Quantitative genetic models, Gene–gene interaction

1 Introduction

Classic models of selection, such as models of directional selection in a single locus two-allele system, inevitably lead to the fixation of the favorable allele, and with that fixation, the loss of genetic variation [1]. In these simple models, the rate of loss changes as a function of the dominance; however, regardless of the rate at which selection proceeds, all else being equal, the ultimate fate of these models is fixation of one allele, and the loss of all genetic variation. In such models, the primary mechanism to preserve genetic variation is stabilizing selection. More complex polygenic models of selection also lead to a monotonic decline in genetic variance and the eventual fixation of the optimal genotype. The exceptions to this are models with infinite numbers of loci such as Fisher's infinitesimal model, which has the paradoxical property that selection does not change gene frequencies [2]. This occurs because, for convenience, Fisher assumed that polygenic traits were determined by infinite loci, each with an infinitesimal effect on the phenotype.

Jason H. Moore and Scott M. Williams (eds.), *Epistasis: Methods and Protocols*, Methods in Molecular Biology, vol. 1253, DOI 10.1007/978-1-4939-2155-3_1, © Springer Science+Business Media New York 2015

As a result, any change in phenotype, regardless of how large, results in an infinitesimal change in gene frequency at any given locus. Interestingly, polygenic models with a finite number of alleles also lead to a loss of genetic variation even when there is stabilizing selection, since stabilizing selection will lead to the fixation of alternative alleles at different loci.

This fixation in polygenic models happens for two reasons. One is that in most cases there will be some combination of alleles at the loci affecting the trait that, when fixed, will give the optimal phenotype. The second is that selection has the effect of reducing the effective population size. This reduction in effective population size occurs because high fitness families will be overrepresented in the next generation. That is, selection will have the effect of increasing the variance in family size. As might be expected, the reduction in effective population size is a function of the heritability of the trait under selection, with high heritabilities causing the greatest reduction in effective population size [3].

The reduction in genetic variation associated with selection is one area where theory and experiment appear to be in conflict. Selection experiments that have lasted a hundred or more generations have been performed. Of all of these cases there appear to be none in which the experiment reached a selection limit due to loss of genetic variation. Typically, selection limits are met either because the trait under selection reaches a natural limit, or because there appears to be strong natural selection opposing the response to selection (e.g., [4]). An example of natural limits is the long-term selection experiment on oil and protein content in corn, which shows an approximately linear response to selection, except in the selection for low oil and protein lines, which show reduced response to selection as the amounts of these compounds approach zero. In this case the selection limit appears to be due to minimum levels of lipids and proteins for proper cell function. An example of natural selection opposing a continued response to selection is found in experiments on long-term selection for pupal weight in *Tribolium* flour beetles. Reverse selection and crosses to known size mutants demonstrated that there was ample genetic variation and no physical limits on size; however, in both the up and down lines there was a strong correlation between fertility and pupal weight that opposed artificial selection, with a positive correlation in the down line and a negative correlation in the up line [5].

This long-term maintenance of additive genetic variance in the face of selection has typically been attributed to ongoing mutation, and the observation that quantitative mutation rates may be quite large. However, Goodnight [6] suggested that selection and gene interaction could interact in a process similar to genetic drift to "convert" nonadditive genetic variance due to gene interaction into additive genetic variance in a process similar to the manner in which genetic drift can drive a similar conversion.

Another common feature of long-term selection experiments is the appearance of long-term intermediate plateaus in the response to selection. These plateaus can last from a few to 20 or more generations. They are typically attributed to waiting for a favorable mutation or recombination event to occur; however, there is no generally accepted explanation for this phenomenon. In this chapter, I will present a model that suggests that gene interaction may be involved in these intermediate plateaus in the response to selection.

2 Genetic Drift and Gene Interaction

The effects of genetic drift in systems with genetic interaction have been well described elsewhere. The interaction between drift and dominance and epistasis has been modeled using an infinite alleles model and a two-locus, two-allele Markov chain model. Both of these approaches give qualitatively similar answers. Goodnight [7, 8] showed that any two-locus, two-allele interaction can be divided into eight orthogonal variance components, additive variance at the two loci, dominance at the two loci, additive-by-additive epistasis, additive-by-dominance epistasis, dominance-by-additive epistasis, and dominance-by-dominance epistasis. As described by Goodnight, these components are only orthogonal at a gene frequency of 0.5, but using appropriate regression methods they can be used to determine the variance components for any gene frequency. By assuming a starting gene frequency of 0.5 at both loci and averaging over a Markov chain model for each generation of inbreeding, it becomes possible to calculate the expected variance partitioning for the "pure" variance components as a function of the inbreeding coefficient (Fig. 1). Figure 1 shows that the additive genetic variance declines linearly as a function of the additive effects, but for all forms of gene interaction the additive genetic variance is maximized at an intermediate inbreeding coefficient. The net result is that in many situations a population bottleneck or period of small population size can lead to an increase in the additive genetic variance rather than the decrease that is predicted by additive theory. As a result, in systems with one of the "pure" forms of gene interactions the additive genetic variance increases in the early stages of inbreeding, followed by a decline as the nonadditive variance is depleted and the population moves towards complete fixation.

It is helpful to have an intuitive understanding of why this "conversion" of nonadditive effects into additive effects occurs. Consider a two-locus, two-allele system with additive-by-additive epistasis (Fig. 2). In these systems the additive genetic variance for a locus is the variance in the marginal values of the genotypes of the locus when averaged over the interacting locus. Thus, the marginal value of the AA genotype is the frequency weighted mean genotypic value of the $A_1A_1B_1B_1$, $A_1A_1B_1B_2$, and $A_1A_1B_2B_2$ genotypes. At a

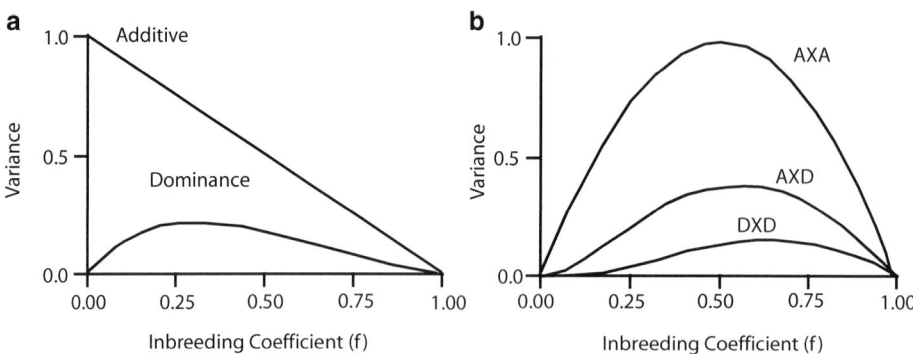

Fig. 1 The additive genetic variance as a function of inbreeding coefficient for the "pure" forms of genetic variance [9]. (**a**) Additive and dominance variance. (**b**) Additive-by-additive epistasis (AXA), additive-by-dominance, dominance-by-additive epistasis (AXD), and dominance-by-dominance epistasis (DXD)

	A₁A₁	A₁A₂	A₂A₂	$p_A=p_B=0.5$ genotypic mean	$p_A=p_B=0.75$ genotypic mean
B₁B₁	1	0	−1	0	0.5
B₁B₂	0	0	0	0	0
B₂B₂	−1	0	1	0	−0.5
$p_A=p_B=0.5$ genotypic mean	0	0	0		
$p_A=p_B=0.75$ genotypic mean	0.5	0	−0.5		

Fig. 2 Additive-by-additive epistasis. The marginal values change as gene frequencies change. For additive-by-additive epistasis, the additive genetic variance is the sum of the variance in the marginal values of the two genotypes; this will not be true if they are interactions involving dominance

frequency of 0.5 the marginal values for all genotypes at both loci are zero, and the additive genetic variance, measured as the variance in marginal values, is also zero. If, however, the gene frequency changes to freq(A) = freq(B) = 0.75, the marginal values will no longer equal zero, and the additive genetic variance will increase. Note that if a metapopulation starts at a gene frequency of 0.5 at both loci in all demes at the start, there will be no additive genetic variance. As genetic drift proceeds, the gene frequencies will change randomly and form a distribution centered around a frequency of 0.5. In this case, any subpopulation that has a gene frequency other than 0.5 for both loci will have an increase in additive genetic variance. Thus, even though the changes in gene frequency

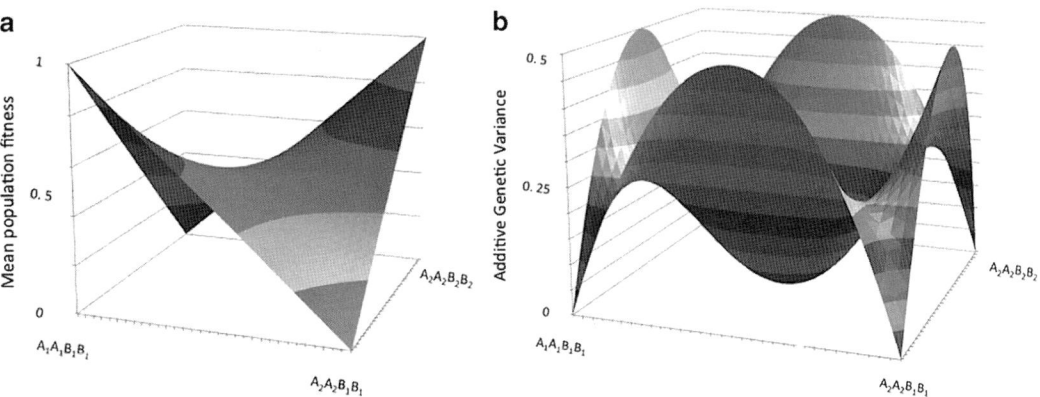

Fig. 3 The fitness surface and additive genetic variance for additive-by-additive epistasis as a function of two-locus gene frequency. (**a**) Mean population fitness on an arbitrary scale. (**b**) Additive genetic variance as a function of two-locus gene frequency. Additive variance is plotted as a proportion of additive-by-additive epistasis measured at a gene frequency of 0.5 for both loci

are random, on average, there will be an increase in the additive genetic variance as a result of genetic drift (Fig. 3).

Another important point that can be seen from Fig. 2 is that, at fixation, the effect of an allele on the phenotype will be different depending on what allele is fixed at the interacting locus. For example, the B_1B_1 genotype confers high fitness in populations fixed for the A_1 allele, and low fitness in populations fixed for the A_2 allele (Fig. 3a). Thus, genetic drift not only causes an increase in the additive genetic variance, it also causes the effects of the genotypes to change. Goodnight [7] defines the "local average effect of an allele to be the average effect of an allelic substitution on the phenotype of an individual in a specific subpopulation measured as a deviation from the metapopulation mean." This definition is similar to Fisher's [2] definition, except that the local average effect of an allele is a property of both the allele and the subpopulation in which it is measured. For quantitative genetic approaches in which specific loci are not identified there is the similar concept of local breeding value, defined as the mean phenotype of the offspring measured as a deviation from the metapopulation mean of an individual mated at random within a specific subpopulation [10].

In general, an increase in additive genetic variance due to genetic drift is always associated with a shift in local average effects of alleles. It is most useful to measure the variance in the allele by deme interaction. That is, the variance in the local average effects with the global average effect of the allele and the effect of the deme averaged across all loci removed (Fig. 4). If this is done in a system with only additive effects, the variance in the local average effects will be zero, regardless of the inbreeding coefficient. On the other hand, if there is gene interaction this variance will be non-zero, and may be quite large.

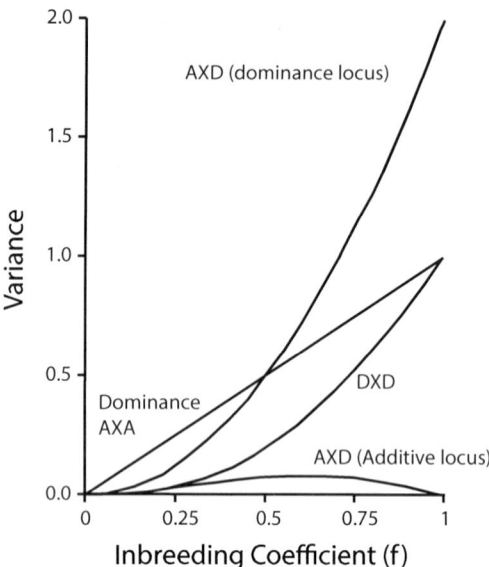

Fig. 4 The variance in local average effects relative to the variance in the ancestral population as a function of the inbreeding coefficient [9]. Additive-by-additive epistasis (AXA), dominance, and dominance-by-dominance epistasis (DXD) are shown. Additive-by-dominance and dominance-by-additive (AXD) epistasis are shown separately for the additive locus and the dominance locus. Additive variance does not contribute to the variance in local average effects, and is not shown. Dominance and additive-by-additive epistasis have the same relative effect on local average effects

The variance in local average effects can be interpreted as a measure of the predictability of the effect of an allele on the phenotype. Consider the situation where the A_1 allele confers a measurable difference in phenotype relative to the A_2 allele. If there are only additive effects and thus the variance in local average effects is zero, then the difference between the effects of the two alleles will be the same regardless of what deme they are measured in. If there is gene interaction, then the variance in local average effects will be nonzero, and the difference in phenotypic effects of the two alleles will vary from deme to deme. The variance in local average effects can be converted into a correlation by dividing the total variance in the phenotypic effect of an allele, that is the local average effect plus the demic effect, by the demic effect. This correlation varies from one for pure additive effects to zero for dominance and dominance-by-dominance epistasis. This correlation is another measure of the predictability of the relative ranking of the different alleles in different populations. In the additive system, if an allele is favored by directional selection in one deme, it will be favored in all demes. For gene interactions involving dominance (i.e., pure dominance, the dominance locus of additive-by-dominance and dominance-by-additive epistasis, and dominance-by-dominance epistasis) the correlation drops to zero. This means that for these pure forms of gene interaction, at fixation, the relative ranking of the two alleles becomes completely unpredictable. For example, in half of the subpopulations the A_1 allele will be superior to the A_2 allele, and in half the populations the opposite will be the case.

This also indicates that there are two distinct types of population differentiation. Most familiar is the "traditional" form of population differentiation, differentiation for population means. However, these models indicate that when there is gene interaction there will also be differentiation for local average effects. Importantly, these two types of population differentiation are not correlated. For example, when there are only additive effects genetic drift will lead to a divergence of population means, but no divergence for local average effects. In contrast, dominance-by-dominance epistasis does not lead to divergence in population means but will lead to large divergences in the local average effects of alleles.

There have been two studies of the variance in local breeding values [11, 12]. Both of these studies found highly significant variances in local breeding values. For example, de Brito et al. [11] crossed standard strains of mice to recombinant inbred lines and measured the variance in local breeding values for 20 traits related to growth and body weight. They found that there were highly significant variances in local breeding values for all of the traits and very low Spearman rank correlations among the different lines, indicating that not only was there substantial gene interaction contributing to the differences among lines, but that the variance resulted in a change in the relative order of the breeding values in the different recombinant inbred lines.

2.1 Simulation Models of Gene Interaction and Selection

Goodnight [6] developed a simulation model of selection in a two-locus, two-allele system. This model assumed two populations of 100 individuals each. At the start every allele was unique, and a random normal deviate from zero. Thus, each population started with 200 unique alleles at each of the two loci. Because of the large number of loci sampled, the two populations started out with similar means. Directional selection was then applied, and the population was followed until fixation occurred (in some treatments it was possible to get overdominance, and when overdominance developed fixation did not occur). Unless there was overdominance, fixation always occurred due to the finite population sizes and the lack of input from mutation or migration.

Although a number of different treatments were examined, I will focus on two. In the first, all of the genes acted additively, that is the phenotype of an individual was determined by the sum of the random normal deviates associated with the four alleles (two at each of the two loci) that made up the individual. The second treatment was identical, except that additive-by-additive epistasis was added. The additive-by-additive epistasis was a random normal deviate that was unique to each allele pair combination. There were four interactions, one for each possible combination of alleles from each locus. For each interacting pair the random normal deviates were multiplied, and their product was used as a seed in a normal random number generator. This gave a repeatable epistatic interaction that was statistically independent from the main effects of the two interacting alleles. In the epistatic case, the phenotype was determined by the sum of the random normal deviates of the individual alleles, plus the sum of the random normal deviates associated with the interactions among pairs of alleles.

For each run of the simulation there were two independent populations. These two populations were both subject to directional selection until fixation. These treatments did not have dominance, so fixation always occurred. In each population the additive genetic variance was measured. In addition, because it was a simulation it was possible to perform tests that are not possible with real organisms. First, to monitor the genetic effects in each population, a set of four standard alleles at one of the loci was inserted into every possible genotype (if the number of possible genotypes was very large a random subsample was used). In this way it was possible to track the local average effect of the test alleles. Second, to monitor the genetic differentiation of populations, every generation 100 F_1 hybrids were created, measured, and discarded (there was no actual gene flow between the two populations). This F_1 population was compared to the two parental populations.

Figure 5 shows the response to selection for representative runs of this simulation. The response to selection was similar in systems with and without epistasis, although the overall response to selection was greater in runs that included epistasis. At one level this is not surprising since there is more overall genetic variance in the system with epistasis. However, it is evidence that, contrary to standard theory, selection is able to act on this nonadditive genetic variance. More important is the F_1 hybrids. In the additive system, the hybrids are always approximately intermediate between the two parental populations. The same is not true for the epistatic system. In these runs, the F_1 hybrid typically started out intermediate between the two parental populations; however, as the populations responded to selection hybrid breakdown developed. At some point the hybrid population dropped below two genetic standard deviations (thin dashed lines in Fig. 5). At this point the hybrids had a lower fitness than approximately 95 % of the individuals within either population, and the two populations can be considered effectively reproductively isolated.

In addition to monitoring the population means, Goodnight [6] also measured the additive genetic variance of each generation (Fig. 6), and the local average effect of standardized test alleles (Fig. 7). In additive systems, the additive genetic variance tended to decline exponentially (Fig. 6, dashed line), whereas in epistatic systems it tended to stay elevated for several generations before declining (Fig. 6, solid line). This suggests that selection is generating new genetic variance by a process similar to the effects of drift on genetic variance.

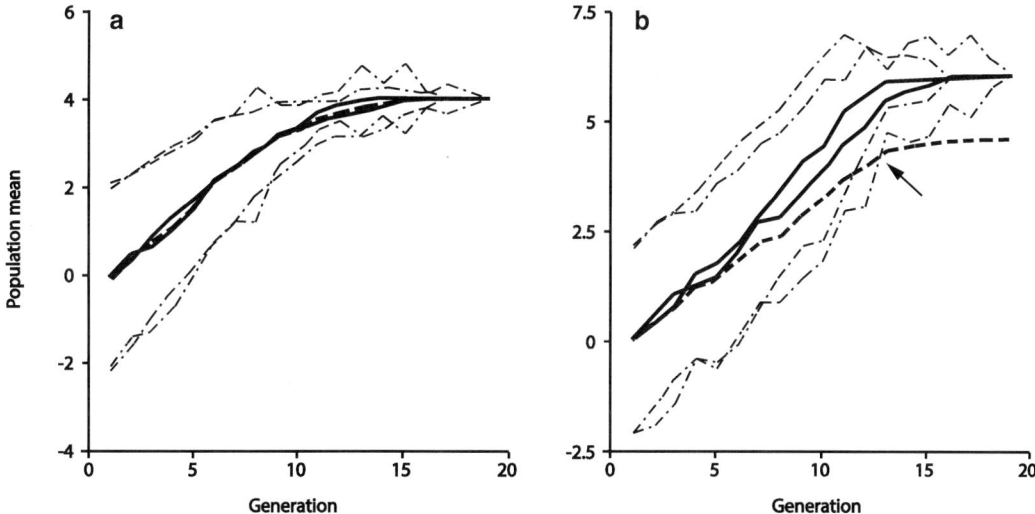

Fig. 5 The response to selection in the simulation study by Goodnight [9]. In both graphs, the *solid lines* are the responses to selection observed in the original populations. The *heavy dashed line* is the F_1 hybrid, and the *light dashed lines* are two genetic standard deviations above and below the population means. (**a**) The response to selection in an additive system. The F_1 hybrid is always approximately intermediate between the two parental strains. (**b**) The response to selection in a system with additive-by-additive epistasis. The F_1 hybrid quickly shows hybrid breakdown as the populations respond to selection. The *arrow* indicates the point where the F_1 hybrid has a fitness that is more than two genetic standard deviations below the fitness in either population. At this point, speciation has occurred. Note that the scale is different in the two graphs

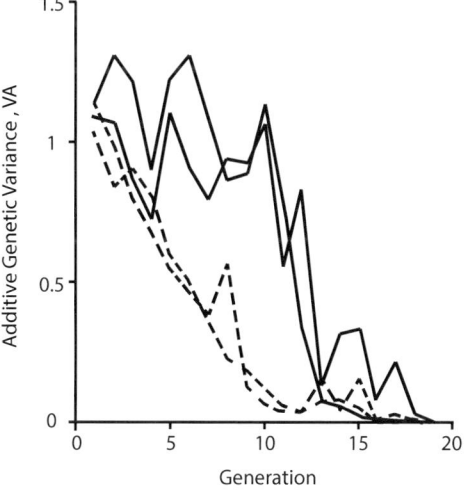

Fig. 6 The additive genetic variance as a function of generation for representative runs of the simulation described in the text [6]. The *dashed line* indicates additive effects only. The *solid line* indicates additive effects plus additive-by-additive epistasis. The measured additive genetic variance was significantly higher in the presence of epistasis

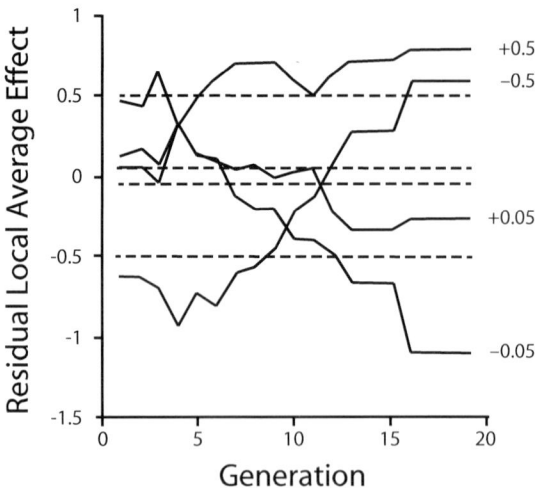

Fig. 7 The local average effects of test alleles in the simulation described in the text. Dashed lines: local average effects expected under an additive model. Solid lines observed local average effects in simulations with epistasis present. Numbers on the right are the assigned main effects for the test alleles [6]

Drift models indicate that increases in additive genetic variance should be associated with changes in the local average effects of the test alleles. In accordance with these predictions, local average effects are constants in the additive model, but change in an apparently random manner in the runs that included epistasis (Fig. 7).

Together these results lead to some interesting insights. In the epistatic system, the response to directional selection is approximately constant for an extended period, and actually for a longer period than would be expected in an additive system. With the exception of the anomaly of the response to selection being longer than expected, the response at the phenotypic level is quite predictable and very similar to what would be predicted in an additive system. Nevertheless this predictability is lost as we move to the genetic level. At this level, selection is converting epistatic interactions into additive variance and, in the process, randomly changes the average effects of alleles. One could imagine a situation in which an allele was of benefit in the early stages of selection, but also led to increased frequency of alleles at other loci that may alter the local average effect and lead to the elimination of the allele that originally got them started. In short, the very regular and predictable behavior at the phenotypic level hides a great deal of apparently random and possibly chaotic changes that take place at the underlying genetic level.

2.2 Analytical Models of Gene Interaction and Selection

I developed a two-locus, two-allele model of selection with arbitrary gene interactions. The basic model allows any set of nine interactions to be used; however, in the example presented here only additive-by-additive epistasis (Fig. 2) will be examined. For simplicity and clarity, I always start from a situation with no linkage disequilibrium with free recombination (both can be varied in the model). Since this is an analytical model there are no stochastic forces such as drift or mutation, thus any effects seen are strictly the results of selection in this simple system.

In this model, fitness is determined by the nine genotypic values, and scaled by a "fitness multiplier." If the fitness multiplier is 1, the lowest fitness genotype has an absolute fitness of 0, and the highest fitness genotype has an absolute fitness of 1, with the fitness of other genotypes being scaled appropriately. If the fitness multiplier has a value of 0, all the genotypes have equal absolute fitness of 0.5. In the examples presented here, the fitness multiplier is 0.25, thus the lowest fitness genotype has an absolute fitness of 0.375, and the highest fitness genotype has an absolute fitness of 0.625. Relative fitness is calculated relative to the genotype frequencies in the population being sampled, which is a function of the gene frequencies at the two loci and the linkage disequilibrium.

Figure 8 is a vector plot where each vector represents the change in gene frequency of a single generation for a population

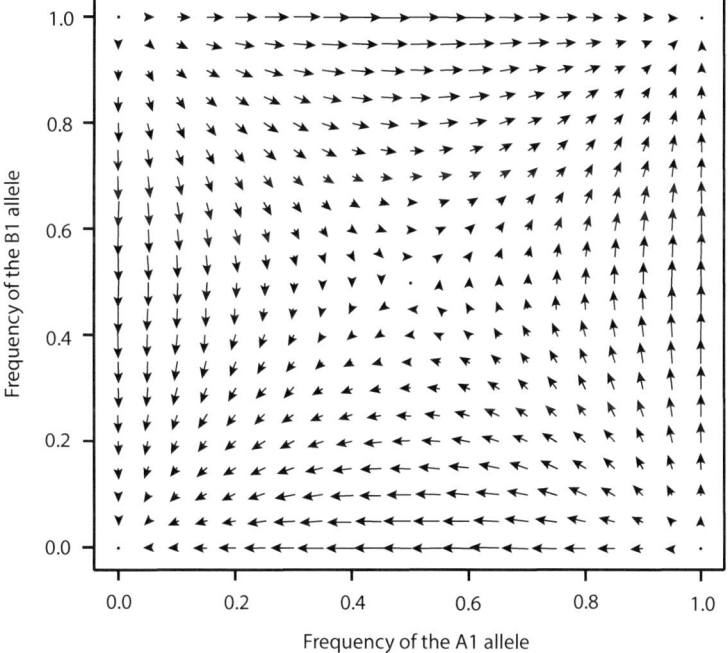

Fig. 8 Vector plot of the change in gene frequency resulting from a single generation of selection in a system with additive-by-additive epistasis as a function of starting gene frequency. Each vector connects the starting and ending gene frequency resulting from the single generation of selection. The response to selection is calculated for gene frequencies from 0 to 1 in increments of 0.05 for both loci. The strength of selection is 0.25

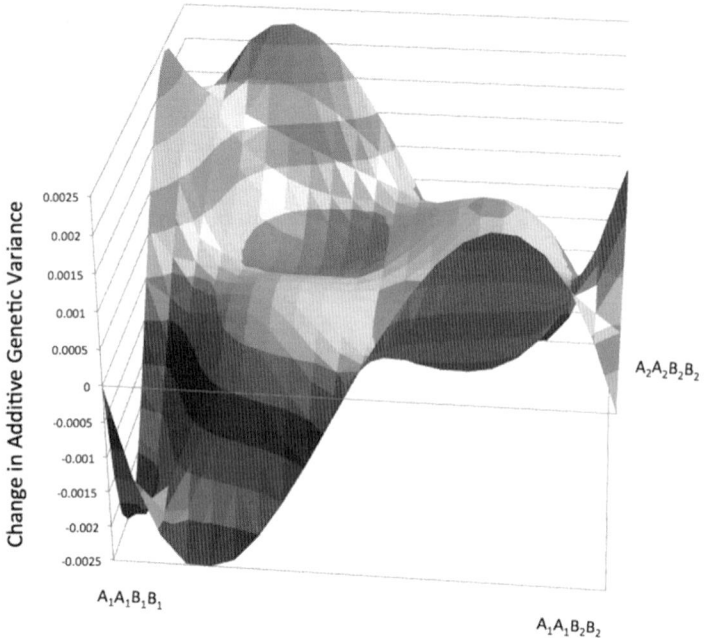

Fig. 9 The change in additive genetic variance associated with a single generation of selection as a function of two-locus gene frequency, with loci in linkage
equilibrium. Additive genetic variance tends to increase in regions where selection is driving the population away from equilibrium and decrease in regions
where selection is driving a population towards equilibrium

in linkage equilibrium. Gene frequencies were sampled between a
frequency of 0 and 1 in 0.05 increments for both loci. As shown
in this figure, there are three unstable equilibria: fixation of the A_1
and B_2 allele, fixation of the A_2 and B_1 allele, and a gene frequency
of 0.5 for both loci, and two stable equilibria: fixation of the A_1
and B_1 alleles and fixation of the A_2 and B_2 alleles. This result is,
not surprisingly, in agreement with the fitness surface for additive-
by-additive epistasis (Fig. 3a); the unstable equilibria correspond
to the fitness low points and the fitness saddle, and the stable equilibria correspond to the fitness high points. Comparison with
Fig. 3b suggests that over at least some of the gene frequency
space, selection will increase, rather than decrease, the additive
genetic variance.

Figure 9 shows the change in additive genetic variance as a
function of gene frequency at both loci. Note that this is not the
additive genetic variance, but rather it is the change in additive
genetic variance. Comparison of Figs. 8 and 9 shows that those
regions where selection is driving the population away from an
unstable equilibrium have the effect of increasing the additive
genetic variance, whereas those regions where selection is driving a
population towards an equilibrium, whether stable or unstable, the
additive genetic variance will tend to decline.

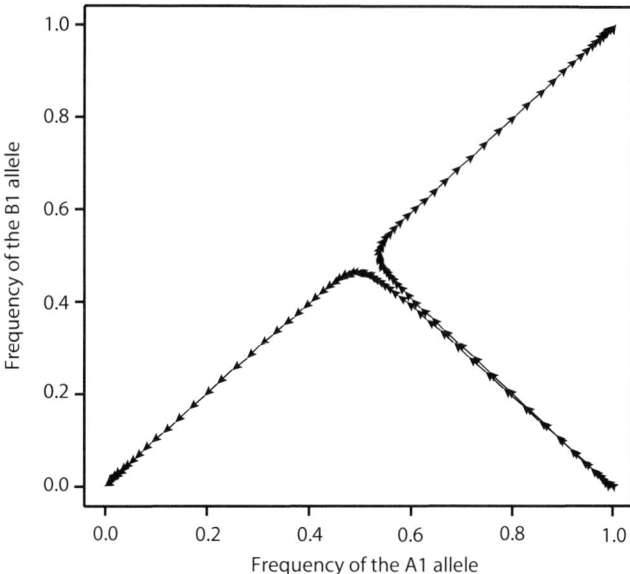

Fig. 10 The response to selection for two similar starting conditions. The initial conditions are the frequency of the A_1 allele equal to 0.99 in both populations, and the frequency of the B_1 allele at 0.0101 in one population and 0.0099 in the other population. This slight difference in initial gene frequency is sufficient to drive the two populations to fixation of opposite alleles. *Arrowheads* indicate generations. Note that the response to selection changes as gene frequencies change

It is unnecessary to examine the full range of possible allele frequency starting conditions, since most of the important points can be illustrated with a single pair of starting gene frequencies. Figure 10 shows the response to selection, measured as a function of population gene frequencies for two populations starting near the unstable equilibrium (fitness low point) of fixation for the A_1 allele and fixation of the B_2 allele. The two starting gene frequencies considered are $p(A_1) = 0.99$, $p(B_1) = 0.0101$, $p(A_1) = 0.99$, and $p(B_1) = 0.0099$. The model is perturbed from the unstable equilibrium because in an analytical model such as this, with no stochasticity, there will be no change for populations starting at the unstable equilibrium. The perturbation is asymmetric, because a symmetrical change in gene frequencies (e.g., $p(A_1) = 0.99$, $p(B_1) = 0.01$) will move directly up the fitness surface (Fig. 3a) and stop at the unstable equilibrium of the saddle point $(p(A_1) = p(B_1) = 0.5)$. Even with starting conditions that differ in gene frequency by only 0.0002, the two populations go to fixation for opposite genotypes. This shows that even in this very simple deterministic model in some regions of the gene frequency space there is sensitivity to initial conditions, and very similar starting gene frequencies coupled with uniform selection pressures can lead to differentiation of populations. In essence this is a simple example of an adaptive topography with multiple selective

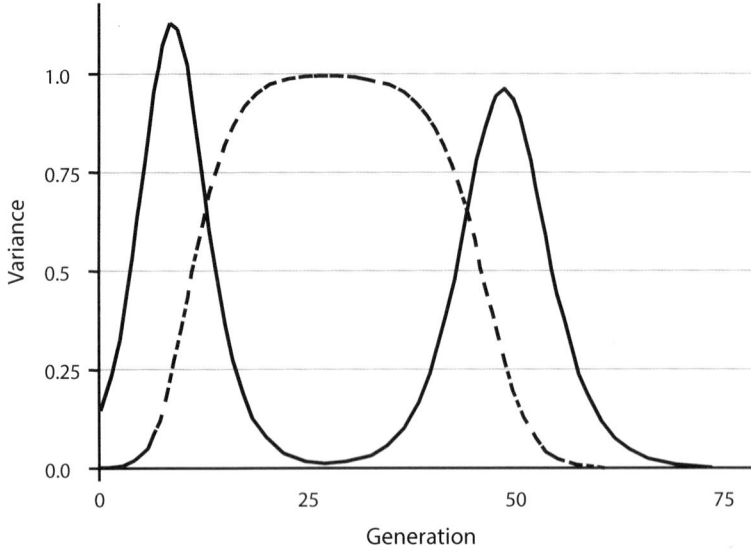

Fig. 11 The change in the additive genetic variance (*solid line*) and the additive-by-additive epistatic variance (*dashed line*) as a function of generation of selection as shown in Fig. 10. Variances are shown relative to the maximum additive-by-additive genetic variance, which occurs at a gene frequency of 0.5 for both loci. Starting conditions: frequency of the A_1 allele: 0.99, frequency of the B_1 allele: 0.0101

peaks, which demonstrates that selection can act as a force magnifying minute differences in initial conditions into dramatically different genetic outcomes.

Note that each arrowhead represents a single generation response to selection, showing that the rate of evolution changes as gene frequencies change, and the response to selection slows down dramatically in the regions near the equilibrium points. Because the selection is constant, the response to selection should be a direct function of the additive genetic variance. This is indeed the case (Fig. 11) with the additive genetic variance first increasing then decreasing and then increasing again, before finally decreasing as the population approaches fixation. A plot of the response to selection as a function of generation (Fig. 12) shows the effects of these changes in the additive genetic variance. It is interesting that in this simple model epistasis generates changes in the additive genetic variance that result in intermediate plateaus in the response to selection that are often seen in long-term selection experiments.

It is useful to measure the local average effects of the alleles. The change in the local average effect for the A locus is shown in Fig. 13. Figure 10 shows that when the starting frequencies are $p(A_1) = 0.99$ and $p(B_1) = 0.0101$ the frequency of the A_1 allele decreases to near 0.5 before eventually increasing to fixation. This is reflected in the local average effect in which the A_1 allele at first has

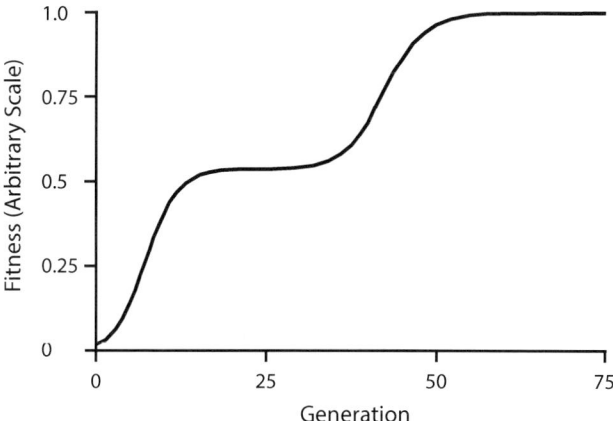

Fig. 12 The response to selection by generation for the model described in the paper. There is an intermediate plateau in the response to selection that corresponds to a set of gene frequencies with low additive genetic variance. Starting conditions: frequency of the A_1 allele: 0.99, frequency of the B_1 allele: 0.0101

a local average effect that is negative. As selection proceeds the local average effects change. The increase in the local average effects starts to occur around generation 40, which is when the gene frequencies are "turning the corner" in Fig. 10 and also where the additive genetic variance is approaching a minimum (Fig. 11).

3 Conclusions and Discussion

Previous work has shown that genetic drift can "convert" nonadditive genetic variance into additive genetic variance. The reason that this occurs can be seen in Fig. 3b, which shows the additive genetic variance as a function of two-locus gene frequency for additive-by-additive epistasis. Additive genetic variance and additive-by-additive epistatic variance are statistical components of the total genetic variance in a system, and the additive-by-additive interactions are only purely epistatic variance when gene frequencies at both loci are 0.5. Deviations from that gene frequency will change the partitioning between additive and epistatic variance. Any force, including selection, that changes gene frequencies will have the effect of changing the partitioning between additive and nonadditive genetic variance.

The models shown here are two-locus, two-allele models, which are the simplest models that can be used to explore the complexities of inter-locus interactions. These simple models show that when there is gene interaction selection can generate "new" additive genetic variance, and with respect to genetic variance components, it acts in many ways like genetic drift. Essentially what both drift and selection do is generate new additive genetic variance by causing

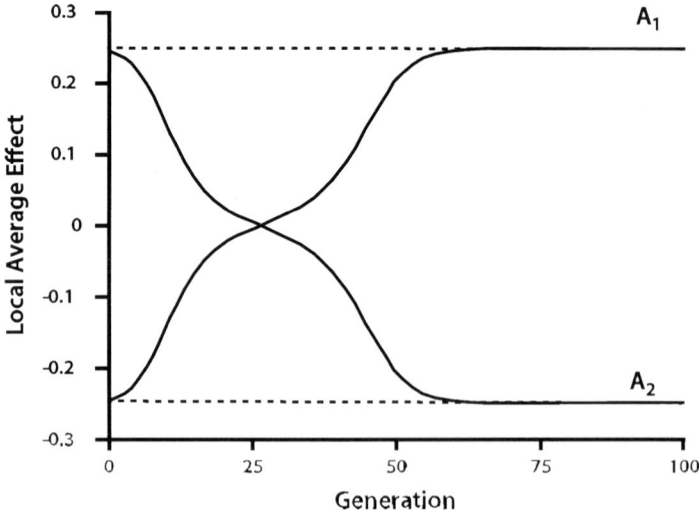

Fig. 13 The local average effects for the two alleles at locus A. These are corrected for the population mean, which in this two-allele case means that the effects will be symmetrical around 0. In the early stages of selection, the A_2 allele has the greater local average effect and increases in frequency. Starting around generation 40 the local average effect of the A_1 allele increases until it eventually becomes positive and the A_1 allele increases in frequency, eventually going to fixation. The *dashed lines* are the local average effects for a comparable system with only additive effects. Starting conditions: frequency of the A_1 allele: 0.99, frequency of the B_1 allele: 0.0101

a shift in the average effects of alleles. Furthermore, in the examples shown in Figs. 8 through 13, it is apparent that because the example starts near fixation there is little additive genetic variance or epistatic variance at the beginning. Thus, in an experimental situation there would be no reason to believe that epistatic variance would complicate the situation. From a phenotypic perspective there is also no sign of complications. Although the rate of advance is not even, the mean fitness of the population increases smoothly as the population responds to selection. Thus, there are no genetic signs of gene interactions at the beginning the "experiment," and the phenotypic response appears to be additive or nearly additive. Temporary plateaus are common in selection experiments (e.g., [4]); thus it is unlikely that this would alert an experimentalist to the possibility that gene interaction was contributing to the response to selection.

Epistasis is probably more easily detected in these simple models than it would be in more complex systems with more loci and more alleles per locus. It is well known that epistatic genetic variance is typically a small component of the genetic variance [13]. This might be expected if most loci have one or a few alleles at high frequency and many others at low frequencies. In this situation the majority of epistatic and dominance interactions will be expressed as additive genetic variance, and the epistatic variance will be very low [7]. Nevertheless, if selection acts to increase the frequency of rare alleles, interactions that were undetectable in the original population may come to the fore as gene frequencies change. Importantly, it is possible that the epistatic variance will never become large, but epistasis will nevertheless contribute to the response to selection. In this case, the only way to detect this contribution will be to perform inter-population crosses and look for the variance in local average effects [8] or local breeding values [10].

As seen from Figs. 7 and 13, the shift in the average effects of alleles is not "adaptive." Selection can generate additive genetic variance, but the variance generated is not due to adaptive shifts in the average effects of alleles. These figures also show a rarely considered problem with developing a phenotype–genotype or genotype–phenotype map. In both of these examples, it is clear that the relationship between the allelic value and its effect on the phenotype is not constant. In Fig. 13, what is originally a "good gene" becomes a "bad gene." In Fig. 7 a set of test alleles do what appears to be a random walk. If such shifts are general, then generating a static relationship between a gene and its affect on the phenotype will be impossible, and a more dynamic view of the relationship between genotype and phenotype will be required.

Selection is generally considered to "use up" genetic variance. As pointed out in the beginning of this chapter, it is rare for this to actually occur. Selection limits are nearly always due to counteracting natural selection rather than a "using up" of variance. The traditional explanation for this is ongoing mutations [14]. The results shown here suggest another important mechanism for the maintenance of variation in the face of selection. That is, alleles at low frequencies that do not make major contributions to any genetic variance components may increase in frequency as a result of selection. As they increase in frequency their interactions with other loci may shift average effects of alleles at these loci generating increases in additive genetic variance well beyond what would be expected from the changes expected from an additive system.

4   Notes

The models presented here are exactly that: models. These two-locus epistatic models demonstrate a much greater richness of possible outcomes than traditional single locus models; however, it is unlikely that they reflect the broad range of outcomes that are possible with all of the complications of real organisms. It is also important to recognize that there are multiple explanations for the results of selection experiments. The continued response to selection experiments without selection limits can be explained by ongoing new mutations, and indeed mutation rates may be very high for polygenic traits. The intermediate selection plateaus can be explained as a period of low genetic variation as the selected population "waits" for the appearance of a favorable mutation or a favorable recombination event between linked loci. Thus, the models presented here should be considered only one of many possible causes for the phenomena observed in selection experiments. Finally, the examples given here mostly involve additive-by-additive epistasis. Other forms of gene interaction, including dominance, have many of the same effects. For example, a nonzero variance in local average effects indicates that there is gene interaction, but does not provide a method for distinguishing between interactions among loci (epistasis) and interactions within loci (dominance). The effects of different forms of gene interaction on the response to selection have not been explored, and their effects are not known at this time.

References

1. Hedrick PW (2005) Genetics of populations, 3rd edn. Jones and Bartlett Publishers, London
2. Fisher RA (1930) The genetical theory of natural selection. Oxford University Press, Oxford
3. Robertson A (1961) Inbreeding in artificial selection programmes. Genet Res Cambr 2:189–194
4. Dudley JW, Lambert RJ (2004) 100 Generations of selection for oil and protein in corn. In: Lamkey KR, Coors JG, Dentine M (eds) Plant breeding reviews, vol 24 pt. 1: long term selection maze. John Wiley and Sons, Hoboken, NJ, pp 79–110
5. Muir WM, Miles D, Bell AE (2004) Long-term selection for pupal weight in Tribolium castaneum. In: Lamkey KR, Coors JG, Dentine M (eds) Plant breeding reviews, vol 24 pt. 2: long term selection maze. John Wiley and Sons, Hoboken, NJ, pp 211–224
6. Goodnight CJ (2004) Gene interaction and selection. In: Lamkey KR, James C, Dentine M (eds) Plant breeding reviews, vol 24 pt. 1. John Wiley & Sons, Hoboken, NJ, pp 269–290
7. Goodnight CJ (2000) Quantitative trait loci and gene interaction: the quantitative genetics of metapopulations. Heredity 84:587–598
8. Goodnight CJ (2000) Modeling gene interaction in structured populations. In: Wolf JB, Brodie ED III, Wade MJ (eds) Epistasis and the evolutionary process. Oxford University Press, New York, NY
9. Goodnight CJ (2004) Metapopulation quantitative genetics: the quantitative genetics of population differentiation. In: Hanski I, Goggiotti OE (eds) Ecology, genetics and evolution of metapopulations. Elsevier Academic Press, Burlington, MA, pp 199–223
10. Goodnight CJ (1995) Epistasis and the increase in additive genetic variance: implications for phase 1 of Wright's shifting balance process. Evolution 49:502–511
11. De Brito RA, Pletscher LS, Cheverud J (2005) The evolution of genetic architecture. I. diversification of genetic backgrounds by genetic drift. Evolution 59:2333–2342
12. Drury DW, Wade MJ (2011) Genetic variation and co-variation for fitness between intra-population and inter-population backgrounds in the red flour beetle, Tribolium castaneum. J Evol Biol 24:168–176
13. Hill WG, Goddard ME, Visscher PM (2008) Data and theory point to mainly additive genetic variance for complex traits. PLoS Genet 4:1–10
14. Keightly PD, Hill W (1992) Quantitative genetic variation in body size of mice from new mutations. Genetics 131:693–700

Chapter 2

Finding the Epistasis Needles in the Genome-Wide Haystack

Marylyn D. Ritchie

Abstract

Genome-wide association studies (GWAS) have dominated the field of human genetics for the past 10 years. This study design allows for an unbiased, dense exploration of the genome and provides researchers with a vast array of SNPs to look for association with their trait or disease of interest. GWAS has been referred to as finding needles in a haystack and while many of these "needles," or SNPs associating with disease, have been identified, there is still a great deal of heritability yet to be explained. The missing or phantom heritability is due, at least in part, to epistasis or gene–gene interactions, which have not been extensively explored in GWAS. Part of the challenge for epistasis analysis in GWAS is the sheer magnitude of the search and the computational complexity associated with it. An exhaustive search for epistasis models is not computationally feasible; thus, alternate approaches must be considered. In this chapter, these approaches will be reviewed briefly, and the incorporation of biological knowledge to guide this process will be further expanded upon. Real biological data examples where this approach has yielded successful identification of epistasis will also be provided. Epistasis has been known to be important since the early 1900s; however, its prevalence in mainstream research has been somewhat overshadowed by molecular technology advances. Due to the increasing evidence of epistasis in complex traits, it continues to emerge as a likely explanation for missing heritability.

Key words Epistasis, Prior knowledge, Missing heritability, Filtering, Enrichment, Pathways

1 Introduction

The search for the missing heritability [1, 2] in genome-wide association studies (GWAS) has become an important focus for the human genetics community – especially as larger and larger sample sizes have resulted in even smaller effect sizes to be identified. The National Human Genome Research Institute (NHGRI) GWAS catalog was developed to store all of the GWAS results in a central database. A few years ago, NHGRI looked at the distribution of GWAS-associated SNPs and found a majority were associated with small effects sizes (odds ratios less than 1.4) [3]. In January 2014, we evaluated the GWAS catalog to see if the trend had changed, and unfortunately, due to increasing sample sizes,

Jason H. Moore and Scott M. Williams (eds.), *Epistasis: Methods and Protocols*, Methods in Molecular Biology,
vol. 1253, DOI 10.1007/978-1-4939-2155-3_2, © Springer Science+Business Media New York 2015

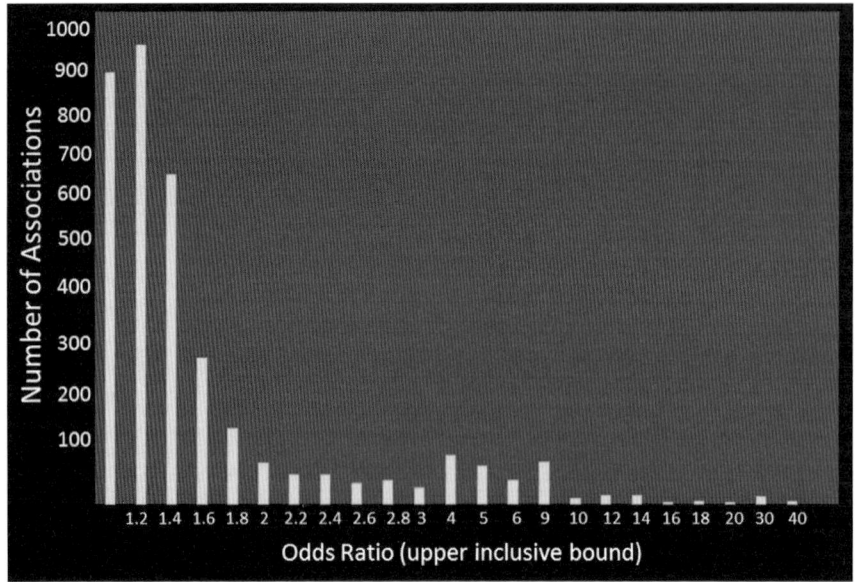

Fig. 1 Distribution of odds ratios (effect size) from the NHGRI GWAS catalog as of January 2014

the effect sizes identified have become even smaller. Figure 1 shows the distribution of these effects; the majority have odds ratios less than 1.2. This leads to much discussion regarding the missing or phantom heritability. Lander and colleagues explain that much of the missing heritability could be due to genetic interactions [4]. This opinion has been shared and emphasized by others in the literature for several years [5–11]. As such, it is believed by many that epistasis is important and should be explored in the context of GWAS; however, specific applications of epistasis analysis in GWAS have been much fewer than single variant main effects analyses. The computational burden of exploring gene–gene interactions in the wealth of data generated in GWAS, along with small to moderate sample sizes, has led to epistasis being an afterthought, rather than a primary focus of GWAS analyses.

In this chapter, we discuss some potential approaches to make epistasis analysis more computationally tractable in a GWAS dataset. A number of alternative approaches are described, but the primary focus is on the use of prior biological knowledge from databases in the public domain to guide the search for epistasis. The manner in which prior knowledge is incorporated into a GWAS study can be done in several different ways and the knowledge can be extracted from a variety of database sources. These approaches will be discussed, and some successful applications will be described. Incorporating biological knowledge is likely to be fruitful in the search for epistasis in large-scale genomic studies of the current state-of-the-art and into the future.

2 The Scope of the Problem

The ultimate goal of any disease gene discovery project is to identify as much of the genomic variation as possible that is relevant to the phenotype (disease or trait) being studied. As molecular technology has advanced, the field has gone from very coarse genomic examination embodied in cytogenetic analyses, to higher resolution linkage analyses, and now to very high-resolution association analyses. Methodological advances in the analysis of large-scale GWAS studies and the ability to integrate results across experiments have simply not kept pace with this flood of genotyping and now sequencing data. It is a central fallacy that simply generating more data and collecting more samples/individuals will solve the problem. Instead, it is this tsunami of data that has made distinguishing true scientific discoveries from the thousands or millions of false discoveries even more challenging. The ultimate success of our monumental investment in data generation will depend largely on the development and use of innovative analytic approaches and intelligent study designs that allow for the detection of gene–gene and gene–environment interactions.

A major hurdle in discovering epistasis, however, is the variable selection problem. Exhaustively evaluating all of the possible combinations of SNPs is not computationally feasible. For example, testing all two-SNP models in genome-wide data including one million SNPs generates 5.00×10^{11} possible two-SNP models; this requires extensive computing resources and produces many statistically significant results. If we consider going beyond pairwise models, one million SNPs generate 1.7×10^{17} three-SNP models, 4.2×10^{22} four-SNP models, 8.3×10^{27} five-SNP models, 1.4×10^{33} six-SNP models, and so on (calculated based on (n/m), where n is the number of SNPs and m is the number of variables in the model). This creates an enormous computing challenge as well as a multiple testing correction issue. It has been shown that using the parallel Multifactor Dimensionality Reduction approach (pMDR), it is possible to scan through an exhaustive search of possible two-SNP models [12]. Steffens et al. also demonstrate a genome-wide interaction analysis (GWIA) and the strategies for data compression, specific data representations, interleaved data organization, and parallelization of the analysis on a multiprocessor system [13]. These strategies, as well as many others that have been developed in the past few years, provide capability to perform an exhaustive pairwise GWIA [14–16]. However, beyond pairwise interactions, exhaustive searching is not tractable.

3 Methods for Data Reduction

It is clear that while the goal of GWAS is to survey the entire genome in an unbiased way, this type of approach simply does not work in the search for epistasis, especially beyond pairwise interactions. A number of filtering approaches have been suggested to reduce the computational burden. First, using statistical evidence of single-SNP effects to prioritize SNPs can be promising and has been shown to have high power [14, 17]. This approach follows from the hierarchical model-building principles of the general linear model whereby interaction terms are tested only after all main effect terms are deemed statistically significant (as some predefined p-value threshold). For example, in a 500,000 SNP GWAS analysis, one might use a threshold of $p < 1 \times 10^{-5}$ based on a chi-square test. As such, it is expected that there would be approximately five SNPs significant by chance alone; presumably additional SNPs will be significant because some of those will be true effects for that particular dataset. If the SNPs that are important for the epistatic model are not among those top hits, the interactions will not be tested. If we select or filter variables based on their main effects, we bias the analysis using statistical information and assume that relevant interactions occur only between markers that independently have some effect on the phenotype alone. Filtering SNPs based on the strength of independent main effects can identify SNP combinations among loci with small to moderate main effects, such as two 2-SNP models identified for Amyotrophic lateral sclerosis (ALS) [18] or multiple sclerosis (MS) [19]. If, however, the genetic variants that are important for disease risk have effects only through their interactions with other genes, this filtering by main effects approach would potentially miss these types of discoveries.

The second approach is to use intrinsic knowledge extracted from the dataset to filter the list of SNPs to test for interactions. Data reduction algorithms that explicitly assess the quality of an SNP in its relationship to the clinical outcome are an alternative to pure statistical or biological filters. A series of Relief algorithms have been explored including Relief, ReliefF, Spatially Uniform ReliefF (SURF), Tuned ReliefF (TuRF), and SURF and TURF [20, 21]. These approaches use a nearest neighbor approach to assess SNP quality to detect attributes associated with disease. In this case, nearest neighbors are individuals in the dataset who are genetically similar at the many SNPs across the genome. Relief uses a single neighbor, ReliefF uses multiple nearest neighbors, and the SURF and TURF are various extensions to the ReliefF filtering. Filtering approaches that use intrinsic properties of the data, such as these ReliefF methods, look like a promising alternative for epistasis in GWAS. According to published studies, they will be successful in removing nonfunctional SNPs while maintaining the SNP–SNP interaction models. This will effectively reduce the number of statistical tests that need to be performed, which relieves computational complexity issues as well as multiple comparisons issues [22].

Third, the use of extrinsic biological knowledge to filter SNPs and then evaluate multi-marker combinations based on biological criteria has been suggested [23–24]. If we filter variables using biological information extrinsic to our dataset—i.e., only examine interactions between SNPs in a common pathway or with a common structure or function based on the literature or information in databases—we bias the analysis in favor of models with an established biological foundation in the literature, and novel interactions between SNPs would be missed. Furthermore, the analysis is conditioned on the quality of the biological information used. However, the interaction models with detectable statistical epistasis will have good evidence for biological epistasis and a high likelihood of being interpretable [22].

Each of these strategies imposes a specific bias into the analysis, and no one strategy will be optimal in all cases (*see* **Note 1**). Each of these has advantages and disadvantages with known biases and limitations. While all of the proposed approaches for filtering have clear strengths and limitations, we propose that filtering based on extrinsic biological knowledge will be a robust approach for the detection of epistasis in large-scale genomic analyses including GWAS as well as next-generation sequencing. While the available biological knowledge is incomplete and always evolving, it provides a framework for exploring epistasis in which models are plausible, more likely to be interpretable, and reduces the computational and statistical burdens. By limiting the search space, we limit the number of statistical tests, multiple comparison burden, as well as computational complexity. The remainder of this chapter will focus on approaches being developed for using biological knowledge to prioritize the search for missing heritability in the epistasis domain and provide some examples where these strategies have been implemented.

4 Methods for Incorporating Biological Knowledge

The incorporation of prior knowledge into GWAS has been proposed and many new tools have been developed to allow for this type of analysis (*see* **Note 2**). While most of them have been utilized and published based on a single-locus test of association, nearly all of them could be used in the search for epistasis. This is certainly a rapidly growing area of research; as such it is not possible to thoroughly describe all of the recent developments. However, in the following sections, a number of approaches will be described with suggestions for how they could guide the search for epistasis in large, genome-wide datasets.

4.1 Protein–Protein Interaction Approaches

Protein–protein interactions can be measured using mass spectrometry, immunoprecipitation, yeast two-hybrids, and affinity pull down followed by mass spectrometry [25]. As discussed in Ritchie [22], a number of protein–protein interaction databases are publicly available, including the database of interacting proteins (DIPs) [26], BioGRID (Biological General Repository for Interaction Datasets) [27], and human protein reference database (HPRD) [28]. As described by Pattin and Moore, a couple of different approaches could be used to incorporate protein–protein interaction data [25]. First, the most straightforward approach includes filtering the full SNP list by the SNPs included in the genes encoding the proteins involved in the interactions [25]. This would reduce the number of SNPs explored for epistasis. However, it would also prevent the identification of models that include novel biology. An alternative and perhaps more promising approach involves developing a metric to score the relative importance of the SNPs such that the full list could be prioritized or weighted, rather than filtered in or out of the dataset. This would allow for novel biology to be discovered, although it would favor models with a priori evidence of support [25]. Scoring systems like this can then be used for filtering as well as for Bayesian priors for analysis.

4.2 Pathway Approaches

As discussed by Ritchie [22], the use of pathway data to look for overrepresentation of genome-wide associated hits has been done in many studies. For example, Perry et al. used Kyoto Encyclopedia of Genes and Genomes (KEGG), BioCarta, and Gene Ontology (GO) to perform a modified gene-set enrichment analysis (GSEA) for type II diabetes [29]. In rheumatoid arthritis (RA), Beyene et al. utilized a selection of prior knowledge from c2 curated gene sets, which are obtained from online pathway databases, citations in PubMed, and domain experts [30]. For GSEA, their final set included 1,900 gene sets collected from canonical pathways, chemical and genetic perturbations, BioCarta pathways, GeneMAPP, and KEGG [30].

Another similar gene set enrichment analysis, the SNP-ratio test (SRT) [31], compares the proportion of statistically significant genes to all SNPs within genes that are part of a specific pathway. An empirical p-value is then calculated based on comparisons in datasets where a permutation test has been performed (i.e., the assignment of case/control status has been randomized). Approaches like this rely largely on single-SNP analysis, but then look for enrichment of sets of SNPs to report interesting findings, with the idea that sets of interacting SNPs/genes would show up in pathway enrichment tests. So while it is not epistasis evidence, they are at least considering polygenic models.

Baranzini and colleagues [18] propose a protein interaction and network-based analysis (PINBPA) for the study of a multiple sclerosis (MS) dataset. An alternative approach was explored by Askland et al., where they used exploratory visual analysis (EVA) to perform a number of pathway-based analyses of bipolar disorder [32]. Another approach, pathway genetic load (PGL), looks for evidence of epistasis between genes confined to a single pathway [33]. This approach dramatically reduces the computational complexity of an epistasis search in GWAS data.

As discussed by Ritchie [22], assessing the statistical significance of pathways is also an important and difficult challenge. It is not enough to simply look for an overrepresentation of hits in a particular pathway or set of pathways. There are reasons unrelated to the associations that can lead to this, such as the selection of SNPs on the genotyping platforms or the choice of pathway annotation for analysis. Large pathways have a greater chance of being statistically significant, and many of the bioinformatics tools used for these types of studies are biased toward detecting large, well-defined pathways [34]. Methods to perform permutation testing in pathway analysis frameworks have been developed to provide increased power and efficiency [35]. Other approaches index pathways using Gene Ontology terms and test for overrepresentation of pathways in a list of hits from a genome-wide association study (such as ALIGATOR-Association LIst Go AnnoTatOR) and successfully identify pathways for complex traits, such as bipolar disorder [36]. It is also important to re-iterate that pathway analysis approaches, in themselves, were not developed for the purpose of detecting epistasis. These methods focus on single-SNP analyses and explore pathways where accumulated single locus associations are detected. However, it is obvious that these pathway approaches will develop hypotheses regarding potential "underground networks" [1], which would be particularly interesting to focus efforts for detection of epistasis.

4.3 Comprehensive Knowledge Approach

Perhaps the most lucrative solution involves a comprehensive knowledge-based approach that includes evidence from pathways, protein–protein interactions, prior association, gene ontology, linkage, or gene expression, etc. Because we have a limited number of known epistatic models in humans, it is currently a challenge to hypothesize what structure models will involve and what relationships between genes we should expect. We can look to the known examples of epistasis in model organisms to point us in the right direction, but until we have more examples in humans, we are merely speculating and may not include all possible types of models.

Ritchie [22] described that one of the major disadvantages of the comprehensive approach is the current inability to accurately evaluate it compared to other approaches in simulated data experiments. Unfortunately, simulation studies, where biological knowledge is concerned, are very difficult to perform. There are two issues. First, if you do the straightforward type of simulation study where you preselect functional SNPs based on biological knowledge, embed them into the simulation, and use that same knowledge to guide the search, the simulation is overly simplistic and really not very interesting. The second issue is that to do it right, we need to have a simulation tool whereby we can simulate pathways and networks, and then create disease models including some of the loci from these networks. This type of tool does not currently exist. So, unfortunately, while a simulation study to compare approaches would be fantastic, it is not currently feasible. Once a body of literature is published demonstrating some of the pathway and network effects we can expect to observe in natural, biological data, we will be able to develop simulation tools to test additional novel analytic methods. After that, we may have a better-detailed critique on the different approaches.

Several approaches have been developed that include a more thorough extraction of prior information from multiple sources. The Biofilter is one such system [37, 38]. Layers of biological machinery exist between genetic variations and the phenotypes they manifest, and imposing this extra dimension of known biological information into statistical analyses may help identify relationships between genetic variants that contribute to common complex disease. The Biofilter is a database system cataloging biological information based on data from BioGRID, dbSNP, NCBI gene, Gene Otology, MINT, NetPath, OregAnno, Pfam, PharmGKB, Reactome, UCSC genome browser, and the NHGRI GWAS catalog [38]. The strategy of Biofilter steps beyond the annotation and grouping of independent SNP effects. The Biofilter uses biological information about gene–gene relationships and gene–disease relationships to construct multi-SNP models before conducting any statistical analyses. Rather than annotating the independent effect of each SNP in a GWAS dataset, the Biofilter allows the explicit detection and modeling of interactions between a set of SNPs preselected by the application. In this manner, the Biofilter process provides a tool to discover significant multi-SNP models with nonsignificant main effects that have established biological plausibility. This approach has the added benefit of reducing both the computational and statistical burden of exhaustively evaluating all possible multi-SNP models. The goal of the Biofilter is to take advantage of what we know, recognizing that there is much more to be discovered [38].

Biofilter uses biological information about gene–gene relationships to construct multi-SNP models that can then be prioritized before conducting any statistical analyses. The key idea behind Biofilter model generation is that any pathway, ontological category, protein family, experimental interaction, or other grouping of genes or proteins implies a relationship between each of those genes or proteins. Thus Biofilter provides a tool to discover significant multi-SNP models with nonsignificant main effects that have established biological plausibility. The Biofilter model generation process thus far has been protein-coding gene-centric, and as such, SNPs from GWAS genotyping platforms must first be assigned to protein-coding regions [38, 39]. Relationships between genes represented by a genotyping platform can then be translated to multi-SNP models. If the same two genes appear together in more than one grouping, they're likely to have an important biological relationship; if they appear in multiple groups from several independent sources, then they're even more likely to be biologically related in some way, and receive a higher implication index. Biofilter has access to thousands of such groupings because of the use of multiple domain sources and can analyze all of them to identify the sets of genes or SNPs appearing together in the greatest number of groupings and the widest array of original data sources. These pairs can then be tested for significance within a research dataset, and, depending on the level of data filtering or application of an implication index cutoff, Biofilter can be used to avoid the prohibitive computational and multiple-testing burden of an exhaustive pairwise analysis. Once multi-SNP models are constructed, they can be evaluated using any relevant analytic method such as logistic regression, multifactor dimensionality reduction (MDR), Bayesian networks, etc.

We have developed the Library of Knowledge Integration (LOKI) database (Fig. 2) and integrated eleven public domain data sources. These sources are combined into one central LOKI database for use in annotation, filtering, and building models of gene–gene interactions for analysis. Biofilter and LOKI are described in detail in Pendergrass et al. [38]. Biofilter has been

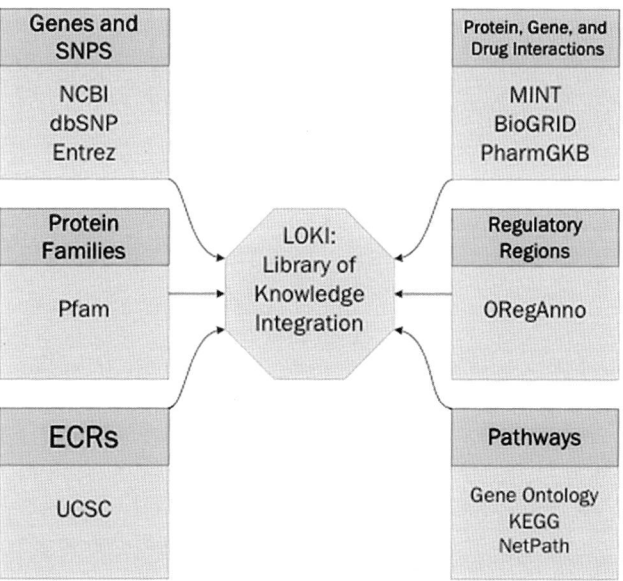

Fig. 2 The current Library of Knowledge Integration (LOKI). LOKI contains information from multiple database repositories, covering multiple domains. From [38]

applied to a number of natural, biological datasets for the discovery of gene–gene interaction models associated with complex traits including Multiple Sclerosis [40], HDL cholesterol [41], HIV Pharmacogenomics [42], and cataract status [43].

Another approach for comprehensive data integration is INTERSNP. INTERSNP is a powerful, flexible approach that implements logistic regression or log-linear models for joint analysis of multiple SNPs [16]. The filtering of SNPs can be done using statistical evidence from single locus statistics, genomic evidence based on genomic location, or biologic relevance based on pathway information from KEGG [16]. Approaches such as these have the greatest potential since they rely on multiple sources and types of information. This is, of course, as long as the analytic strategy is implemented in such a way that the incorporation of incorrect knowledge does not impede the ability to detect the correct models. Using prior knowledge can be an incredibly powerful tool, but we should be careful to use it in an efficient manner (*see* **Note 4**).

5   Real World Example: HDL Cholesterol

Plasma concentrations of low-density lipoprotein (LDL) cholesterol, high-density lipoprotein (HDL) cholesterol, triglycerides (TRI), and total cholesterol are among the most important risk factors for coronary artery disease (CAD). Lipid traits have been well studied in genome-wide association studies with between 100,000 [44] and over 188,000 individuals [45] included. While there have been over 150 loci identified for association with lipid traits, the proportion of heritability explained is still modest with ~25–30 % of the genetic variance for each lipid trait [44, 45]. Due to this missing heritability, several groups have embarked on explorations of epistasis or gene–gene interactions in lipid traits.

For example, Turner et al. looked for epistasis associated with HDL cholesterol [41] using a biological knowledge-driven filtering method. Here, the Biofilter [37, 38] was used to decrease the number of SNP–SNP models evaluated from genome-wide genotype data. Through the application of the Biofilter, eleven significant GxG models were in the discovery Biobank cohort, eight of which show evidence of replication in a second biobank cohort [41]. The strongest predictive model included a pairwise interaction between *LPL* (which modulates the incorporation of triglyceride into HDL) and *ABCA1* (which modulates the incorporation of free cholesterol into HDL) [41]. The authors required that any GxG interactions in the discovery cohort ($n = 3,740$ participants) showed evidence of replication in the de-identified EMRs of a second cohort ($n = 1,858$ participants). This resulted in replicated GxG interactions associated with variation in HDL-C, all of which have potential biological relevance.

A similar approach was taken by Ma et al.; they used prior knowledge from established genome-wide association study (GWAS) hits, protein–protein interactions, as well as pathway information to guide their gene–gene interaction analyses [46]. These results were further followed up through the evaluation of gene-based interaction analysis [47] as well as potential eQTLs involved in gene–gene interactions [48]. These results demonstrated that gene–gene interactions modulate complex human traits, including HDL cholesterol, and the use of prior biological knowledge can increase power to identify biologically interesting and relevant models (*see* **Note 3**).

6 Notes

1. The search space for enumerating all possible epistasis models in genome-wide datasets is computationally prohibitive; thus numerous data reduction or filtering strategies have been employed to reduce the SNP set for epistasis modeling including:

   - Statistical filtering using single-SNP statistics (such as the chi-square test).

     – Advantage: simple, unbiased with respect to the biologist.

     – Disadvantage: relies on all important genes having independent main effects.

   - Intrinsic filtering using statistical or computational data-driven approaches (such as ReliefF).

     – Advantage: unbiased with respect to the biologist; uses the data.

     – Disadvantage: complicated; models may not have biological relevance.

   - Extrinsic filtering using biological knowledge (such as Biofilter or pathway analysis).

     – Advantage: results are biologically relevant.

     – Disadvantage: limited by current state of biology; biased toward genes we know something about as a field.

2. Biological knowledge-based epistasis methods are emerging as powerful strategies for epistasis analyses.

3. Real data applications have been deemed successful finding evidence of epistasis replicating across multiple datasets.

4. Many methods for incorporation of biological knowledge into epistasis analysis exist and continue to be developed.

References

1. Maher B (2008) Personal genomes: the case of the missing heritability. Nature 456:18–21. doi:10.1038/456018a

2. Manolio TA, Collins FS, Cox NJ, Goldstein DB, Hindorff LA, Hunter DJ, McCarthy MI, Ramos EM, Cardon LR, Chakravarti A, Cho JH, Guttmacher AE, Kong A, Kruglyak L, Mardis E, Rotimi CN, Slatkin M, Valle D, Whittemore AS, Boehnke M, Clark AG, Eichler EE, Gibson G, Haines JL, Mackay TFC, McCarroll SA, Visscher PM (2009) Finding the missing heritability of complex diseases. Nature 461:747–753. doi:10.1038/nature08494

3. Hindorff LA, Sethupathy P, Junkins HA, Ramos EM, Mehta JP, Collins FS, Manolio TA (2009) Potential etiologic and functional implications of genome-wide association loci for human diseases and traits. Proc Natl Acad Sci U S A 106:9362–9367. doi:10.1073/pnas.0903103106

4. Zuk O, Hechter E, Sunyaev SR, Lander ES (2012) The mystery of missing heritability: genetic interactions create phantom heritability. Proc Natl Acad Sci U S A 109(4):1193–1198, 201119675. doi: 10.1073/pnas.1119675109

5. Moore JH (2003) The ubiquitous nature of epistasis in determining susceptibility to common human diseases. Hum Hered 56:73–82

6. Moore JH, Williams SM (2005) Traversing the conceptual divide between biological and statistical epistasis: systems biology and a more modern synthesis. Bioessays 27:637–646

7. Cordell HJ (2009) Detecting gene-gene interactions that underlie human diseases. Nat Rev Genet 10:392–404. doi:10.1038/nrg2579

8. Templeton AR (2000) Epistasis and complex traits. Epistasis and the evolutionary process. Oxford University Press, New York, pp 41–57

9. Gibson G (1996) Epistasis and pleiotropy as natural properties of transcriptional regulation. Theor Popul Biol 49:58–89

10. Moore JH (2005) A global view of epistasis. Nat Genet 37:13–14. doi:10.1038/ng0105-13

11. McKinney BA, Pajewski NM (2011) Six degrees of epistasis: statistical network models for GWAS. Front Genet 2:109. doi:10.3389/fgene.2011.00109

12. Bush WS, Dudek SM, Ritchie MD (2006) Parallel multifactor dimensionality reduction: a tool for the large-scale analysis of gene-gene interactions. Bioinformatics 22:2173–2174

13. Steffens M, Becker T, Sander T, Fimmers R, Herold C, Holler DA, Leu C, Herms S, Cichon S, Bohn B, Gerstner T, Griebel M, Nöthen MM, Wienker TF, Baur MP (2010) Feasible and successful: genome-wide interaction analysis involving all 1.9 × 1011 pair-wise interaction tests. Hum Hered 69:268–284. doi:10.1159/000295896

14. Evans DM, Marchini J, Morris AP, Cardon LR (2006) Two-stage two-locus models in genome-wide association. PLoS Genet 2:e157. doi:10.1371/journal.pgen.0020157

15. Ueki M, Cordell HJ (2012) Improved statistics for genome-wide interaction analysis. PLoS Genet 8:e1002625. doi:10.1371/journal.pgen.1002625

16. Herold C, Steffens M, Brockschmidt FF, Baur MP, Becker T (2009) INTERSNP: genome-wide interaction analysis guided by a priori information. Bioinform Oxf Engl 25:3275–3281. doi:10.1093/bioinformatics/btp596

17. Kooperberg C, Leblanc M (2008) Increasing the power of identifying gene x gene interactions in genome-wide association studies. Genet Epidemiol 32:255–263. doi:10.1002/gepi.20300

18. Sha Q1, Zhang Z, Schymick JC, Traynor BJ, Zhang S. Genome-wide association reveals three SNPs associated with sporadic amyotrophic lateral sclerosis through a two-locus analysis. BMC Med Genet. 2009 Sep 9;10:86

19. Baranzini SE, Galwey NW, Wang J, Khankhanian P, Lindberg R, Pelletier D, Wu W, Uitdehaag BMJ, Kappos L, GeneMSA Consortium, Polman CH, Matthews PM, Hauser SL, Gibson RA, Oksenberg JR, Barnes MR (2009) Pathway and network-based analysis of genome-wide association studies in multiple sclerosis. Hum Mol Genet 18:2078–2090. doi:10.1093/hmg/ddp120

20. Greene CS, Penrod NM, Kiralis J, Moore JH (2009) Spatially uniform relieff (SURF) for computationally-efficient filtering of gene-gene interactions. BioData Min 2:5. doi:10.1186/1756-0381-2-5

21. Moore JH, White BC (2007) Tuning relieff for genome-wide genetic analysis. In: Moore JH, Rajapakse JC, Marchiori E (eds) Evolutionary computation, machine learning and data mining, bioinformatics. Springer, Berlin, pp 166–175

22. Ritchie MD (2011) Using biological knowledge to uncover the mystery in the search for epistasis in genome-wide association studies. Ann Hum Genet 75:172–182. doi:10.1111/j.1469-1809.2010.00630.x

23. Carlson CS, Eberle MA, Kruglyak L, Nickerson DA (2004) Mapping complex disease loci in whole-genome association studies. Nature 429:446–452

24. Sun X, Lu Q, Mukheerjee S, Crane PK, Elston R, Ritchie MD (2014) Analysis pipeline for the

epistasis search – statistical versus biological filtering. Front Genet 5:106. doi:10.3389/fgene.2014.00106

25. Pattin KA, Moore JH (2008) Exploiting the proteome to improve the genome-wide genetic analysis of epistasis in common human diseases. Hum Genet 124:19–29. doi:10.1007/s00439-008-0522-8

26. Salwinski L, Miller CS, Smith AJ, Pettit FK, Bowie JU, Eisenberg D (2004) The database of interacting proteins: 2004 update. Nucleic Acids Res 32:D449–D451. doi:10.1093/nar/gkh086

27. Breitkreutz B-J, Stark C, Reguly T, Boucher L, Breitkreutz A, Livstone M, Oughtred R, Lackner DH, Bahler J, Wood V, Dolinski K, Tyers M (2008) The BioGRID interaction database: 2008 update. Nucleic Acids Res 36:D637–D640. doi:10.1093/nar/gkm1001

28. Mishra GR, Suresh M, Kumaran K, Kannabiran N, Suresh S, Bala P, Shivakumar K, Anuradha N, Reddy R, Raghavan TM, Menon S, Hanumanthu G, Gupta M, Upendran S, Gupta S, Mahesh M, Jacob B, Mathew P, Chatterjee P, Arun KS, Sharma S, Chandrika KN, Deshpande N, Palvankar K, Raghavnath R, Krishnakanth R, Karathia H, Rekha B, Nayak R, Vishnupriya G, Kumar HGM, Nagini M, Kumar GSS, Jose R, Deepthi P, Mohan SS, Gandhi TKB, Harsha HC, Deshpande KS, Sarker M, Prasad TSK, Pandey A (2006) Human protein reference database – 2006 update. Nucleic Acids Res 34:D411–D414. doi:10.1093/nar/gkj141

29. Perry JRB, McCarthy MI, Hattersley AT, Zeggini E, Wellcome Trust Case Control Consortium, Weedon MN, Frayling TM (2009) Interrogating type 2 diabetes genome-wide association data using a biological pathway-based approach. Diabetes 58:1463–1467. doi:10.2337/db08-1378

30. Beyene J, Hu P, Hamid JS, Parkhomenko E, Paterson AD, Tritchler D (2009) Pathway-based analysis of a genome-wide case-control association study of rheumatoid arthritis. BMC Proc 3(Suppl 7):S128

31. O'Dushlaine C, Kenny E, Heron EA, Segurado R, Gill M, Morris DW, Corvin A (2009) The SNP ratio test: pathway analysis of genome-wide association datasets. Bioinform Oxf Engl 25:2762–2763. doi:10.1093/bioinformatics/btp448

32. Askland K, Read C, Moore J (2009) Pathways-based analyses of whole-genome association study data in bipolar disorder reveal genes mediating ion channel activity and synaptic neurotransmission. Hum Genet 125:63–79. doi:10.1007/s00439-008-0600-y

33. Huebinger RM, Garner HR, Barber RC (2010) Pathway genetic load allows simultaneous evaluation of multiple genetic associations. Burns 36:787–792. doi:10.1016/j.burns.2010.02.001

34. Elbers CC, van Eijk KR, Franke L, Mulder F, van der Schouw YT, Wijmenga C, Onland-Moret NC (2009) Using genome-wide pathway analysis to unravel the etiology of complex diseases. Genet Epidemiol 33:419–431. doi:10.1002/gepi.20395

35. Guo Y-F, Li J, Chen Y, Zhang L-S, Deng H-W (2009) A new permutation strategy of pathway-based approach for genome-wide association study. BMC Bioinformatics 10:429. doi:10.1186/1471-2105-10-429

36. Holmans P, Green EK, Pahwa JS, Ferreira MAR, Purcell SM, Sklar P, Owen MJ, O'Donovan MC, Craddock N (2009) Gene ontology analysis of GWA study data sets provides insights into the biology of bipolar disorder. Am J Hum Genet 85:13–24. doi:10.1016/j.ajhg.2009.05.011

37. Bush WS, Dudek SM, Ritchie MD (2009) Biofilter: a knowledge-integration system for the multi-locus analysis of genome-wide association studies. Pac Symp Biocomput 368–379

38. Pendergrass SA, Frase AT, Wallace JR, Wolfe D, Katiyar N, Moore C, Ritchie MD (2013) Genomic analyses with biofilter 20: knowledge driven filtering, annotation, and model development. BioData Min 6(1):25

39. Bush WS, Chen G, Torstenson ES, Ritchie MD (2009) LD-spline: mapping SNPs on genotyping platforms to genomic regions using patterns of linkage disequilibrium. BioData Min 2:7. doi:10.1186/1756-0381-2-7

40. Bush WS, McCauley JL, DeJager PL, Dudek SM, Hafler DA, Gibson RA, Matthews PM, Kappos L, Naegelin Y, Polman CH, Hauser SL, Oksenberg J, Haines JL, Ritchie MD (2011) A knowledge-driven interaction analysis reveals potential neurodegenerative mechanism of multiple sclerosis susceptibility. Genes Immun 12:335–340. doi:10.1038/gene.2011.3

41. Turner SD, Berg RL, Linneman JG, Peissig PL, Crawford DC, Denny JC, Roden DM, McCarty CA, Ritchie MD, Wilke RA (2011) Knowledge-driven multi-locus analysis reveals gene-gene interactions influencing HDL cholesterol level in two independent EMR-linked biobanks. PLoS One 6:e19586. doi:10.1371/journal.pone.0019586

42. Grady BJ, Torstenson ES, McLaren PJ, De Bakker PIW, Haas DW, Robbins GK, Gulick RM, Haubrich R, Ribaudo H, Ritchie MD (2011) Use of biological knowledge to inform the analysis of gene-gene interactions involved

in modulating virologic failure with efavirenz-containing treatment regimens in art-naïve actg clinical trials participants. Pac Symp Biocomput 2011:253–264

43. Pendergrass SA, Verma SS, Holzinger ER, Moore CB, Wallace J, Dudek SM, Huggins W, Kitchner T, Waudby C, Berg R, McCarty CA, Ritchie MD (2013) Next-generation analysis of cataracts: determining knowledge driven gene-gene interactions using Biofilter, and gene-environment interactions using the PhenX Toolkit. Pac Symp Biocomput 147–158

44. Teslovich TM, Musunuru K, Smith AV, Edmondson AC, Stylianou IM, Koseki M, Pirruccello JP, Ripatti S, Chasman DI, Willer CJ, Johansen CT, Fouchier SW, Isaacs A, Peloso GM, Barbalic M, Ricketts SL, Bis JC, Aulchenko YS, Thorleifsson G, Feitosa MF, Chambers J, Orho-Melander M, Melander O, Johnson T, Li X, Guo X, Li M, Shin Cho Y, Jin Go M, Jin Kim Y, Lee J-Y, Park T, Kim K, Sim X, Twee-Hee Ong R, Croteau-Chonka DC, Lange LA, Smith JD, Song K, Hua Zhao J, Yuan X, Luan J, Lamina C, Ziegler A, Zhang W, Zee RYL, Wright AF, Witteman JCM, Wilson JF, Willemsen G, Wichmann H-E, Whitfield JB, Waterworth DM, Wareham NJ, Waeber G, Vollenweider P, Voight BF, Vitart V, Uitterlinden AG, Uda M, Tuomilehto J, Thompson JR, Tanaka T, Surakka I, Stringham HM, Spector TD, Soranzo N, Smit JH, Sinisalo J, Silander K, Sijbrands EJG, Scuteri A, Scott J, Schlessinger D, Sanna S, Salomaa V, Saharinen J, Sabatti C, Ruokonen A, Rudan I, Rose LM, Roberts R, Rieder M, Psaty BM, Pramstaller PP, Pichler I, Perola M, Penninx BWJH, Pedersen NL, Pattaro C, Parker AN, Pare G, Oostra BA, O'Donnell CJ, Nieminen MS, Nickerson DA, Montgomery GW, Meitinger T, McPherson R, McCarthy MI, McArdle W, Masson D, Martin NG, Marroni F, Mangino M, Magnusson PKE, Lucas G, Luben R, Loos RJF, Lokki M-L, Lettre G, Langenberg C, Launer LJ, Lakatta EG, Laaksonen R, Kyvik KO, Kronenberg F, König IR, Khaw K-T, Kaprio J, Kaplan LM, Johansson A, Jarvelin M-R, Janssens ACJW, Ingelsson E, Igl W, Kees Hovingh G, Hottenga J-J, Hofman A, Hicks AA, Hengstenberg C, Heid IM, Hayward C, Havulinna AS, Hastie ND, Harris TB, Haritunians T, Hall AS, Gyllensten U, Guiducci C, Groop LC, Gonzalez E, Gieger C, Freimer NB, Ferrucci L, Erdmann J, Elliott P, Ejebe KG, Döring A, Dominiczak AF, Demissie S, Deloukas P, de Geus EJC, de Faire U, Crawford G, Collins FS, Chen YI, Caulfield MJ, Campbell H, Burtt NP, Bonnycastle LL, Boomsma DI, Boekholdt SM, Bergman RN, Barroso I, Bandinelli S, Ballantyne CM, Assimes TL, Quertermous T, Altshuler D, Seielstad M, Wong TY, Tai E-S, Feranil AB, Kuzawa CW, Adair LS, Taylor HA Jr, Borecki IB, Gabriel SB, Wilson JG, Holm H, Thorsteinsdottir U, Gudnason V, Krauss RM, Mohlke KL, Ordovas JM, Munroe PB, Kooner JS, Tall AR, Hegele RA, Kastelein JJP, Schadt EE, Rotter JI, Boerwinkle E, Strachan DP, Mooser V, Stefansson K, Reilly MP, Samani NJ, Schunkert H, Cupples LA, Sandhu MS, Ridker PM, Rader DJ, van Duijn CM, Peltonen L, Abecasis GR, Boehnke M, Kathiresan S (2010) Biological, clinical and population relevance of 95 loci for blood lipids. Nature 466:707–713. doi:10.1038/nature09270

45. Global Lipids Genetics Consortium, Willer CJ, Schmidt EM, Sengupta S, Peloso GM, Gustafsson S, Kanoni S, Ganna A, Chen J, Buchkovich ML, Mora S, Beckmann JS, Bragg-Gresham JL, Chang H-Y, Demirkan A, Den Hertog HM, Do R, Donnelly LA, Ehret GB, Esko T, Feitosa MF, Ferreira T, Fischer K, Fontanillas P, Fraser RM, Freitag DF, Gurdasani D, Heikkilä K, Hyppönen E, Isaacs A, Jackson AU, Johansson A, Johnson T, Kaakinen M, Kettunen J, Kleber ME, Li X, Luan J, Lyytikäinen L-P, Magnusson PKE, Mangino M, Mihailov E, Montasser ME, Müller-Nurasyid M, Nolte IM, O'Connell JR, Palmer CD, Perola M, Petersen A-K, Sanna S, Saxena R, Service SK, Shah S, Shungin D, Sidore C, Song C, Strawbridge RJ, Surakka I, Tanaka T, Teslovich TM, Thorleifsson G, Van den Herik EG, Voight BF, Volcik KA, Waite LL, Wong A, Wu Y, Zhang W, Absher D, Asiki G, Barroso I, Been LF, Bolton JL, Bonnycastle LL, Brambilla P, Burnett MS, Cesana G, Dimitriou M, Doney ASF, Döring A, Elliott P, Epstein SE, Eyjolfsson GI, Gigante B, Goodarzi MO, Grallert H, Gravito ML, Groves CJ, Hallmans G, Hartikainen A-L, Hayward C, Hernandez D, Hicks AA, Holm H, Hung Y-J, Illig T, Jones MR, Kaleebu P, Kastelein JJP, Khaw K-T, Kim E, Klopp N, Komulainen P, Kumari M, Langenberg C, Lehtimäki T, Lin S-Y, Lindström J, Loos RJF, Mach F, McArdle WL, Meisinger C, Mitchell BD, Müller G, Nagaraja R, Narisu N, Nieminen TVM, Nsubuga RN, Olafsson I, Ong KK, Palotie A, Papamarkou T, Pomilla C, Pouta A, Rader DJ, Reilly MP, Ridker PM, Rivadeneira F, Rudan I, Ruokonen A, Samani N, Scharnagl H, Seeley J, Silander K, Stancáková A, Stirrups K, Swift AJ, Tiret L, Uitterlinden AG, van Pelt LJ, Vedantam S, Wainwright N, Wijmenga C, Wild SH, Willemsen G, Wilsgaard T, Wilson JF, Young EH, Zhao JH, Adair LS, Arveiler D, Assimes TL, Bandinelli S, Bennett F, Bochud M,

Boehm BO, Boomsma DI, Borecki IB, Bornstein SR, Bovet P, Burnier M, Campbell H, Chakravarti A, Chambers JC, Chen Y-DI, Collins FS, Cooper RS, Danesh J, Dedoussis G, de Faire U, Feranil AB, Ferrières J, Ferrucci L, Freimer NB, Gieger C, Groop LC, Gudnason V, Gyllensten U, Hamsten A, Harris TB, Hingorani A, Hirschhorn JN, Hofman A, Hovingh GK, Hsiung CA, Humphries SE, Hunt SC, Hveem K, Iribarren C, Järvelin M-R, Jula A, Kähönen M, Kaprio J, Kesäniemi A, Kivimaki M, Kooner JS, Koudstaal PJ, Krauss RM, Kuh D, Kuusisto J, Kyvik KO, Laakso M, Lakka TA, Lind L, Lindgren CM, Martin NG, März W, McCarthy MI, McKenzie CA, Meneton P, Metspalu A, Moilanen L, Morris AD, Munroe PB, Njølstad I, Pedersen NL, Power C, Pramstaller PP, Price JF, Psaty BM, Quertermous T, Rauramaa R, Saleheen D, Salomaa V, Sanghera DK, Saramies J, Schwarz PEH, Sheu WH-H, Shuldiner AR, Siegbahn A, Spector TD, Stefansson K, Strachan DP, Tayo BO, Tremoli E, Tuomilehto J, Uusitupa M, van Duijn CM, Vollenweider P, Wallentin L, Wareham NJ, Whitfield JB, Wolffenbuttel BHR, Ordovas JM, Boerwinkle E, Palmer CNA, Thorsteinsdottir U, Chasman DI, Rotter JI, Franks PW, Ripatti S, Cupples LA, Sandhu MS, Rich SS, Boehnke M, Deloukas P, Kathiresan S, Mohlke KL, Ingelsson E, Abecasis GR (2013) Discovery and refinement of loci associated with lipid levels. Nat Genet 45:1274–1283. doi:10.1038/ng.2797

46. Ma L, Brautbar A, Boerwinkle E, Sing CF, Clark AG, Keinan A (2012) Knowledge-driven analysis identifies a gene-gene interaction affecting high-density lipoprotein cholesterol levels in multi-ethnic populations. PLoS Genet 8:e1002714. doi:10.1371/journal.pgen.1002714

47. Ma L, Clark AG, Keinan A (2013) Gene-based testing of interactions in association studies of quantitative traits. PLoS Genet 9:e1003321. doi:10.1371/journal.pgen.1003321

48. Ma L, Ballantyne C, Brautbar A, Keinan A (2014) Analysis of multiple association studies provides evidence of an expression QTL hub in gene-gene interaction network affecting HDL cholesterol levels. PLoS One 9:e92469. doi:10.1371/journal.pone.0092469

Chapter 3

Biological Knowledge-Driven Analysis of Epistasis in Human GWAS with Application to Lipid Traits

Li Ma, Alon Keinan, and Andrew G. Clark

Abstract

While the importance of epistasis is well established, specific gene–gene interactions have rarely been identified in human genome-wide association studies (GWAS), mainly due to low power associated with such interaction tests. In this chapter, we integrate biological knowledge and human GWAS data to reveal epistatic interactions underlying quantitative lipid traits, which are major risk factors for coronary artery disease. To increase power to detect interactions, we only tested pairs of SNPs filtered by prior biological knowledge, including GWAS results, protein–protein interactions (PPIs), and pathway information. Using published GWAS and 9,713 European Americans (EA) from the Atherosclerosis Risk in Communities (ARIC) study, we identified an interaction between *HMGCR* and *LIPC* affecting high-density lipoprotein cholesterol (HDL-C) levels. We then validated this interaction in additional multiethnic cohorts from ARIC, the Framingham Heart Study, and the Multi-Ethnic Study of Atherosclerosis. Both *HMGCR* and *LIPC* are involved in the metabolism of lipids and lipoproteins, and *LIPC* itself has been marginally associated with HDL-C. Furthermore, no significant interaction was detected using PPI and pathway information, mainly due to the stringent significance level required after correcting for the large number of tests conducted. These results suggest the potential of biological knowledge-driven approaches to detect epistatic interactions in human GWAS, which may hold the key to exploring the role gene–gene interactions play in connecting genotypes and complex phenotypes in future GWAS.

Key words Epistasis, Gene–Gene Interaction, Biological Knowledge, GWAS, Lipid

1 Introduction

As of June 2013, over 1,638 publications and 10,859 single nucleotide polymorphisms (SNPs) have been included in the catalog of human genome-wide association studies (GWAS) [1]. Although these SNPs are significantly associated with human diseases and traits, most of them have small effects that, in aggregate, account for only a moderate proportion of heritable variance [2–6]. Four lipid traits, total cholesterol (TC), low-density lipoprotein cholesterol (LDL-C), triglyceride (TG), and high-density lipoprotein cholesterol (HDL-C) levels, are among the most important risk factors for coronary heart disease. Recently, meta-analyses of many

Jason H. Moore and Scott M. Williams (eds.), *Epistasis: Methods and Protocols*, Methods in Molecular Biology,
vol. 1253, DOI 10.1007/978-1-4939-2155-3_3, © Springer Science+Business Media New York 2015

GWAS, with a combined sample size of up to 100,000, have detected hundreds of loci associated with levels of these four lipids [7, 8]. However, these loci collectively only explain 25–30 % of heritable variance of each lipid [7, 8]. Many hypotheses have been offered to explain the missing heritability, including rare and structural variants, gene–environment interactions, epigenetics, and complex inheritance [2–5]. The missing heritability may also be partially attributed to epistatic or gene–gene interactions [9–11]. Here, we seek to identify examples of pairwise SNP by SNP interaction effects on any of the four lipid levels in this study.

Since Bateson's first discovery in 1905 that some genes suppress the effects of other genes [12], researchers have been investigating the effect of epistasis to better understand the complex relationships between genotypes and phenotypes. Studies of model organisms suggested epistasis or gene–gene interactions are a common phenomenon [13–16], and a number of gene–gene interactions have been reported in gene mapping studies in animals, plants, and other model organisms [17–20]. However, gene–gene interactions have proven difficult to find in humans [21, 22], mainly due to low statistical power caused by the small effect size, the low minor genotype frequency of the multiple-SNP combinations, the large combinatorial number of interaction tests required [14, 23], and the lack of control over environmental conditions. Hence, to improve power and enable detection of gene–gene interactions in human GWAS, many approaches have been developed to prioritize candidate genes or SNPs using biological knowledge from established GWAS hits [6, 24], protein–protein interactions (PPIs) [25, 26], and pathway information [27].

Tests of gene–gene interactions are not as powerful as tests of single-marker association, so a judicious strategy is essential for successful epistasis analyses in human GWAS [11, 14, 15]. One fundamental limitation to the analysis of epistasis is that the statistical power of each particular test can be easily eroded by the performance of vast numbers of pairwise or higher-order interaction tests. In order to achieve similar success for interaction analyses as one obtains with single-marker tests in GWAS, we are limited to performing a similar number of tests (~1 million), assuming the nominal per-test power to detect interaction is the same as that of a single-marker association test. This limitation prevents us from conducting an inclusive all-by-all interaction analysis in current human GWAS, which typically examines millions of SNPs. If one were to attempt all-by-all epistasis tests, the result is such reduced power that only the most exceptionally strong interactions could be detected. Therefore, we recommend epistasis analysis be done on a reduced number of SNPs, using prior biological knowledge as the simplest way to restrict SNP numbers. The selection of candidate genes or SNPs can be based on any factor(s) that the biologist or medical practitioner chooses. So long as the gene choice is not based on the tests of interactions itself, winnowing down the gene set will not inflate the type I error and can increase the power when the underlying interactions are enriched between candidate genes or SNPs. Criteria such as SNP density, local linkage disequilibrium (LD), and data quality can further supplement primary knowledge of gene function, pathway position, connectedness to networks, etc. in selecting genes.

We illustrate the biological knowledge-driven approach with examples from analyses of large consortium cohort studies of cardiovascular disease [6]. We tested for pairwise SNP by SNP epistatic interactions affecting the level of four lipids, TC, LDL-C, TG, and HDL-C, based on prior knowledge of published GWAS hits, PPIs, and pathway information. Based on GWAS hits, we detected an interaction between *HMGCR* and a locus upstream of *LIPC* in their effect on HDL-C levels in the discovery data set from the Atherosclerosis Risk in Communities (ARIC) study. Using a locus-based replication procedure, we validated this interaction in cohorts from the Framingham Heart Study (FHS) and from the Multi-Ethnic Study of Atherosclerosis (MESA). In summary, a biological, knowledge-driven approach might be crucial to the detection of epistatic interactions underlying complex traits and diseases in human GWAS.

2 Methods and Results

2.1 Descriptions of GWAS Data

Data sets from three GWAS were considered here, the Atherosclerosis Risk in Communities (ARIC) study, the Framingham Heart Study (FHS), and the Multi-Ethnic Study of Atherosclerosis (MESA). The ARIC study is a multicenter prospective investigation of atherosclerotic disease [28]. This analysis included 9,713 European Americans (EAs) in the discovery study and 3,207 African American (AA) individuals in the validation study. The FHS is a prospective cohort study to evaluate cardiovascular disease (CVD) risk factors, which has been described in detail previously [29]. This analysis included 6,575 EA subjects in the validation study, while accounting for familial relatedness (*see* Subheading 2.4 below). MESA is a prospective cohort study of individuals aged 45–84 years without clinical CVD recruited from 6 US centers [30], which is designed to study the characteristics of subclinical CVD and its progression. Participants of MESA self-reported their race or ethnicity group as EA, AA, Chinese American, or Hispanic American (HA). In total 2,685 EA, 2,588 AA, and 2,174 HA individuals were included in the validation. All of the included subjects from the three GWAS studies have both genotypic and phenotypic (lipid) measurements available, as described in the following sections.

2.2 Genotype Data and Quality Control (QC)

Genotyping of samples was obtained from the ARIC and MESA studies by Affymetrix 6.0 SNP arrays. Affymetrix 6.0 SNP array genotyping of MESA samples and Affymetrix 500K SNP array genotyping of FHS samples were obtained from dbGaP (MESA SHARe, downloaded in May 2011; Framingham Cohort, downloaded in April 2010) [31]. Standard quality control (QC) filters were applied across each of the samples and SNPs, including (1) exclusion of subjects with >10 % missing data; (2) removing SNPs with call rates <90 %; (3) removing SNPs with minor allele frequencies (MAF) ≤1 %; and (4) removing SNPs with Hardy–Weinberg Equilibrium (HWE) test with $P < 10^{-6}$. For the SNP pairs to be tested, we also required (1) sample size of each of the nine two-SNP combinations greater than 20 in the discovery analysis and greater than 10 in the validation study; and (2) LD measure of $r^2 < 0.1$ between the two candidate SNPs.

When necessary, untyped SNPs were imputed using IMPUTE2 [32] with HapMap3 [33] and 1,000 Genomes [34] reference haplotypes, which resulted in about the same set of SNPs across the three studies. No imputation was performed in MESA HA samples due to lack of appropriate reference haplotype panels. SNPs with information scores <0.6 were excluded, and each genotype was imputed to be the genotype with the highest posterior probability. When the highest posterior probability is less than 0.8, the genotype was treated as missing.

2.3 Phenotype Data and Covariates

Four quantitative lipid measurements, total cholesterol (TC), LDL cholesterol (LDL-C), triglyceride (TG), and HDL cholesterol (HDL-C), were considered in the analysis. Each lipid level is measured at multiple time points and the average level per individual was used in all studies. A log transformation was applied to TG levels to normalize its distribution because of the skewness in the original distribution. Individuals self-reported to be taking lipid-lowering medications were excluded. Sex, age, age squared, body mass index (BMI) were included as covariates in all analyses. The average values for age, age squared, and BMI were also used whenever multiple measurements were available. Plate number is also included as a covariate factor in the ARIC data due to its correlation with some of the lipid levels (known as "plate effect").

2.4 Test of Statistical Interactions

Statistical interactions between pairs of SNPs were tested on a quantitative trait. For each individual, let Y denote the trait of interest and G_i denote the genotype of the ith SNP ($i = 1, 2$). G_i is the number of copies of the reference allele, thus with possible values 0, 1, or 2. Two indicator variables, x_i and z_i, were defined for each of the two SNPs as,

$$
x_i = \begin{cases} 1, & G_i = 0 \\ 0, & G_i = 1 \\ -1 & G_i = 2 \end{cases} \quad z_i = \begin{cases} -0.5, & G_i = 0 \\ 0.5, & G_i = 1 \\ -0.5, & G_i = 2 \end{cases}
$$

Two linear models were fitted and compared for testing of interaction. The first model (M1) included additive and dominance effect terms for each of the two SNPs without including any interaction effects between the two SNPs. The second model (M2), on top of M1, allows for four classic forms of epistatic interactions (additive × additive, additive × dominance, dominance × additive, and dominance × dominance), as follows:

$$\Upsilon = Z_0\beta_0 + x_1 a_1 + z_1 d_1 + x_2 a_2 + z_2 d_2 + \varepsilon \qquad (M_1)$$

$$\Upsilon = Z_0\beta_0 + x_1 a_1 + z_1 d_1 + x_2 a_2 + z_2 d_2 + x_1 x_2 i_{aa} + x_1 z_2 i_{ad} + z_1 x_2 i_{da} + z_1 z_2 i_{dd} + \varepsilon \qquad (M_2)$$

Here, β_0 is a vector of the intercept and possible nongenetic covariates; a_i and d_i are the additive and dominance effects of the ith SNP; and i_{aa}, i_{ad}, i_{da}, and i_{dd} denote the four classic interaction effects between the two SNPs. The existence of an epistatic interaction of any combination of the four types was tested by an F-test comparing M1 and M2 with four degrees of freedom [6, 35]. This classical test of statistical interactions is similar to the "--epistasis" option in PLINK [36], except that only additive effects and additive × additive interactions are considered, such that an F-test with one degree of freedom is performed in PLINK. We note that the power of this test is strongly impacted by the marginal SNP allele frequencies.

Potential population stratification was corrected using the principal component (PC) approach and the familial relationship in FHS was accounted for using a mixed model approach [22, 37]. PC analysis was conducted using EIGENSOFT [37] and the top ten PCs were included in the analysis as covariates to account for population stratification in ARIC and MESA samples. For FHS, a mixed model approach was first applied to account for familial relatedness and then pairwise interaction was tested on the residuals from the mixed models [22].

2.5 Locus-Based Validation of Interactions

We sought to validate (replicate) the interactions detected in the discovery study using data from FHS, MESA, and another AA sample from the ARIC study. "Validate" rather than "replicate" was used here because linked and proximate SNPs were included in the validation in addition to the original SNPs, as follows. For a significant interaction between two SNPs (e.g., A and B) in the discovery study, we performed the following possible stages of tests in the validation study: (1) Test for interaction directly between SNP A and SNP B; (2) If the interaction is not significant in stage (i), test for interactions between SNP A and each SNP less than 200 kb away from SNP B, and, similarly, between SNP B and each SNP surrounding SNP A; (3) If no test in stage (ii) is significant following multiple-testing correction, test for interactions between each SNP less than 100 kb away from A and each SNP less than 100 kb away from B. Assume there are n_1 and n_2 SNPs, respectively, surrounding SNPs A and B and the number of statistical tests to be conducted is 1, $n_1 + n_2$, and n^2 in the three validation stages, respectively. To reduce the number of tests and the cost of multiple-testing correction on power, the validation process proceeds sequentially and stops at any stage where significant results were found after multiple-testing correction.

2.6 Prioritize SNP Pairs Using Biological Knowledge

Though only pairwise interactions are considered, the total number of possible interaction tests across 2.5 million SNPs is still huge, at more than 3×10^{12} tests. Due to the severe reduction in power entailed by stringent multiple-testing correction for such a large number of tests, it is crucial to restrict the total number of tests for the whole study. Therefore, through the following three strategies, we aimed to reduce the total number of interaction tests and to enrich interaction signals in the tests considered.

2.6.1 Lipid GWAS Results

Recently, 95 genetic loci were associated with TC, LDL-C, TG, or HDL-C in a GWAS meta-analysis [7]. We considered 125 SNPs, previously associated with lipid levels [7], in the 95 loci and tested each pair of SNPs on TC, LDL-C, TG, and HDL-C in the discovery sample of 9,713 EAs from the ARIC study. Using this approach, we tested ~7,748 pairwise interactions for each trait. We identified one significant interaction affecting LDL-C levels and one interaction on HDL-C. The interaction on LDL-C was between rs2247056 and rs1030431 ($P_c = 0.03$ after Bonferroni correction). Both rs2247056 and rs1030431 were marginally associated with LDL-C, TG, and TC [7, 38]. We then performed fine mapping analyses on the loci surrounding the two SNPs to find the most significant interaction between rs2853928 and rs1993453 ($P_c = 0.01$). We tried to validate the interaction on LDL-C in additional replication samples from ARIC, FHS, and MESA, but had no success.

The interaction on HDL-C was between rs12916 and rs1532085 ($P_c = 0.008$) in the discovery study and was successfully validated in the replication samples (Table 1). To further explore the interaction between the two loci, we tested interactions between each SNP surrounding rs12916 and each SNP surrounding rs1532085 within 100 kb. While many of these SNP pairs show significant interactions due to LD, the strongest signal was between rs3846662 and rs2043085 ($P_c = 0.002$). SNP rs3846662, located in the intron of gene *HMGCR*, has been previously associated with TC and LDL-C [7, 39], but not with HDL-C. Rs3846662 has also been associated with a 2.2-fold change in *HMGCR* expression in vitro [40] and alternative splicing of exon 13 [41]. The other SNP, rs2043085, has been found to be marginally associated with HDL-C [7]. Rs2043085 is located upstream of *LIPC* and is associated with the expression of the gene [6]. On top of marginal effects, interaction between the two SNPs affects HDL-C twice as much as the effect of *LIPC* alone.

Table 1
Significant interactions affecting HDL-C levels validated in multiple population cohorts (Regenerated from ref. 6)

Test stage	Cohort	SNP 1				SNP 2				P_c^b
		rsID	Chr	Pos[a]	Gene	rsID	Chr	Pos[a]	Gene	
Discovery	ARIC EA	rs12916	5	74656539	*HMGCR* (3' UTR)	rs1532085	15	58683366	40.8 k U *LIPC*	0.008
Fine mapping	ARIC EA	rs3846662	5	74651084	*HMGCR* (Intron)	rs2043085	15	58680954	43.2 k U *LIPC*	0.002
Validation	MESA EA	rs3846662	5	74651084	*HMGCR* (Intron)	rs1973688	15	58582540	141.6 k U *LIPC*	0.006
Validation	FHS EA	rs55727654	5	74651864	*HMGCR* (Intron)	rs473422	15	58666341	57.8 k U *LIPC*	0.002
Validation	MESA HA	rs1423527	5	74602699	30.3 k U *HMGCR*	rs7163280	15	58718340	5.8 k U *LIPC*	0.04
Validation	ARIC AA	rs3761743	5	74685520	27.6 k D *HMGCR*	rs567838	15	58736623	*LIPC* (Intron)	0.004

U upstream of, *D* downstream of
[a]Build 37.1 (GRCh37)
[b]*P*-value after Bonferroni correction

On average, individuals with the TT genotype at rs2043085 show an increase of 2.63 mg/dl in HDL-C; subjects with TT at rs2043085 and AA at rs3846662 exhibit an average increase of 5.72 mg/dl. The linear model with these two SNPs and their interaction has an r^2 value of 0.8 %, meaning that these two SNPs and their interaction combined to explain 0.8 % of phenotypic variation in HDL-C. Using the locus-based validation procedure, the interaction between rs3846662 and rs2043085 on HDL-C was successfully validated in stage (ii) for MESA EA samples and in stage (iii) for FHS EA, MESA HA, and ARIC AA cohorts (Table 1). It did not validate after multiple-testing correction for the MESA AA sample, due to smaller sample size.

2.6.2 Protein–Protein Interactions (PPIs)

Over 3,000 high-confidence human PPIs were carefully assembled [42]. For each gene pair indicated by a PPI, we exhaustively tested all pairwise interactions between each SNP in the first gene and each SNP in the second gene. To map SNPs to genes, gene information (hg18) was obtained from the UCSC genome browser [43] and we considered all SNPs located between 5 kb upstream and downstream of a gene. Suppose there are n_1 and n_2 SNPs in the first and second genes, respectively, the number of possible pairwise interactions is $n_1 \times n_2$. For each of the four lipid traits, we performed ~6 million pairwise interaction tests according to the 3,000 PPIs, which resulted in no significant interactions following multiple-testing correction.

2.6.3 Lipid Pathway Information

We hypothesized that possible gene–gene interactions are enriched between genes in the lipid-related pathways. To test this hypothesis, we used the metabolic pathway of lipids and lipoproteins as an example. There are a total of 228 genes in this pathway [44] and 12,716 SNPs are mapped to the 228 genes. We tested all pairwise interactions among the 12,716 SNPs, resulting in a total of ~27 million interaction tests for each lipid trait. Hence, it is not surprising that we found nothing significant after multiple-testing correction. However, there is a deviation in the QQ plot of the P values for interactions underlying TC levels. The interaction between rs4804546 and rs914196 is the strongest, though not significant following correction for the ~27 million tests ($P_c = 0.14$). The interaction is between two genes, *CARM1* and *AGPAT3*, from the metabolism of lipids and lipoproteins pathway. Gene *AGPAT3* has been associated with phospholipid levels [45], and *CARM1* has not been associated with any lipid levels.

3 Notes

To ensure accuracy and power, quality control process is of crucial importance in genetic association studies, as well as in studies of epistasis [46, 47]. Non-normality distribution and outliers of the quantitative phenotype can potentially lead to false positive results with very small P values, particularly when individuals carrying the minor multi-SNP genotype incidentally are also outliers of the phenotype. When both outliers and low-frequency SNPs exist in the data, the chance of false positives can be inflated. Therefore, a minor multi-SNP genotype frequency filter (20 in discovery and 10 in validation in this study) is necessary when testing for interactions, much like the MAF filtering in a typical GWAS quality control.

While the interaction we identified was replicated in multiethnic populations, it still has a moderate effect size, which is approximately the same magnitude as the marginal effect discovered in GWAS for lipid traits. If this is the case for general epistatic interactions, we will have a lower power for detecting epistasis than marginal associations, which also explains why so few interactions have been detected and replicated in human GWAS. Moreover, unfortunately, epistatic interactions with such small effect sizes may only explain a minor proportion of the missing heritability, unless as some argue, there are a great many weakly interacting pairs.

Note that the interaction we detected in this study was validated in part by proximate SNPs, thus indicating the power of integrating information from genomic regions surrounding target SNPs, like a gene-based test for marginal associations. Recently, a series of gene-based interaction testing methods have been developed in the literature [48–52], which can be employed to increase power of detecting and replicating gene–gene interactions.

In summary, we detected significant interactions after multiple-testing correction only in <10k tests guided by GWAS, but found nothing significant in 6 and 27 million tests using PPI and pathway information, respectively. By noticing that the study is already powerful with a sample size of over 10,000, multiple measurements of phenotypes, and a genome-wide one million SNPs, we only afford to perform <10k tests to be able to identify the significant gene–gene interactions from false positives. In conclusion, a small-scale, biological, knowledge-driven study with higher enrichment of putative signals may hold the key to identifying gene–gene interactions underlying complex diseases or traits in current and future association studies.

References

1. Hindorff LA, Sethupathy P, Junkins HA, Ramos EM, Mehta JP et al (2009) Potential etiologic and functional implications of genome-wide association loci for human diseases and traits. Proc Natl Acad Sci U S A 106:9362–9367

2. Manolio TA, Collins FS, Cox NJ, Goldstein DB, Hindorff LA et al (2009) Finding the missing heritability of complex diseases. Nature 461:747–753

3. Frazer KA, Murray SS, Schork NJ, Topol EJ (2009) Human genetic variation and its contribution to complex traits. Nat Rev Genet 10: 241–251

4. Maher B (2008) Personal genomes: the case of the missing heritability. Nature 456:18–21

5. Eichler EE, Flint J, Gibson G, Kong A, Leal SM et al (2010) Missing heritability and strategies for finding the underlying causes of complex disease. Nat Rev Genet 11:446–450

6. Ma L, Ballantyne CM, Belmont JW, Keinan A, Brautbar A (2012) Interaction between SNPs in the RXRA and near ANGPTL3 gene region inhibit apolipoprotein B reduction following statin-fenofibric acid therapy in individuals with mixed dyslipidemia. J Lipid Res 53(11): 2425–2428

7. Teslovich TM, Musunuru K, Smith AV, Edmondson AC, Stylianou IM et al (2010) Biological, clinical and population relevance of 95 loci for blood lipids. Nature 466:707–713

8. Asselbergs FW, Guo YR, van Iperen EPA, Sivapalaratnam S, Tragante V et al (2012) Large-scale gene-centric meta-analysis across 32 studies identifies multiple lipid loci. Am J Hum Genet 91:823–838

9. Cheverud JM, Routman EJ (1995) Epistasis and its contribution to genetic variance components. Genetics 139:1455–1461

10. Cockerham CC (1954) An extension of the concept of partitioning hereditary variance for analysis of covariances among relatives when epistasis is present. Genetics 39:859–882

11. Zuk O, Hechter E, Sunyaev SR, Lander ES (2012) The mystery of missing heritability: genetic interactions create phantom heritability. Proc Natl Acad Sci 109:1193–1198

12. Bateson W, Saunders ER, Punnett RC, Hurst CC (eds) (1905) Reports to the Evolution Committee of the Royal Society, report II. Harrison and Sons, London

13. Carlborg O, Haley CS (2004) Epistasis: too often neglected in complex trait studies? Nat Rev Genet 5:618–625

14. Cordell HJ (2009) Detecting gene-gene interactions that underlie human diseases. Nat Rev Genet 10:392–404

15. Moore JH, Williams SM (2009) Epistasis and its implications for personal genetics. Am J Hum Genet 85:309–320

16. Gao H, Granka JM, Feldman MW (2010) On the classification of epistatic interactions. Genetics 184:827–837

17. Shimomura K, Low-Zeddies SS, King DP, Steeves TDL, Whiteley A et al (2001) Genome-wide epistatic interaction analysis reveals complex genetic determinants of circadian behavior in mice. Genome Res 11:959–980

18. Carlborg Ö, Kerje S, Schütz K, Jacobsson L, Jensen P et al (2003) A global search reveals epistatic interaction between QTL for early growth in the chicken. Genome Res 13:413–421

19. Caicedo AL, Stinchcombe JR, Olsen KM, Schmitt J, Purugganan MD (2004) Epistatic interaction between Arabidopsis FRI and FLC flowering time genes generates a latitudinal cline in a life history trait. Proc Natl Acad Sci U S A 101:15670

20. Clark AG, Doane WW (1984) Interactions between the amylase and adipose chromosomal regions of Drosophila melanogaster. Evolution 957–982

21. Ma L, Dvorkin D, Garbe J, Da Y (2007) Genome-wide analysis of single-locus and epistasis single-nucleotide polymorphism effects on anti-cyclic citrullinated peptide as a measure of rheumatoid arthritis. BMC Proc 1:S127

22. Ma L, Yang J, Runesha HB, Tanaka T, Ferrucci L et al (2010) Genome-wide association analysis of total cholesterol and high-density lipoprotein cholesterol levels using the Framingham Heart Study data. BMC Med Genet 11:55

23. Ma L, Runesha HB, Dvorkin D, Garbe JR, Da Y (2008) Parallel and serial computing tools for testing single-locus and epistatic SNP effects of quantitative traits in genome-wide association studies. BMC Bioinformatics 9:315

24. Marchini J, Donnelly P, Cardon LR (2005) Genome-wide strategies for detecting multiple loci that influence complex diseases. Locus 2:0.0

25. Jia P, Zheng S, Long J, Zheng W, Zhao Z (2011) DmGWAS: dense module searching for genome-wide association studies in protein–protein interaction networks. Bioinformatics 27:95

26. Sun YV, Kardia SLR (2010) Identification of epistatic effects using a protein–protein interaction database. Hum Mol Genet 19:4345

27. Wu X, Dong H, Luo L, Zhu Y, Peng G et al (2010) A novel statistic for genome-wide interaction analysis. PLoS Genet 6:e1001131

28. Williams OD (1989) The atherosclerosis risk in communities (ARIC) study – design and objectives. Am J Epidemiol 129:687–702

29. Dawber TR, Meadors GF, Moore FE (1951) Epidemiological approaches to heart disease: the Framingham study. Am J Public Health Nations Health 41:279–286

30. Bild DE, Bluemke DA, Burke GL, Detrano R, Roux AVD et al (2002) Multi-ethnic study of atherosclerosis: objectives and design. Am J Epidemiol 156:871–881

31. Mailman MD, Feolo M, Jin Y, Kimura M, Tryka K et al (2007) The NCBI dbGaP database of genotypes and phenotypes. Nat Genet 39:1181–1186

32. Howie BN, Donnelly P, Marchini J (2009) A flexible and accurate genotype imputation method for the next generation of genome-wide association studies. PLoS Genet 5: e1000529

33. Altshuler DM, Gibbs RA, Peltonen L, Dermitzakis E, Schaffner SF et al (2010) Integrating common and rare genetic variation in diverse human populations. Nature 467: 52–58

34. Altshuler DL, Durbin RM, Abecasis GR, Bentley DR, Chakravarti A et al (2010) A map of human genome variation from population-scale sequencing. Nature 467:1061–1073

35. Cordell HJ (2002) Epistasis: what it means, what it doesn't mean, and statistical methods to detect it in humans. Hum Mol Genet 11:2463–2468

36. Purcell S, Neale B, Todd-Brown K, Thomas L, Ferreira MAR et al (2007) PLINK: a tool set for whole-genome association and population-based linkage analyses. Am J Hum Genet 81:559–575

37. Price AL, Patterson NJ, Plenge RM, Weinblatt ME, Shadick NA et al (2006) Principal components analysis corrects for stratification in genome-wide association studies. Nat Genet 38:904–909

38. Haas BE, Horvath S, Pietilainen KH, Cantor RM, Nikkola E et al (2012) Adipose co-expression networks across Finns and Mexicans identify novel triglyceride-associated genes. BMC Med Genomics 5:61. doi:10.1186/1755-8794-1185-1161

39. Aulchenko YS, Ripatti S, Lindqvist I, Boomsma D, Heid IM et al (2009) Loci influencing lipid levels and coronary heart disease risk in 16 European population cohorts. Nat Genet 41: 47–55

40. Burkhardt R, Kenny EE, Lowe JK, Birkeland A, Josowitz R et al (2008) Common SNPs in HMGCR in Micronesians and Whites associated with LDL-cholesterol levels affect alternative splicing of exon13. Arterioscler Thromb Vasc Biol 28:U2078–U2332

41. Burkhardt R, Kenny EE, Lowe JK, Birkeland A, Josowitz R et al (2008) Common SNPs in HMGCR in micronesians and whites associated with LDL-cholesterol levels affect alternative splicing of exon13. Arterioscler Thromb Vasc Biol 28:2078–2084

42. Das J, Yu H (2012) HINT: high-quality protein interactomes and their applications in understanding human disease. BMC Syst Biol 6:92

43. Kent WJ, Sugnet CW, Furey TS, Roskin KM, Pringle TH et al (2002) The human genome browser at UCSC. Genome Res 12:996–1006

44. Matthews L, Gopinath G, Gillespie M, Caudy M, Croft D et al (2009) Reactome knowledge-base of human biological pathways and processes. Nucleic Acids Res 37:D619–D622

45. Lemaitre RN, Tanaka T, Tang WH, Manichaikul A, Foy M et al (2011) Genetic loci associated with plasma phospholipid n-3 fatty acids: a meta-analysis of genome-wide association studies from the CHARGE consortium. PLoS Genet 7

46. Lambert CG, Black LJ (2012) Learning from our GWAS mistakes: from experimental design to scientific method. Biostatistics 13:195–203

47. Burton PR, Clayton DG, Cardon LR, Craddock N, Deloukas P et al (2007) Genome-wide association study of 14,000 cases of seven common diseases and 3,000 shared controls. Nature 447:661–678

48. He J, Wang K, Edmondson AC, Rader DJ, Li C et al (2011) Gene-based interaction analysis by incorporating external linkage disequilibrium information. Eur J Hum Genet 19: 164–172

49. Oh S, Lee J, Kwon M-S, Weir B, Ha K et al (2012) A novel method to identify high order gene-gene interactions in genome-wide association studies: gene-based MDR. BMC Bioinformatics 13:S5

50. Ma L, Clark AG, Keinan A (2013) Gene-based testing of interactions in association studies of quantitative traits. PLoS Genet 9:e1003321

51. Li SY, Cui YH (2012) Gene-centric gene-gene interaction: a model-based Kernel machine method. Ann Appl Stat 6:1134–1161

52. Rajapakse I, Perlman MD, Martin PJ, Hansen JA, Kooperberg C (2012) Multivariate detection of gene-gene interactions. Genet Epidemiol 36:622–630

Chapter 4

Epistasis for Quantitative Traits in *Drosophila*

Trudy F.C. Mackay

Abstract

The role of gene–gene interactions in the genetic architecture of quantitative traits is controversial, despite the biological plausibility of nonlinear molecular interactions underpinning variation in quantitative traits. In strictly outbreeding populations, genetic architecture is inferred indirectly by estimating variance components; however, failure to detect epistatic variance does not mean lack of epistatic gene action and is even consistent with pervasive epistasis. In *Drosophila*, more focused approaches to detecting epistatic gene action are possible, based on the ability to create de novo mutations and perform crosses among them; to construct inbred lines, artificial selection lines, and chromosome substitution lines; to map quantitative trait loci affecting complex traits by linkage and association; and to evaluate effects of induced mutations on multiple wild-derived backgrounds. Here, I review evidence for epistasis in *Drosophila* from the application of these methods, and conclude that additivity is an emergent property of underlying epistatic gene action for *Drosophila* quantitative traits. Such studies can be used to infer novel, highly interconnected genetic networks that are enriched for gene ontology categories and metabolic and cellular pathways. The consequence of epistasis is that the main effects of each of the interacting loci depend on allele frequency, which negatively impacts the predictive ability of additive models. Finally, epistasis results in hidden quantitative genetic variation in natural populations (genetic canalization) and the potential for rapid evolution of Dobzhansky–Muller incompatibilities (speciation).

Key words Genetic interaction, *P*-Element mutagenesis, Diallel cross, QTL mapping, Introgression analysis, *Drosophila* Genetic Reference Panel, Cryptic genetic variation, Canalization

1 Introduction

The role of epistasis (interactions between segregating loci) in the genetic architecture of quantitative traits has been controversial since early formulations of quantitative genetic theory, with Fisher's [1] emphasis on genetic architecture and response to natural selection attributable to alleles with small additive effects at many loci and Wright's [2] explicit consideration of epistatic interactions and their key roles in his "shifting balance" theory of evolution. This controversy continues today [3, 4]. Biologically, we know that developmental, neural, transcriptional, metabolic, and biochemical networks are highly dynamic, interconnected, and consist of nonlinear positive

Jason H. Moore and Scott M. Williams (eds.), *Epistasis: Methods and Protocols*, Methods in Molecular Biology, vol. 1253, DOI 10.1007/978-1-4939-2155-3_4, © Springer Science+Business Media New York 2015

and negative feedback loops. Consequently, genetic systems that regulate these networks are complex and contain large numbers of interacting molecules [5]. Quantitative genetic variation in phenotypes results from multifactorial genetic perturbations of these networks. Thus, from first principles one can argue that interactions among genetic variants affecting quantitative traits are likely to be important aspects of genetic architecture. Further, for any n loci affecting variation in a trait, there are n additive effects, but on the order of n^2 two-way and n^3 three-way epistatic interactions [6], therefore the summed effects of individually small epistatic interactions may be large.

Given the biological plausibility and likelihood of the importance of epistatic interactions in determining quantitative trait phenotypes, why does the controversy exist? At the heart of the argument is the "central dogma" of quantitative genetics, which states that the relationship between relatives, predictive ability, and responses to natural and artificial selection are all driven by the additive component of genetic variance [6, 7]. Quantitative genetics theory defines additive (V_A), dominance (V_D), and epistatic (V_{AA}, V_{AD}, and V_{DD} for all possible pairwise interactions) genetic variance in terms of additive, dominance, and epistatic effects and allele frequencies. Additive genetic variance accounts for most of the total genetic variance for a wide range of allele frequencies, even if gene action is purely dominant and/or epistatic [4, 6–8]. This is because dominant and epistatic loci have nonzero marginal (additive) effects in most cases. Further, all estimates of additive genetic variance from resemblance among relatives include fractions of the interaction variance components [7]. Thus, one cannot infer underlying modes of gene action from relative magnitudes of variance components. Small and even negligible amounts of epistatic variance are not inconsistent with pervasive epistatic gene action.

While predicting short-term population responses to selection or genomic prediction of breeding values using dense molecular marker genotypes [9, 10] utilizes only additive genetic variance, understanding the underlying biology of quantitative traits can be greatly informed by deriving genetic interaction networks from epistatic interactions between loci [11]. Epistatic gene action is the genetic mechanism underlying canalization (genetic homeostasis) [12, 13] and speciation (Dobzhansky–Muller incompatibilities) [14, 15]; therefore identifying interacting loci segregating in natural populations is relevant to understanding both evolutionary stasis and change. Finally, knowledge of interacting loci will improve predictions of long-term responses to selection and inbreeding depression (and its converse, heterosis) in agricultural animal and crop species, and individual disease risk in humans. Since classical quantitative genetic methods for partitioning genetic variance components in outbreeding populations yield no insight regarding the underlying gene action of causal alleles, how can we assess whether epistasis is an important feature of the genetic architecture of quantitative traits? Genetically tractable model organisms afford the opportunity to utilize many different experimental designs to evaluate the contribution of epistatic gene action to quantitative trait phenotypes. In this chapter, I review methods for detecting epistasis in *Drosophila melanogaster*, summarize empirical results showing that epistasis is pervasive, and discuss the implications of these observations.

2   Defining Epistasis

In classical Mendelian genetics, epistasis between loci with alleles of large effect refers to the masking of genotypic effects at one locus by the genotype of another, as reflected by distorted segregation ratios in a di-hybrid cross [11]. The usage of the term in quantitative genetics is much broader, where it refers to any statistical interaction between genotypes at two (or more) loci [6, 7, 16–19]. Epistasis can refer to a modification of the homozygous and/or heterozygous effects of the interacting loci (Fig. 1). Two-locus epistasis can be easily visualized by plotting the phenotypes of the nine different genotypes (Fig. 2). In the absence of epistasis the estimates of additive (a) and dominance (d) effects at each locus are the same regardless of the genotype of the other locus (Fig. 2a). In the presence of epistasis, the effect of one locus depends on the genotype at the interacting locus (Fig. 2b–e), and estimates of additive and dominance effects at a locus will vary depending on the allele frequency of the interacting locus. Thus, if allele frequencies of interacting loci vary among populations, estimates of additive and dominance effects can be significant in one population but not another, or even of opposite sign, if both loci are not included in the model to estimate the main (additive) effects. Epistatic interactions can take a wide variety of forms for quantitative traits, but fall into two general categories. The first is a change of the magnitude of effects (Fig. 2b, c), in which the phenotype of one locus is enhanced or suppressed by the genotype at another locus. The second is a change of direction of effects (Fig. 2d, e).

2.1   Epistasis Between Induced Mutations

Single *P*-transposable element insertional mutations can easily be induced in *D. melanogaster* by simple crosses. Since the exact genomic location of a *P*-element insertion can be readily determined, the *Drosophila* community has adopted this technique to achieve the goal of disrupting all *D. melanogaster* genes [20]. *P*-element insertional mutagenesis is also a highly efficient method for identifying loci affecting quantitative traits, provided the crossing scheme ensures the mutations are generated in a common homozygous genetic background [21–24]. Such screens have been performed for a large number of quantitative traits, including

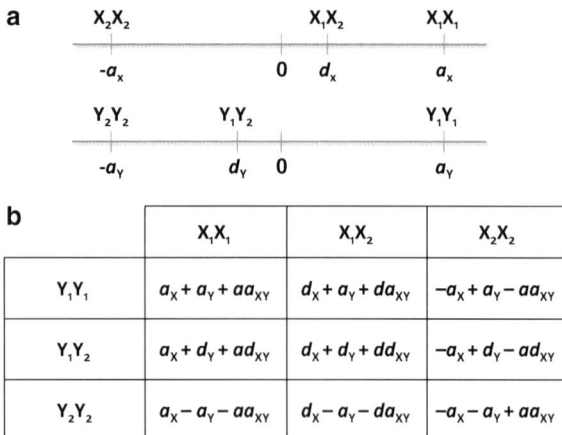

Fig. 1 Quantitative genetic genotypic effects. (**a**) Three genotypes for two loci, X and Y, each of which have two alleles (X_1, X_2, Y_1, Y_2). By convention, the X_1 and Y_1 alleles are associated with the higher phenotypic values. For each locus, the additive effect, *a*, is one half the difference in mean phenotype between the two homozygous genotypes. The dominance effect, *d*, is the difference between the mean phenotype of heterozygous individuals and the average phenotype of the homozygous genotypes. If $d = 0$, then there is additive gene action at the locus. Values of *d* between $\pm a$ denote partial dominance: $d = \pm a$ is the case of complete dominance of one of the alleles and $d < -a$ or $d > +a$ are cases of overdominance. The *subscript* denotes the locus. (**b**) Genotypic values for nine two-locus genotypes. The first two terms for each genotype denote the additive combination of single locus additive and dominance effects. With epistasis, the genotypic values cannot be predicted from the single locus additive and dominance effects alone; additional terms reflecting additive by additive (aa_{XY}), additive by dominance (ad_{XY}, da_{XY}), and dominance by dominance (dd_{XY}) contribute to the genotype value. (**a**) Originally published in Falconer DS, Mackay TFC (1996) Introduction to Quantitative Genetics. Longman, Harlow, Essex [7]. (**b**) Reproduced from [81] with permission from the author

sensory bristle number [21, 23, 24]; metabolic phenotypes, including fat and glycogen contents, enzyme activities, total protein and body weight [22]; olfactory [25, 26], locomotor [27, 28], and aggressive behaviors [29]; alcohol sensitivity [30]; starvation stress resistance [31]; and life span [32]. All studies reveal a large mutational target size for each quantitative trait. Since the mutations have been induced in the same homozygous background, they can be crossed to create genotypes that can be used to estimate the magnitude and nature of epistatic interactions. The advantage of assessing epistatic interactions between single mutations is that the interacting partners are known, but the disadvantage is that these methods are difficult to apply to large numbers of mutations.

2.1.1 Pairwise Epistatic Interactions

It is easy to construct all nine two-locus genotypes (Fig. 1b) between pairs of mutations, particularly if they are on different chromosomes. In this design, the phenotype of interest is assessed

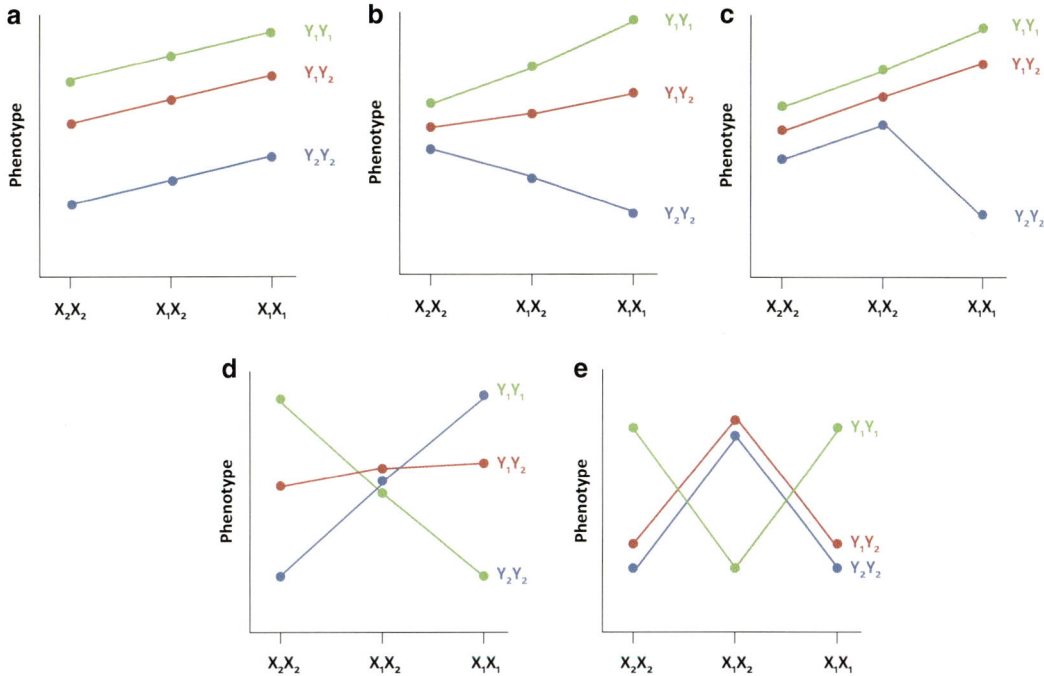

Fig. 2 Graphical representations of genotypic effects at two biallelic loci as reaction norms. (**a**) In this panel, gene action at the X locus is purely additive, with the mean heterozygote phenotype the same as the average phenotype of the two homozygous genotypes. At the Y locus, Y_1 is partially dominant over Y_2. There is no epistasis between the X and Y loci, indicated by the parallel reaction norms. The additive and dominance effects at the X locus are the same regardless of the Y locus genotype, and vice versa. (**b**, **c**) Epistasis occurs when the reaction norms are not parallel. In these panels, the additive effect of the Y locus is much greater in the X_1X_1 than the X_2X_2 genetic background, and the additive effects at the X locus are opposite in the Y_1Y_1 and Y_2Y_2 genetic backgrounds. (**d**) An epistatic interaction in which the additive effects at the X locus are in opposite directions in the Y_1Y_1 and Y_2Y_2 backgrounds, and the additive effects of the Y locus are in opposite directions in the X_1X_1 and X_2X_2 genetic backgrounds. The additive effect of the X locus is very small in the Y_1Y_2 genetic background, and the additive effect of the Y locus is very small in the X_1X_2 genetic background. If both loci are at intermediate frequency, no variation would be attributable to either locus with this form of interaction. (**e**) Epistasis involving dominance by dominance and dominance by additive interactions. (**a–c**) Reprinted from [81] with permission from the author

for all nine genotypes, and analyzed by two-way analyses of variance (ANOVA) of general form $Y = \mu + X + Y + X \times Y + \varepsilon$, where Y is the phenotype, μ is the overall mean, X and Y are the main effects of the two loci, $X \times Y$ is the epistatic interaction between the two loci, and ε is the variance within genotypes. There are eight effects to be estimated: a_X, a_Y, d_X, d_Y, aa_{XY}, ad_{XY}, da_{XY}, and dd_{XY} (Fig. 1b), which represent additive and dominance effects at each locus as well as additive by additive, additive by dominance, dominance by additive, and dominance by dominance epistatic effects. Clark and Wang [33] did this for eight pairs of second and third chromosome *P*-element mutations chosen at random, and scored 16 quantitative traits (live weight, total soluble protein, lipid and glycogen contents, and the activities of 12 enzymes in intermediary metabolism) for each genotype. A total of 35 of the 128 tests for epistasis (27 %) were significant. Additive by dominance and dominance by dominance epistasis effects were more common than additive by additive effects, and when present, epistatic effects were of the same order as the main additive and dominance effects. One caveat for applying this method for *P*-element mutations is that the *P*-element transgenes contain an eye-color marker, and the wild-type strain contains a null mutation of this gene. Thus, the inferred epistatic interactions could potentially be confounded by differences in copy number of the selectable marker in homozygous wild type (0 copies), single heterozygotes (1 or 3 copies), double heterozygotes (2 copies), and homozygous mutations (4 copies). This confounding effect is easy to detect if multiple pairs of mutations are tested, since variation in copy number of the selectable marker that affects the phenotype will result in similar estimates of all genetic parameters for all mutant pairs. This was not the case in the Clark and Wang [33] study.

2.1.2 Diallel Cross Analyses

In a full diallel cross analysis, all n^2 F_1 genotypes between n autosomal *P*-element insertion lines of interest are generated. In a half diallel cross analysis, the $n(n-1)/2$ crosses excluding reciprocal crosses and parental lines are generated. The latter is more appropriate for detecting epistasis with *P*-element insertions since it avoids the issue of variation in transgene copy number, because all F_1 genotypes contain two copies of the *P*-element. The first step of the analysis is an ANOVA of all F_1 genotypes to test for phenotypic differences among them. If differences are detected, the next step is to partition the phenotypic variation among F_1 genotypes into sources attributable to general combining ability (*GCA*) and specific combining ability (*SCA*) [34]. The *GCA* effect of each mutation is estimated as $GCA_i = T_i/(n-2) - \Sigma T/n(n-2)$, where T_i is the sum of the mean phenotypic values of F_1 genotypes with the ith mutation, ΣT is twice the sum of the mean phenotypic values of all F_1 genotypes, and n is the number of mutations [7, 35]. The *GCA* of a mutation is thus an estimate of its average heterozygous effect relative to the heterozygous effects of the other mutations. The *SCA* effect of each F_1 genotype is estimated as $SCA_{ij} = x_{ij} - (T_i + T_j)/(n-2) + \Sigma T/(n-1)(n-2)$, where x_{ij} is the phenotype of the F_1 genotype from the ith and jth mutations [35]. *SCA* effects in this experimental design can only be caused by dominance by dominance epistatic interactions.

Federowicz et al. [36] pioneered the use of a half-diallel cross design to detect epistasis between *P*-element mutations generated in a common homozygous background, and applied it to data on olfactory behavior for all 66 doubly heterozygous genotypes resulting from crossing 12 autosomal *smell-impaired* [25] *P*-element mutants. There were 9 (13.6 %) significant *SCA* interactions, involving 10 of the 12 mutations assessed, and one interaction involving an additional mutation approached formal significance. These results indicate that epistatic interactions must be extensive. Further, the interactions among eight of the mutations formed a genetic interaction network of enhancing and suppressing effects. Similar studies subsequently confirmed the prevalence of epistasis and defined new genetic interaction networks for olfactory behavior [26], temperature-sensitive loss of coordinated movement [27], locomotor startle response [28], life span [32], aggressive behavior, brain morphology, and gene expression [37]. Perhaps surprisingly, strong epistasis can occur between mutations without significant main effects on a trait [37]. Further, the interaction networks are not stable, and vary according to the environmental conditions under which the phenotypes are assessed [26], between males and females [32], and in the presence or absence of an additional interacting mutation [27].

2.1.3 Mutations in Different Genetic Backgrounds

P-elements do not insert randomly in the *Drosophila* genome, but have particular "hot" and "cold" spots for insertion [20]. Further, the *P*-element insertional mutations generated as part of the *Drosophila* Gene Disruption Project were induced in one of six different co-isogenic Canton S backgrounds (A–F), and were obtained by isogenizing chromosomes in a segregating Canton S stock using balancer chromosomes [20, 24]. The different Canton S genetic backgrounds vary phenotypically for quantitative traits, and thus care must be taken to ensure the correct background control is used to evaluate the effects of *P*-element insertions on quantitative traits and also to perform epistasis analyses only between mutations induced in the same background. However, cases in which the *P*-element has inserted at the same nucleotide position in different genetic backgrounds can be used to infer whether epistatic modifier loci differ in the different backgrounds. For example, Rollmann et al. [38] observed that two *P*-element insertions in the same orientation at the same intergenic location between *Tre1* and *Gr5a* had different effects on longevity, starvation stress resistance, and heat stress resistance. In a screen of 1,332 *P*-element mutations for effects on life span, Magwire et al. [32] observed 25 independent sites at which *P*-elements had inserted in different genetic backgrounds. Eight of these insertion sites were not associated with an effect on life span in any genetic background. All others had effects on life span that differed according to the genetic background, a sample of which is depicted in Fig. 3. Although based on the serendipitous occurrence of mutations at the same site in different genetic backgrounds and the fact that the nature of the interacting loci is unknown, these observations speak to the prevalence of epistasis for *Drosophila* quantitative traits.

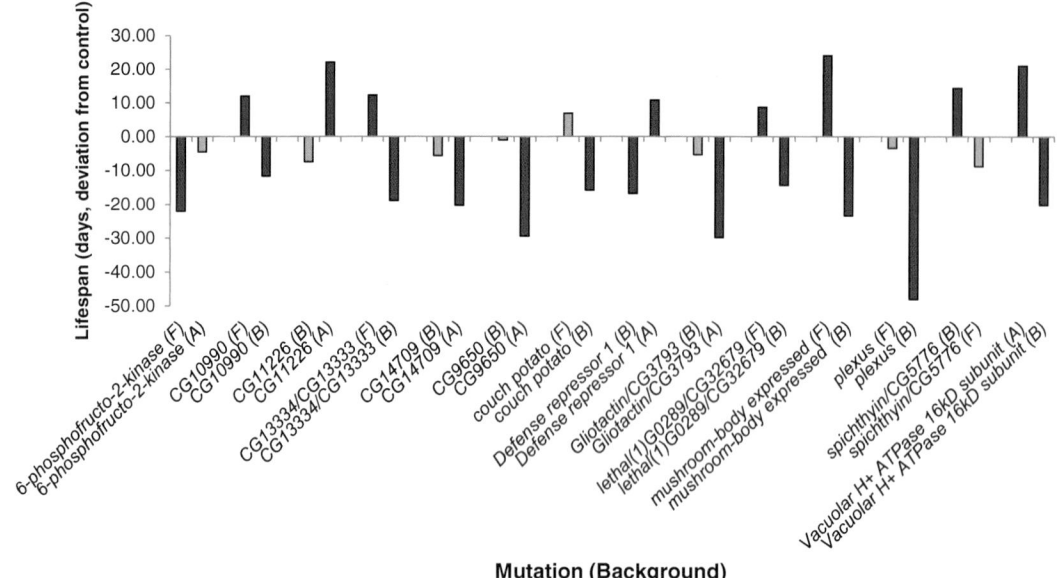

Fig. 3 Mean life span for pairs of homozygous *P*-element mutations inserted in identical locations in one of three different (A, B, F) Canton S homozygous genetic backgrounds, expressed as the deviation from the appropriate co-isogenic control (Magwire et al. [32]). *Red* and *grey bars* denote significant and nonsignificant differences in mean life span from the control line, respectively

2.1.4 A Matter of Scale   A comprehensive evaluation of only the pairwise epistatic interactions between n mutations requires the generation of n^2 genotypes. Given that different insertion sites in the same gene are typically associated with different phenotypes for the same quantitative trait [24, 32, 38], and epistasis can occur between mutations without main effects [37], the number of crosses necessary to comprehensively explore interaction space is astronomical. Further, accurate measurement of quantitative trait phenotypes for a single genotype is labor intensive. Methods that can prioritize mutations for testing epistatic interactions are thus extremely valuable. One such method posits that de novo mutations do not only affect organismal phenotypes, but also have pleiotropic effects on gene expression traits [39]. The genes for which expression is altered in the mutant genetic background are thus candidate genes affecting the same organismal phenotype as the focal mutation, and are also candidate genes participating in a genetic interaction network affecting the phenotype. Several genome-wide expression studies of single *P*-element mutations and co-isogenic control lines have confirmed that the mutations are typically associated with up- or downregulation of hundreds of gene expression traits [32, 37–41]. Six of eight candidate genes implicated in the gene expression analysis of mutations affecting aggressive behavior (75 %) affected aggression [37]. Two studies identified interactions between mutations in genes with altered expression in the mutant background with the focal mutation (both of these studies were for olfactory behavior). Between 37.5 % [40] and 67 % [39] of these tests indeed showed significant epistasis, attesting to the power of using mutational perturbations to reverse engineer networks. However, the large numbers of candidate genes implicated by gene expression profiling still indicate that interaction space is very large—and this is only for pairwise interactions.

2.2 Epistasis Between Naturally Occurring Variants

To what extent does the extensive epistasis implicated by analysis of induced mutations translate to epistatic interactions in natural populations? The ability to construct inbred lines, artificial selection lines, and chromosome substitution lines, as well as map quantitative trait loci (QTLs) affecting complex traits by linkage and association, facilitate analyses of epistasis between naturally occurring variants in *Drosophila*.

2.2.1 Epistasis Inferred from Broad and Narrow Sense Heritability

The response to a single generation of artificial selection for a quantitative trait is given by the breeder's equation, $R = h^2 S$. R is the difference in mean between the parental and offspring generation, h^2 is the realized heritability $[(V_A + \frac{1}{2}V_{AA})/V_P$, where V_A is the additive and V_{AA} is the additive by additive genetic variance (ignoring higher order epistatic interactions for simplicity)], and V_P is the phenotypic variance ($V_A + V_{AA} + V_E$, where V_E is the environmental variance). S, the selection differential, is the difference between the mean of the parental population and the mean of the selected group [7]. The realized heritability is thus $h^2 = R/S$. If selection is conducted over several generations, the realized heritability can be estimated from the regression of the cumulated response on the cumulated selection differential ($h^2 = \Sigma R / \Sigma S$). Realized heritabilities for several behavioral traits determined from long-term replicated divergent selection from a Raleigh, NC derived base population ranged from 7 to 16 % (Table 1). In contrast, broad sense heritabilities determined from variation among completely homozygous inbred lines (ignoring higher order additive by additive epistatic interactions) are $H^2 = (2V_A + 4V_{AA})/V_P$, where V_P, the phenotypic variance, is $2V_A + 4V_{AA} + V_E$ [42]. If all variation is additive ($V_{AA} = 0$), then the broad sense heritability among inbred lines is related to the realized heritability from artificial selection of the outbred populations from which the inbred lines were derived as $H^2 = 2h^2/(1 + h^2)$ [43]. Broad sense heritabilities for the same behavioral traits were estimated from inbred lines derived from the Raleigh, NC population from which the selection lines were derived. In all cases the estimates were high and greater than expected from strictly additive variance (Table 1). Thus, epistatic variance contributes to the genetic architecture of these traits, but analysis at the level of phenotypic variation gives no indication about the individual loci affecting the traits.

Table 1
Evidence for epistasis from realized heritability (h^2) and broad sense heritability (H^2)

Trait	Observed h^2	H^2	Expected $H^2 = 2\,h^2/(1 + h^2)$
Copulation latency	0.07 (ref. 74)	0.25 (ref. 75)	0.13
Startle response	0.16 (ref. 76)	0.58 (ref. 75)	0.28
Aggressive behavior	0.09 (ref. 77)	0.78 (ref. 78)	0.17
Ethanol knock-down time	0.08 (ref. 79)	0.24 (ref. 80)	0.15

2.2.2 Epistasis Inferred from Linkage Mapping

In *Drosophila*, mapping QTLs by linkage to molecular markers is usually done by line cross analysis. Beginning with a pair of inbred strains that differ genetically for the trait of interest, one creates a segregating, mapping population (backcross, F_2, recombinant inbred lines, or advanced intercross, Fig. 4a), and obtains molecular marker genotypes and trait phenotypes for members of the mapping population. If there is a significant difference in phenotype between marker genotypes, the locus affecting the trait is linked to the marker. One then steps along each chromosome performing these tests, constructing a classical QTL plot of position (*x*-axis) versus *P*-value (*y*-axis) for each test. Since there are many tests, and markers in close physical proximity are correlated, correction for multiple tests is usually done by permutation. The most likely location of the QTL is given by the peak of the likelihood profile (Fig. 4b). The precision of QTL mapping depends on the density of markers and the number of recombination events, which in turn depends on sample size. Early mapping experiments in *Drosophila* relied on highly polymorphic insertion sites of transposable elements as molecular markers. These genotypes were determined by destructive cytogenetic analyses of third instar larvae; therefore, inbred lines rather than individuals were used [44, 45]. Early mapping populations were not large (on the order of 100 lines), limiting the ability to precisely localize QTLs, but benefiting from the ability to measure multiple individuals per line to get an accurate measure of genotypic value, which increases the accuracy of estimated effects.

Epistasis between QTLs is estimated in the same way as described above for mutations, although with inbred lines only additive by additive epistasis is examined. The power to detect epistasis between QTLs in mapping populations derived from inbred lines is maximal, because all polymorphic alleles have frequencies of 0.5. However, small mapping populations mean that the number of lines within any one of the four two-locus genotype classes is not very large, increasing the variance in the mean value of the trait within each class, and other segregating

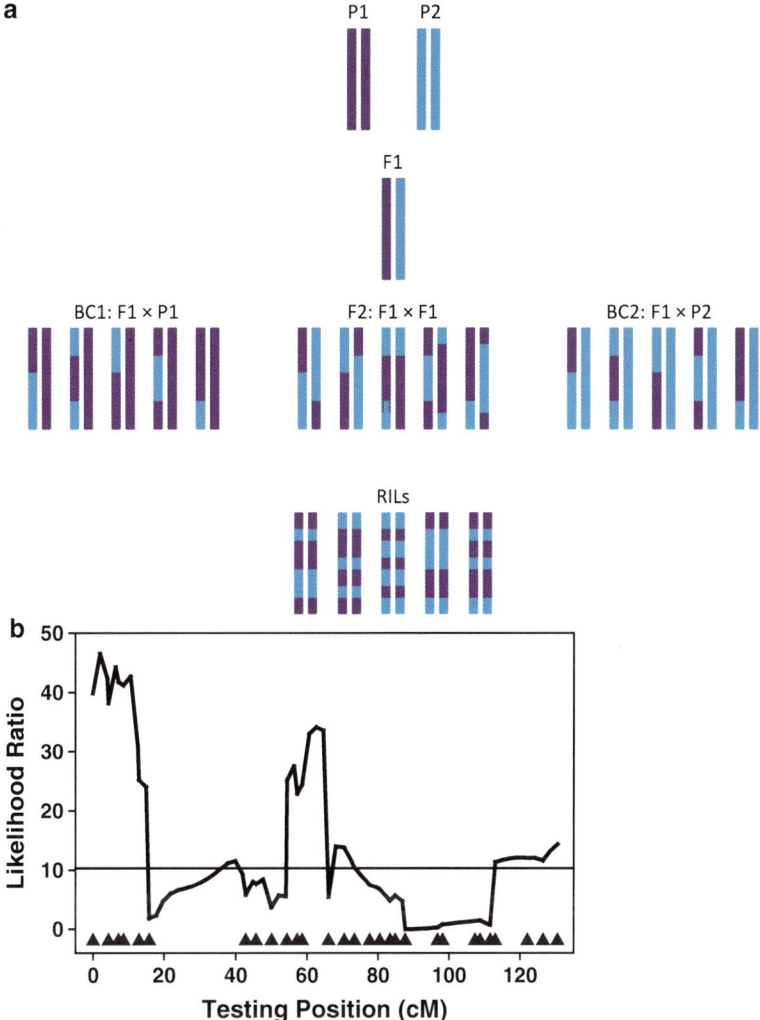

Fig. 4 Linkage mapping of QTLs. (**a**) Line cross analysis begins with two (or more) inbred lines that differ genetically for a quantitative trait (P1, P2). The lines are crossed to produce an F1 generation, and segregating generations are derived by backcrossing the F1 (BC) to one or the other parental line, crossing F1 individuals together to produce an F2 generation, or inbreeding the progeny of BC or F2 individuals for 20 generations to create recombinant inbred lines (RILs). (**b**) Molecular marker genotypes (*triangles*, *x*-axis) and trait phenotypes are obtained for all individuals in the mapping population. Interval mapping analysis evaluates the statistical likelihood that there is a QTL between each set of adjacent markers (*y*-axis), while simultaneously controlling for the segregation of unlinked QTLs, resulting in a QTL likelihood profile. The *horizontal line* is the experiment-wide 5 % significance threshold obtained by permutation. QTLs above this threshold are considered significant, with the most likely location given by the marker position corresponding to the likelihood peaks. (**a**) Reproduced from Anholt RRH, Mackay TFC (2010) *Principles of Behavioral Genetics*. Elsevier, Inc. [43] with permission from Elsevier Inc

QTLs can confound the estimate of epistasis for the tested pair of loci. These factors and the severe multiple testing penalties for pairwise epistasis screens conspire against the ability to detect all but extremely strong interactions. Given these inherent biases against detecting epistasis, most studies report only additive QTL effects. Perhaps surprisingly, when epistasis is evaluated, it is often found. Epistatic interaction effects can be as large as main effects as well as between QTLs that were not individually significant. Epistasis has been reported between QTLs affecting the quintessential additive traits, abdominal [44, 46], and sternopleural [46, 47], bristle number; wing shape [48, 49]; longevity [50, 51]; and enzyme activity, metabolic rate, and flight velocity [52].

One powerful method for detecting epistasis is to construct overlapping segmental introgression lines between a pair of parental lines with different trait phenotypes. Each segmental introgression line contains a fragment of one genome (A) in the background of the other (B). Ideally, the introgressed genomic segments are overlapping, and together the collection of lines comprises the genome of the introgressed line. Depending on the number of segments and the extent of overlap, such lines can be used to map QTLs as well as infer epistasis. First, one performs t-tests for each introgressed genomic fragment of line A to determine whether the mean phenotype of individuals from this line is different from the mean phenotype of the parental line, B. If not, then under an additive model there are no QTLs affecting the trait in the interval encompassed by the introgressed fragment. If so, the fragment does harbor one or more QTLs that affect the trait. If the mean phenotype of the two overlapping introgression lines is different from that of the recurrent parent, but not from each other, then the most parsimonious location of the QTL is in the region of overlap. Thus, a relatively small number of introgression lines can be used to map QTLs with high precision. Epistasis is indicated if the sum of the effects of the introgressed fragments is significantly greater than, or significantly less than, the mean difference in phenotype between the two parental strains. In addition, the location of QTLs participating in an epistatic interaction can be inferred if there are departures from additive expectations from overlapping introgressions, or if one introgression has an effect that greatly exceeds the difference between the parental lines (Fig. 5). This approach was used to map QTLs for aggressive behavior [53]. Of the five QTLs detected, at least three and possibly all five showed evidence of epistasis between QTLs. However, this design is generally not informative with respect to mapping the interacting loci.

2.2.3 Epistasis Inferred from Association Mapping

Association mapping of QTLs differs from linkage mapping in that it relies on correlations in allele frequency between polymorphic markers segregating in a natural population (linkage disequilibrium, LD) to map QTLs. The principle of mapping remains the same. One obtains a sample of individuals from a natural population, and determines the molecular marker genotypes and quantitative trait phenotypes of these individuals. If there is a difference in phenotype associated with the marker genotype, then a QTL affecting the trait is in LD with the marker locus. The advantages of association

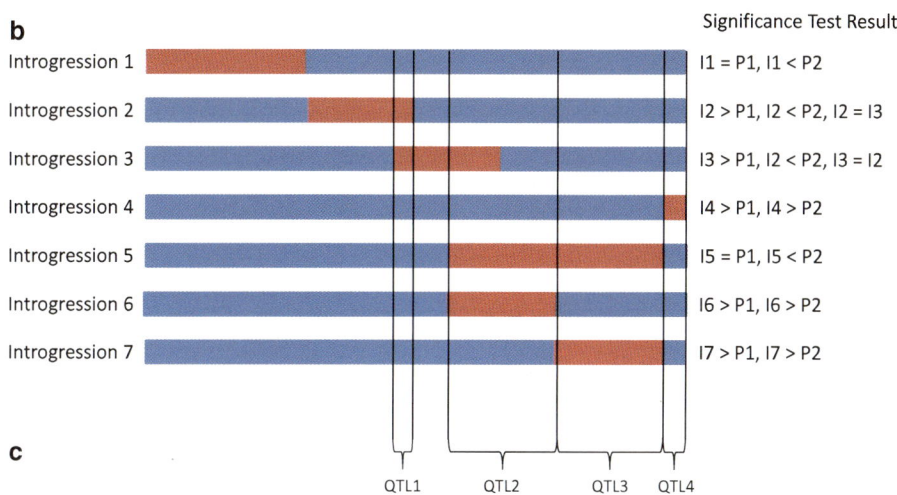

Fig. 5 Introgression mapping of QTLs. (**a**) As for QTL linkage mapping, introgression analyses begin with two inbred lines that differ genetically for a quantitative trait. In this example, the mean phenotype of Parent 2 (*red*) is greater than that of Parent 1 (*blue*). (**b**) *Seven lines* in which genomic fragments of the Parent 2 genome have been introgressed into the Parent 1 genome are shown, as are the results of statistical tests for differences in phenotype of the introgression lines (I1–I7) from the parental lines (P1, P2) or each other. (**c**) I1 has the same phenotype as P1. The most parsimonious interpretation is that there is no QTL in the region of this introgression. I2 and I3 have the same phenotype intermediate between the two parental lines, which is consistent with a QTL affecting the trait in both lines. The most parsimonious interpretation is that the QTL is located in the region of overlap between the two introgressions. The phenotype of the I4 introgression exceeds that of P2. This is not possible with strictly additive inheritance. Thus, at least one epistatically interacting QTL occurs in this small genomic region, the effect of which is suppressed by another locus or loci in the P2 genome. I5 has the same effect as I1, indicating no QTL in this region. However, both I6 and I7, which are contained within I5, have large effects similar to I4. This is not consistent with additive gene action, and we conclude there are at least two epistatically interacting QTLs in this region

mapping over linkage mapping are that more genetic diversity is sampled (as opposed to the two genotypes sampled in a traditional line cross analysis), and the precision of mapping is increased relative to a linkage mapping population of the same size because of the reduction in LD over many generations of historical recombination in the natural population [6, 7]. On the other hand, the power to detect QTLs in an association mapping population is less than a linkage mapping population of the same size for loci at less than intermediate frequencies, and LD can be caused by population structure, which must be accounted for should it exist. Association mapping can be performed for single genes (the candidate gene approach) or genome wide (genome-wide association, GWA).

In an association mapping population, the precision with which a QTL can be localized depends on the scale of LD in that population. LD depends on recombination rates and population history. LD is low when recombination rates are high, and in large equilibrium populations. LD is high when recombination rates are low and the population size is small, as could happen with inbreeding, selection, or a recent population size bottleneck. If LD is low, molecular markers in close physical proximity are statistically independent, enabling precise localization of QTLs in large mapping populations, but requiring high-density genotyping. If LD is high, blocks of adjacent markers will be highly correlated, such that any one marker can serve as a proxy for others in the LD (or haplotype) block, thus reducing genotyping costs and the precision of mapping to the level of a haplotype block. *D. melanogaster* is at the high recombination and large population end of the LD spectrum [54, 55], thus high-density genotyping, ideally using complete DNA sequences, is needed to map QTLs causally associated with quantitative traits in this species.

The small size of individual *Drosophila* and the need to genotype large numbers of molecular markers have restricted association mapping in fruit flies to inbred lines, which have the same advantages as inbred line designs for linkage mapping of QTLs. The *D. melanogaster* Genetic Reference Panel (DGRP) is a population of approximately 200 inbred lines derived from the Raleigh, NC population for which complete DNA sequences have been obtained [55]. The DGRP harbors substantial molecular polymorphisms, with approximately 2.5 million single nucleotide polymorphisms (SNPs) at intermediate frequency that are amenable to association mapping analysis. LD in the DGRP on average declines rapidly with physical distance, and there is no global population structure [55]. The DGRP varies phenotypically for every quantitative trait examined to date. These features make the DGRP a valuable community resource for genome-wide association mapping in *Drosophila*.

With 2.5×10^6 single marker tests for association and 3.1×10^{12} possible tests for pairwise epistatic interactions in the DGRP, only associations with very large effects will exceed a Bonferroni threshold for multiple tests. However, we can take advantage of the short generation interval and the ability to rear large populations of *Drosophila* to go beyond the small size of the DGRP population by creating large outbred populations derived from DGRP lines for testing predictions of multi-locus associations. Huang et al. [56] constructed a large outbred advanced intercross population (Flyland) from 40 DGRP lines. After 70 generations of random mating, 2,000 Flyland individuals were scored for each of three quantitative traits (starvation resistance, chill coma recovery time, and startle response), and two pools of 300 extreme-scoring individuals were deeply sequenced for each trait. This design enabled

GWA mapping of QTLs for each trait by significant differences in allele frequencies between the two pools of sequenced flies. The expectation was that common SNPs with true positive associations in the DGRP would be confirmed, false positives rejected, and rare alleles not amenable to association mapping in the DGRP would be detected in the Flyland population. Large numbers of SNPs were significant at Bonferroni-corrected thresholds for each trait in the Flyland population, as expected from GWA analysis in the DGRP population. Perhaps surprisingly, rare alleles did not disproportionately contribute to the genetic architecture of the three traits assessed. Even more surprisingly, there was no concordance at all between effects of common SNPs in the DGRP and the Flyland populations. The only viable explanation for this result is that epistasis dominates the genetic architecture of these traits, and changes in allele frequency between the DGRP and Flyland populations were responsible for the different estimates of marginal effects of common SNPs. Huang et al. [56] used SNPs with significant main effects in either of the two association analyses as focal SNPs for pairwise GWA analyses for each trait in the DGRP population, and found that 50–60 % of these SNPs participated in at least one epistatic interaction. Despite apparent additive effects at distinct SNPs in the two populations, the epistatic interactions perturbed common, biologically plausible, and highly connected genetic networks [56]. A similar result was obtained from GWA analyses for olfactory behavior in the DGRP and an advanced intercross population derived from two extreme DGRP lines [57]. Although these analyses point to the importance of epistatic gene action in predicting individual trait phenotypes, the problem of inferring which of the statistically predicted interactions are biologically significant remains.

2.2.4 Epistasis at the Adh Locus

One of the first association mapping analyses in *Drosophila* was for *Alcohol dehydrogenase* (*Adh*) enzyme activity and protein level. A total of 13 polymorphisms differentiate typical Fast (F) and Slow (S) electrophoretic alleles and account for a 1.5-fold difference in Adh protein level between these alleles [58]. To determine which of these were functionally associated with this difference, Stam and Laurie [58] divided a minimal *Adh* construct that could recapitulate this difference into three domains, A, B, and C. All eight combinations of F and S alleles were engineered ($A^FB^FC^F$, $A^FB^FC^S$, $A^FB^SC^F$, $A^FB^SC^S$, $A^SB^FC^F$, $A^SB^FC^S$, $A^SB^SC^F$, $A^SB^SC^S$), inserted in a *P*-element transformation vector, and transformed into a common *Adh*-null genetic background to produce an average of 20 independent transformants per construct. Protein levels were measured for all 164 transformant genotypes. Not only did each of the three domains contribute to the difference in Adh protein level between the F and S alleles, but there was a significant epistatic interaction between a variant in the A (intronic) domain (likely a complex insertion/deletion allele) with a variant in the C 3′ UTR domain. This interaction would have been difficult to detect by analyses of all pairwise interactions at *Adh* due to the strong LD across this locus. This classic study illustrates not only intragenic epistasis but the power of the *Drosophila* model to dissect causal interacting factors.

2.3 Epistasis Between Mutations and Naturally Occurring Variants

Analysis of epistasis between induced mutations does not scale well to large numbers of mutations, but has the advantage that the interacting partners are specified. Analysis of epistasis between QTLs in an association mapping context has the advantage that interactions among large numbers of polymorphisms and genes can be evaluated, but the severe multiple testing penalty means that there will be large numbers of false positive associations among the top interactions. One strategy for performing a one-dimensional screen for epistasis is to evaluate the phenotypic effects of a known mutation in different genetic backgrounds and to map epistatic modifiers by linkage or association. Although these designs have not been implemented on a large scale, there are many studies indicating they will be very powerful.

Waddington [12] first commented on the large effects and phenotypic variability of mutations versus the apparent stability of wild-type strains, despite exposure to naturally occurring genetic and environmental perturbations. He coined the term "canalization" to refer to the buffering of natural variation against such perturbations. In modern parlance, genetic canalization refers to epistatic interactions between naturally segregating variants. To the extent that these interactions occur between different genetic loci, one can probe the nature and magnitude of the naturally occurring epistatic modifier loci by asking to what extent they modify the effects of a mutant allele. One of the first experiments to demonstrate this phenomenon was Rendel's [59] introgression of a *scute* (*sc*) mutation into a wild-derived background. *D. melanogaster* has four scutellar macrochaetae (bristles) on the dorsal thorax, and this number is invariant in nature. Mutations at *sc* reduce this number to an average of one or fewer in a sex-specific manner. However, 22 generations of artificial selection for scutellar bristle number in a stock in which *sc* and *sc⁺* alleles segregate raised the number of scutellar bristles in *sc* hemizygous males and homozygous females to 2.8 and 3.4, respectively, with many flies having the wild-type number of four in each population. Further, *sc⁺* males and *sc/sc⁺* females from these selected lines had individuals with five and even six scutellar bristles. These results are consistent with the selection of epistatic modifiers of *sc* that were segregating in the initial population and that suppress the mutant *sc* phenotype. However, the genetic backgrounds used in this experiment were not well defined.

One experimental method for determining whether naturally segregating alleles interact epistatically with a mutant allele is to introgress the mutation for ten generations or more into a sample of wild-derived backgrounds. If there is variation among the wild-derived lines for a normally invariant trait, it can only be attributable to epistasis. One caveat is that each introgression will have 10 cM, on average, of introgressed genome on either side of the mutant allele from the line harboring the mutation after ten generations of introgression. Therefore, one cannot unequivocally exclude interactions with loci in this region as the cause of the variation. Several studies have demonstrated highly significant variation outside the normal wild-type range using this design: variation in haltere size and shape as well as haltere bristle number in the presence of a heterozygous *Ultrabithorax* (Ubx^1) allele [60, 61], variation in the antenna to leg transformation homeotic phenotype of an *Antennapedia* ($Antp^{73b}$) allele [61], variation in eye roughness and size on introgression of a *Sevenless* ($Sev^{S11.1}$) allele [62], and variation in expressivity of the *scalloped* (sd^{E3}) wing morphology phenotype [63]. The epistatic effects ranged from complete suppression to enhancement of the mutant phenotype. An advantage of this design is that the epistatic modifiers can, in principle, be mapped [61, 63].

A variant of this design is to evaluate phenotypes of a mutant allele in F_1 genotypes between lines with the mutation and wild-derived lines. The advantage of this method is that it is much easier to implement than constructing introgression lines. The disadvantage is that any phenotypic variation cannot be unambiguously attributed to allelic complementation (dominance effects) or nonallelic complementation (epistasis). Crosses of a dominant gain of function mutation of the *Epidermal Growth Factor Receptor* (*Ellipse*, $Egfr^{E1}$) to wild-derived lines give a range of eye roughness phenotypes [62, 64], at least partially attributable to variation in *Egfr* itself [64]. Approximately 1–2 % of F_1 progeny from crosses of *Drosophila* with mutant alleles of the major heat shock protein *Hsp90* (*Hsp83*) to outbred strains had a wide variety of morphological abnormalities, suggesting that alleles at *Hsp90* normally suppress alleles affecting multiple phenotypes [65]. These results indicate that populations harbor a hidden reservoir of genetic variation for invariant traits that is only revealed in the "decanalizing" background of the mutation. This variation has been called "potential variance" or "cryptic genetic variation" [66].

Introgression experimental designs need to be slightly modified to assess the effects of naturally segregating epistatic modifiers of mutations that affect traits exhibiting quantitative phenotypic variation in natural populations, as opposed to phenotypically invariant traits. The question to be addressed in this case is whether the additive effect of a mutation affecting a quantitative trait that has been characterized in a particular genetic background varies or not when the mutation is evaluated in a range of different genetic backgrounds, which also display quantitative variation for the trait. This requires that separate introgressions are constructed for the mutant allele as well as the wild-type allele from the strain in which the mutant was induced (Fig. 6). All lines are then evaluated for the quantitative trait phenotype, and a two-way ANOVA of form $Y = \mu + G + L + G \times L + \varepsilon$ is performed, where G denotes the difference between the mutant and wild-type allele at the introgressed locus, L indicates the panel of wild-derived lines, and the $G \times L$ interaction is the epistatic interaction term. This design has been utilized by substituting the co-isogenic wild-type chromosome or chromosomes bearing P-element mutations affecting startle response [67], olfactory behavior, and sleep traits [68] into a panel of DGRP lines. In all cases the $G \times L$ term was significant. The nature of the epistatic effects was that variants segregating in the DGRP lines suppressed the mutant phenotypes. In addition, there was variation in the magnitude of epistasis among the DGRP lines. This design can thus be used in the future to map chromosomally unlinked modifiers of defined mutations in a one-dimensional screen.

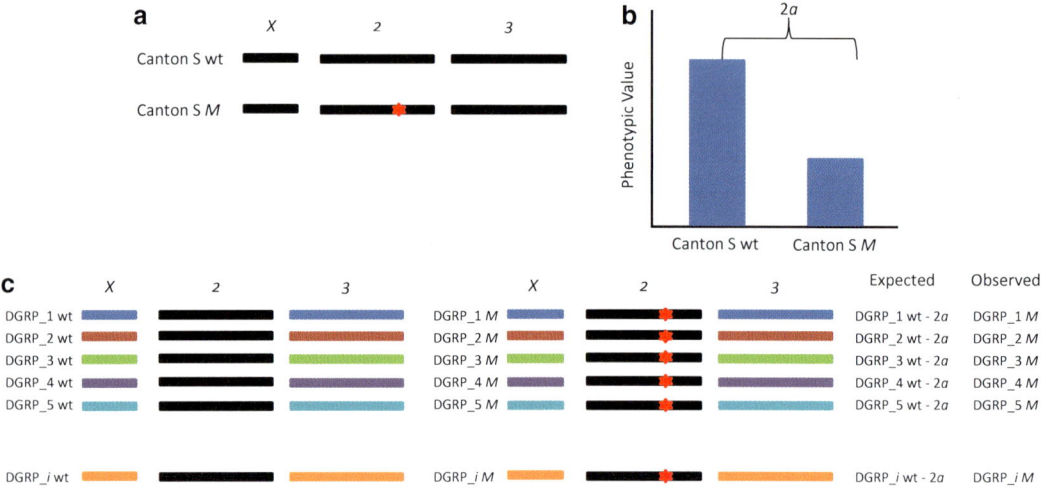

Fig. 6 Effects of naturally occurring variation on induced mutations. (**a**) Schematic diagram of a single *P*-element induced mutation (*red star*) in a homozygous background. The *black bars* represent the three major *Drosophila* chromosomes (*X*, *2*, and *3*). The mutation has occurred on chromosome *2* and has a significant effect on a quantitative trait. (**b**) The difference between the homozygous wild type and homozygous mutant alleles is 2a in the Canton S background in which the mutation was induced. (**c**) Co-isogenic chromosomes containing either the wild type or mutant allele are substituted into different wild-derived inbred lines (DGRP_*i*), and the phenotypes of all lines are assessed. The additive expectation is that the effect of the mutation will be the same in each DGRP line background. If that is not the case, the difference between the observed and expected is due to epistasis. (**c**) Reproduced from [81] with permission from the author

**2.4   Implications
of Pervasive Epistasis**

The studies reviewed here indicate that there is extensive epistatic gene action for *Drosophila* quantitative traits, and by extension, quantitative traits in other organisms, including humans. The epistatic interactions define novel, highly interconnected genetic networks that are enriched for biologically plausible gene ontology categories and metabolic and cellular pathways. Analysis of epistasis reveals that much quantitative genetic variation is hidden and not apparent from analyses of the main effects of causal variants, and that additivity is an emergent property of underlying epistatic networks. Observations of cryptic genetic variation for invariant phenotypes in natural populations that can only be observed in the presence of a decanalizing mutation, as well as naturally segregating variation that largely suppresses the effects of induced mutations for quantitative traits, suggest that natural populations have evolved suppressing epistatic interactions as homeostatic, or canalizing mechanisms.

This realization is paradigm-shifting. Rather than perceiving phenotypic variation for quantitative traits in natural populations as highly variable, it may be more accurate to wonder why there is so little variation in organismal phenotypes, given the large amount of segregating molecular genetic polymorphism. All GWA analyses of *Drosophila* quantitative traits to date [55, 57, 69, 70] have observed an inverse relationship between minor allele frequency and additive effect, such that more rare alleles are associated with larger effects than common alleles. Statistically, rare alleles must have larger effects than common alleles to be detected in a mapping population of the same size. The puzzle is why so few common alleles of large effect are found segregating within a population. One possible explanation is that rare alleles have large effects because they are relatively new mutations, and epistatic modifiers that ameliorate their effects have not yet occurred in the population. Common alleles are presumably older, and could achieve an intermediate frequency due to a modifier mutation at another locus that suppresses the effect of the polymorphism. On the other hand, QTLs affecting *Drosophila* quantitative traits detected by linkage mapping in populations derived from crosses of inbred lines typically have moderately large effects [71]. It is possible that these loci were not common in the populations from which the parental inbred lines were derived. However, it is also possible that the lines that survived inbreeding were enriched for compatible epistatic interactions that were decanalized by crossing to a different genetic background, as exemplified by mapping QTLs affecting aggressive behavior [53].

The consequence of epistasis is that the main effects of each of the interacting loci depend on allele frequency. Thus, even if the true genetic architecture underlying a quantitative trait is the same between two populations with different allele frequencies, estimates of additive effects of causal alleles will be population-specific.

Thus, efforts to validate predicted associations derived from one population in another population may be unsuccessful [56], and genomic prediction based on cross-validation may be poor if the frequency of causal alleles varies between the training and test populations [72]. With epistasis, the genetic architecture of the response to natural selection will be different in different populations, potentially increasing the likelihood of evolution of Dobzhansky–Muller incompatibilities and consequent speciation events. Similarly, in the presence of epistasis the genetic architecture of the response to artificial selection from the same base population could differ among replicate lines as well as within the same line over time, due to allele frequency drift and changes in frequency of causal alleles due to selection.

GWA analyses in human populations typically find that many loci with small effects affect each trait, but together they only account for a small fraction of the total heritability [73]. With suppressing epistasis, additive effects of common interacting loci will be small. In humans, estimates of true narrow sense heritability (V_A/V_P) are obtained from twice the difference in the correlation of monozygotic and dizygotic twins [7], and are biased upward by $(1.5\,V_D + 1.5\,V_{AA} + 1.75\,V_{AD} + 1.875\,V_{DD})/V_P$ in the presence of dominance and epistasis. Thus, suppressing epistasis could potentially account for high heritability estimates coupled with little additive genetic variation from mapped loci in human populations. If true, then individual risk prediction will require knowledge of interacting loci.

3   Conclusions and Future Prospects

Mapping epistatic interactions is statistically and experimentally challenging. However, much progress in understanding and predicting genetic interaction networks affecting *Drosophila* quantitative traits can be made in the future, by taking advantage of the unique biology and resources available for this genetically tractable model organism. The large libraries of mutations that have been induced in a common background, the DGRP resource of naturally occurring genetic variation, and the ability to readily construct advanced intercross outbred populations specific for any quantitative trait as well as chromosome substitution and introgression lines can be used individually and in combination to not only predict but also test the mechanistic basis of genetic interactions.

4   Notes

- A major challenge of contemporary biology is to understand how naturally occurring genetic variation causes phenotypic variation in quantitative traits. Despite the biological plausibility that genetic variation affects nonlinear networks at multiple levels of biological organization, most efforts to explain the relationship between genetic and phenotypic variation concentrate on additive effects of single loci and ignore epistatic interactions. Empirical studies in *D. melanogaster* show that epistasis is a pervasive feature of the genetic architecture of quantitative traits.

• Epistasis for a quantitative trait refers to a statistical interaction between genotypes at two or more loci. The additive or dominance effects of one locus depend on the genotype at the interacting locus, and thus will vary depending on the allele frequency of the interacting locus.

• In *D. melanogaster*, epistatic interactions for many quantitative traits have been demonstrated between *P*-element mutations generated in a common isogenic background, either by constructing all nine possible two locus genotypes for each pair of *P*-element mutations tested, by performing diallel crosses, or by generating the same mutations in different genetic backgrounds. The advantage of this approach is that the interacting partners are known and can be used to derive genetic interaction networks, but the disadvantage is that such studies are not feasible for large numbers of mutations.

• Because outbred populations of *D. melanogaster* can be subjected to artificial selection as well as inbred to homozygosity, epistasis between naturally occurring variants can be inferred when the total genetic variance among inbred lines is greater than twice the additive variance estimated from selection response. Epistasis between QTLs has also been detected by pairwise interaction screens in linkage or association mapping populations and by analysis of segmental introgression lines. However, such studies do not have the resolution to identify interacting genes and are underpowered for realistic sizes of the mapping populations.

• Analysis of epistasis between induced mutations does not scale well to large numbers of mutations, but has the advantage that the interacting partners are specified. Analysis of epistasis between QTLs in an association mapping context has the advantage that interactions among large numbers of polymorphisms and genes can be evaluated, but the severe multiple testing penalty means that there will be large numbers of false positive associations among the top interactions. A powerful strategy for performing a one-dimensional screen for epistasis is to evaluate the phenotypic effects of a known mutation in different genetic backgrounds and to map epistatic modifiers by linkage or association. Such studies in *D. melanogaster* have revealed an impressive reservoir of cryptic genetic variation only observable in the presence of a decanalizing mutation, as well as natural variation that suppresses (canalizes) the effects of induced mutations.

• With pervasive epistasis, we expect that estimates of additive effects of causal alleles will vary between populations with different allele frequencies of interacting loci. This will lead to lack of replication of additive genetic architecture between populations. With epistasis, the genetic architecture of the response to natural selection will be different in different populations, potentially increasing the likelihood of evolution of Dobzhansky–Muller incompatibilities and consequent speciation events. Similarly, the genetic architecture of the response to artificial selection from the same base population could differ among replicate lines as well as within the same line over time in the presence of epistasis, due to allele frequency drift and changes in frequency of causal alleles due to selection. These phenomena are unlikely to be restricted to *Drosophila*. Pervasive epistasis has consequences for plant and animal breeding, evolutionary biology, and human genetics.

References

1. Fisher RA (1930) The genetical theory of natural selection. Clarendon, Oxford
2. Wright S (1931) Evolution in Mendelian populations. Genetics 16:97–159
3. Carlborg O, Haley CS (2004) Epistasis: too often neglected in complex trait studies? Nat Rev Genet 5:618–625
4. Hill WG, Goddard ME, Visscher PM (2008) Data and theory point to mainly additive genetic variance for complex traits. PLoS Genet 4:e1000008
5. Kauffman SA (1993) The origins of order. Oxford University Press, New York
6. Lynch M, Walsh JB (1998) Genetics and analysis of quantitative traits. Sinauer Associates, Sunderland, MA
7. Falconer DS, Mackay TFC (1996) Introduction to quantitative genetics. Harlow Longman, Essex
8. Crow JF, Kimura M (1970) An introduction to population genetics theory. Harper and Row, New York
9. Meuwissen TH, Hayes BJ, Goddard ME (2001) Prediction of total genetic value using genome-wide dense marker maps. Genetics 157:1819–1829
10. Zhou X, Carbonetto P, Stephens M (2013) Polygenic modeling with Bayesian sparse linear mixed models. PLoS Genet 9:e1003264
11. Phillips PC (2008) Epistasis - the essential role of gene interactions in the structure and evolution of genetic systems. Nat Rev Genet 9: 855–867

12. Waddington CH (1942) Canalization of development and the inheritance of acquired characters. Nature 150:563–565
13. Waddington CH (1957) The strategy of genes. George Allen and Unwin, London
14. Dobzhansky T (1937) Genetics and the origin of species. Columbia University Press, New York
15. Muller HJ (1940) Bearing of the Drosophila work on systematics. In: Huxley JS (ed) The new systematics. Clarendon, Oxford
16. Cockerham CC (1954) An extension of the concept of partitioning hereditary variance for analysis of covariances among relatives when epistasis is present. Genetics 39:859–881
17. Kempthorne O (1954) The correlation between relatives in a random mating population. Proc R Soc Lond B 143:168–174
18. Mather K, Jinks JL (1977) Introduction to biometrical genetics. Chapman and Hall, London
19. Cheverud JM, Routman EJ (1995) Epistasis and its contribution to genetic variance components. Genetics 139:1455–1461
20. Bellen HJ, Levis RW, Liao G et al (2004) The BDGP gene disruption project: single transposon insertions associated with 40 % of Drosophila genes. Genetics 167:761–781
21. Mackay TFC, Lyman RF, Jackson MS (1992) Effects of *P* element insertions on quantitative traits in *Drosophila melanogaster*. Genetics 130: 315–332

22. Clark AG, Wang L, Hulleberg T (1995) P-Element-induced variation in metabolic regulation in Drosophila. Genetics 139:337–348

23. Lyman RF, Lawrence F, Nuzhdin SV et al (1996) Effects of single *P* element insertions on bristle number and viability in *Drosophila melanogaster*. Genetics 143:277–292

24. Norga KK, Gurganus MC, Dilda CL et al (2003) Quantitative analysis of bristle number in Drosophila mutants identifies genes involved in neural development. Curr Biol 13:1388–1397

25. Anholt RRH, Lyman RF, Mackay TFC (1996) Effects of single *P* element insertions on olfactory behavior in *Drosophila melanogaster*. Genetics 143:293–301

26. Sambandan D, Yamamoto A, Fanara JJ et al (2006) Dynamic genetic interactions determine odor-guided behavior in *Drosophila melanogaster*. Genetics 174:1349–1363

27. van Swinderen B, Greenspan RJ (2005) Flexibility in a gene network affecting a simple behavior in *Drosophila melanogaster*. Genetics 169:2151–2163

28. Yamamoto A, Zwarts L, Callaerts P et al (2008) Neurogenetic networks for startle-induced locomotion in *Drosophila melanogaster*. Proc Natl Acad Sci U S A 105:12393–12398

29. Edwards AC, Zwarts L, Yamamoto A et al (2009) Mutations in many genes subtly affect aggressive behavior in *Drosophila melanogaster*. BMC Biol 7:29

30. Morozova TV, Mackay TFC, Anholt RRH (2011) Transcriptional networks for alcohol sensitivity in *Drosophila melanogaster*. Genetics 187:1193–1205

31. Harbison ST, Yamamoto AH, Fanara JJ et al (2004) Quantitative trait loci affecting starvation resistance in *Drosophila melanogaster*. Genetics 166:1807–1823

32. Magwire MM, Yamamoto A, Carbone MA et al (2010) Quantitative and molecular genetic analyses of mutations increasing *Drosophila* life span. PLoS Genet 6:e1001037

33. Clark AG, Wang L (1997) Epistasis in measured genotypes: Drosophila *P*-element insertions. Genetics 147:157–163

34. Sprague GF, Tatum LA (1942) General *vs.* specific combining ability in single crosses of corn. J Am Soc Agron 34:923–932

35. Griffing B (1956) Concept of general and specific combining ability in relation to diallel crossing systems. Aust J Biol Sci 9:463–493

36. Fedorowicz GM, Fry JD, Anholt RRH et al (1998) Epistatic interactions between *smell-impaired* loci in *Drosophila melanogaster*. Genetics 148:1885–1891

37. Zwarts L, Magwire MM, Carbone MA et al (2011) Complex genetic architecture of *Drosophila* aggressive behavior. Proc Natl Acad Sci U S A 108:17070–17075

38. Rollmann SM, Magwire MM, Morgan TJ et al (2006) Pleiotropic fitness effects of the *Tre1/Gr5a* region in *Drosophila*. Nat Genet 38:824–829

39. Anholt RRH, Dilda CL, Chang S et al (2003) The genetic architecture of odor-guided behavior in Drosophila: epistasis and the transcriptome. Nat Genet 35:180–184

40. Rollmann SM, Yamamoto A, Goossens T et al (2007) The early neurodevelopmental gene *Semaphorin 5c* is essential for olfactory behavior in adult *Drosophila*. Genetics 176:947–956

41. Rollmann SM, Edwards AC, Yamamoto A et al (2008) Pleiotropic effects of *Drosophila neuralized* on complex behaviors and brain structure. Genetics 179:1327–1336

42. Bulmer MG (1985) The mathematical theory of quantitative genetics. Clarendon Press, Oxford

43. Anholt RRH, Mackay TFC (2010) Principles of behavioral genetics. Elsevier, Burlington, MA

44. Long AD, Mullaney SL, Reid LA et al (1995) High resolution mapping of genetic factors affecting abdominal bristle number in *Drosophila melanogaster*. Genetics 139:1273–1291

45. Nuzhdin SV, Pasyukova EG, Dilda C et al (1997) Sex-specific quantitative trait loci affecting longevity in *Drosophila melanogaster*. Proc Natl Acad Sci U S A 94:9734–9739

46. Dilda CL, Mackay TFC (2002) The genetic architecture of *Drosophila* sensory bristle number. Genetics 162:1655–1674

47. Gurganus MC, Nuzhdin SV, Leips JW et al (1999) High resolution mapping of quantitative trait loci affecting sternopleural bristle number in *Drosophila melanogaster*. Genetics 152:1585–1604

48. Weber K, Eisman R, Morey L et al (1999) An analysis of polygenes affecting wing shape on chromosome *3* in *Drosophila melanogaster*. Genetics 153:773–786

49. Weber K, Eisman R, Higgins S et al (2001) An analysis of polygenes affecting wing shape on chromosome *2* in *Drosophila melanogaster*. Genetics 159:1045–1057

50. Leips J, Mackay TFC (2000) Quantitative trait loci for lifespan in *Drosophila melanogaster*: interactions with genetic background and larval density. Genetics 155:1773–1788

51. Mackay TFC, Roshina NV, Leips JW et al (2006) Complex genetic architecture of Drosophila longevity. In: Masaro EJ, Austad SN (eds) Handbook of the biology of aging, 6th edn. Elsevier, Burlington, MA

52. Montooth KL, Marden JH, Clark AG (2003) Mapping determinants of variation in energy metabolism, respiration and flight in Drosophila. Genetics 165:623–635

53. Edwards AC, Mackay TFC (2009) Quantitative trait loci for aggressive behavior in *Drosophila melanogaster*. Genetics 182:889–897

54. Long AD, Lyman RF, Langley CH et al (1998) Two sites in the *Delta* gene region contribute to naturally occurring variation in bristle number in *Drosophila melanogaster*. Genetics 149:999–1017

55. Mackay TFC, Richards S, Stone EA et al (2012) The *Drosophila melanogaster* genetic reference panel. Nature 482:173–178

56. Huang W, Richards S, Carbone MA et al (2012) Epistasis dominates the genetic architecture of *Drosophila* quantitative traits. Proc Natl Acad Sci U S A 109:15553–15559

57. Swarup S, Huang W, Mackay TFC et al (2013) Analysis of natural variation reveals neurogenetic networks for *Drosophila* olfactory behavior. Proc Natl Acad Sci U S A 110:1017–1022

58. Stam LF, Laurie CC (1996) Molecular dissection of a major gene effect on a quantitative trait: the level of alcohol dehydrogenase expression in *Drosophila melanogaster*. Genetics 144:1559–1564

59. Rendel JM (1959) Canalization of the *scute* phenotype of Drosophila. Evolution 13:425–439

60. Gibson G, van Helden S (1997) Is function of the Drosophila homeotic gene *Ultrabithorax* canalized? Genetics 147:1155–1168

61. Gibson G, Wemple M, van Helden S (1999) Potential variance affecting homeotic *Ultrabithorax* and *Antennapedia* phenotypes in *Drosophila melanogaster*. Genetics 151:1081–1091

62. Polaczyk PJ, Gasperini R, Gibson G (1998) Naturally occurring genetic variation affects Drosophila photoreceptor determination. Dev Genes Evol 207:462–470

63. Dworkin I, Kennerly E, Tack D et al (2009) Genomic consequences of background effects on *scalloped* mutant expressivity in the wing of *Drosophila melanogaster*. Genetics 181:1065–1076

64. Dworkin I, Palsson A, Birdsall K et al (2003) Evidence that *Egfr* contributes to cryptic genetic variation for photoreceptor determination in natural populations of *Drosophila melanogaster*. Curr Biol 13:1888–1893

65. Rutherford SL, Lindquist S (1998) Hsp90 as a capacitor for morphological evolution. Nature 396:336–342

66. Gibson G, Dworkin I (2004) Uncovering cryptic genetic variation. Nat Rev Genet 5:681–690

67. Yamamoto A, Anholt RRH, Mackay TFC (2009) Epistatic interactions attenuate mutations affecting startle behaviour in *Drosophila melanogaster*. Genet Res 91:373–382

68. Swarup S, Harbison ST, Hahn LE et al (2012) Extensive epistasis for olfactory behavior, sleep and waking activity in *Drosophila melanogaster*. Genet Res 94:9–20

69. Weber AL, Khan GF, Magwire MM et al (2012) Genome-wide association for oxidative stress resistance in *Drosophila melanogaster*. PLoS One 7:e34745

70. Jordan KW, Craver KL, Magwire MM et al (2012) Genome wide association for sensitivity to chronic oxidative stress in *Drosophila melanogaster*. PLoS One 7:e38722

71. Mackay TFC (2001) The genetic architecture of quantitative traits. Annu Rev Genet 35:303–339

72. Ober U, Ayroles JF, Stone EA et al (2012) Using whole genome sequence data to predict phenotypic traits in *Drosophila melanogaster*. PLoS Genet 8:e1002685

73. Manolio TA, Collins FS, Cox NJ et al (2009) Finding the missing heritability of complex diseases. Nature 461:747–753

74. Mackay TFC, Heinsohn SL, Lyman RF et al (2005) Genetics and genomics of *Drosophila* mating behavior. Proc Natl Acad Sci U S A 102:6622–6629

75. Ayroles JF, Carbone MA, Stone EA et al (2009) Systems genetics of complex traits in *Drosophila melanogaster*. Nat Genet 41:299–307

76. Jordan KW, Carbone MA, Yamamoto A et al (2007) Quantitative genomics of locomotor behavior in *Drosophila melanogaster*. Genome Biol 8:R172

77. Edwards AC, Rollmann SM, Morgan TJ et al (2006) Quantitative genomics of aggressive behavior in *Drosophila melanogaster*. PLoS Genet 2:e154

78. Edwards AC, Ayroles JF, Stone EA et al (2009) A transcriptional network associated with natural variation in *Drosophila* aggressive behavior. Genome Biol 10:R76

79. Morozova TV, Anholt RRH, Mackay TFC (2007) Phenotypic and transcriptional response to selection for alcohol sensitivity in *Drosophila melanogaster*. Genome Biol 8:R231

80. Morozova TV, Ayroles JF, Jordan KW et al (2009) Alcohol sensitivity in Drosophila: translational potential of systems genetics. Genetics 83:733–745

81. Mackay TFC (2014) Epistasis and quantitative traits: using model organisms to study gene-gene interactions. Nat Rev Genet 15:22–33

Chapter 5

Epistasis in the Risk of Human Neuropsychiatric Disease

Scott M. Williams

Abstract

Neuropsychiatric disease represents the ideal class of disease to assess the role of epistasis, as more genes are expressed in the brain than in any other tissue. In this chapter, two well-studied neuropsychiatric diseases are examined, Alzheimer's disease (AD) and schizophrenia, which have been shown to have multiple and, often, replicated interactions that associate with clinical endpoints or related phenotypes. In each case, a single gene is represented in a plurality of epistatic interactions, apolipoprotein E (APOE) for AD and catechol-O-methyltransferase for schizophrenia. Interestingly, of the two, only *APOE* has clear-cut and consistent evidence for a marginal association. Unraveling the underlying reasons is important in understanding both genetic etiology and architecture as well as how to use genetics to provide better personalized treatments.

Key words Epistasis and Alzheimer's disease, Epistasis and brain, Schizophrenia, Epistasis and pleiotropy

1 Introduction

Despite the identification of many genetic variants that affect the risk of complex human disease in both agnostic and hypothesis-driven studies, there remains a large gap between the total impact of the specific genes identified and the estimated overall genetic effects that underlie each of the examined phenotypes. This has been phrased in terms of missing heritability [1], and several ideas have been proposed to explain why we have under-detected genetic effects with current experimental designs [2]. These explanations include rare variants with large effects not assessed in genome-wide association study (GWAS) designs that focused only on common variants, structural and/or copy number variants, gene by environment interactions, and epigenetics. A common thread in many of the explanations revolves around our limited understanding of the genetic architectures of most diseases studied. One genetic architecture, which is both theoretically and experimentally validated in many systems but has not played, as of yet, a significant role in

Jason H. Moore and Scott M. Williams (eds.), *Epistasis: Methods and Protocols*, Methods in Molecular Biology, vol. 1253, DOI 10.1007/978-1-4939-2155-3_5, © Springer Science+Business Media New York 2015

studies of the genetics of human disease, is the likelihood that effects of single gene variants are dependent on the presence of variants in other genes, or epistasis; the focus of this book.

Epistasis, or nonlinear genetic context dependency, stands in stark contrast to most human geneticists' assessment of the average effect of single alleles across all backgrounds in which that allele occurs. The assumption that the average effect is relevant to the underlying genetic model of disease is simply that, an assumption of convenience that may not be true in many cases. In fact, evolutionary theory predicts that for many situations evolution will favor complex genetic models that involve epistasis to maintain robust or canalized phenotypes [3–5]. Such robustness of phenotype to perturbations is another way of viewing the "healthy" phenotype; that is, evolution has favored genetic architectures that are robust such that disease does not frequently occur. Hence, if health is a selected phenotype, then it is reasonable to expect that epistasis is a major and not a minor component of the genetics of health and hence risk of disease. However, the major assumption of most human genetic association studies to date has been the opposite, namely that effects of single genetic variants are essentially independent of other variants in the genome. This book and this chapter explicitly challenge this assumption and ask what the evidence for nonindependence is?

One reason that research has often ignored the role of epistasis is not that it is denied by most geneticists, but rather with large-scale studies the number of additional tests needed to assess epistasis produces a multiple testing burden that cannot reasonably be overcome with typical sample sizes (*see* **Note 1**). This would be true under certain assumptions, but many of the key assumptions have not been questioned or tested appropriately. For example, power, as typically estimated, is a function of the average effect size, and average effects have been consistently small in GWAS studies. Therefore, with the numerous additional tests required in models of epistasis many investigators assume that the presumed smaller effects due to interactions are undetectable. This implicitly makes a standard statistical assumption—i.e., that interaction effects are, in general, smaller than marginal effects. A typical statistical approach would argue this, but a genetics approach might argue otherwise as the models of genetic disease are often not based on simple mathematical relationships but ones that have evolved to be inherently complex and robust. Hence, it may be essential to first question this assumption. Any nonlinearity in the true genetic model can in fact minimize marginal effects, and the marginal effects may also be extremely sensitive to changes in allele frequencies at interacting loci [6].

It also may be necessary to assess whether agnosticism is the most valid genetic approach to studying human disease. Or viewed alternatively, have we prematurely discarded the concept of candidate studies, especially in light of the role of epistasis in disease phenotypes? Even if multiple testing for epistasis produces an undue multiple testing burden, is it possible to address this using prior biological knowledge? Knowing physiology and biochemistry certainly can help direct analyses such that tests for association are performed on more reasonable scales.

The goal of this chapter is not to argue that epistasis is the only factor that drives missing heritability. For example, multiple rare variants of larger effect will play roles in some common diseases even using derivatives of current marginal analysis methodology, but that if one carefully considers prior biological knowledge and how the evolutionary process shapes genomes epistasis must play an important role in shaping complex phenotypes, such as health and disease. Clearly, using model organisms to direct our thinking about human disease, it is imperative to seriously consider epistasis, as it is such a well-documented biological phenomenon [7] that is too often dismissed prematurely without appropriate investigation.

2 Empirical Evidence for Epistasis in Human Disease: The Literature

Clearly, epistasis in human disease is, to date, understudied. A search in PubMed for the terms "Epistasis" AND "human disease" yielded just under 1,100 entries (as of April 2014), but many of these do not relate directly to results in humans but model systems and/or methodological studies to improve the detection of epistasis. However, a substantial number do. In this chapter, I will focus on one specific class of disease—neuropsychiatric disease. I do this for two reasons. (1) The vast majority of genes are expressed in the human brain in at least one compartment and/or during one developmental time point [8]; hence the opportunity for epistasis is substantial (*see* **Note 2**). (2) A disproportionate number of the citations in the PubMed search relate to this disease class, providing ample opportunity to evaluate the degree of epistasis detected in human disease. Because of the overrepresentation of epistasis in diseases such as Alzheimer's, it affords the kind of validation generally taken as the gold standard for genetic epidemiological studies, replication [9].

3 The Role of Epistasis in Neuropsychiatric Disease Risk

Several neuropsychiatric phenotypes have been carefully studied for epistasis over the past few years. Among them, Alzheimer's disease (AD), schizophrenia, and their related endophenotypes have been particularity well studied. I will focus on AD and schizophrenia in this chapter, recognizing that this does not represent an exhaustive catalogue of the effects that have been found. In addition, much of the work has included not only clinical diagnoses as analytical endpoints for these phenotypes, but imaging data or biochemical data as well as animal models to support relationships to clinical phenotypes and epistasis, especially for schizophrenia. Finally, several of the genes shown to play roles in these diseases are pleiotropic in that variants in them can affect more than one disease, often through interactions with different sets of interacting genes. It has been argued that this kind of pattern—pleiotropy—is driven by the same genes interacting with other players to affect multiple phenotypes [10].

3.1 Alzheimer's Disease

AD has been the focus of genetic research for more than two decades with late onset AD (LOAD) having been shown to strongly associate with apolipoprotein E (*APOE*) variants. Variant *APOE4* has been shown to have a consistent and large effect on LOAD, increasing risk from approximately threefold in heterozygotes with the major allele, *APOE3*, to as much as tenfold in *APOE4/4* homozygotes. Examining the relationship of *APOE* synergistically with other genes may reveal how a complex underlying genetic architecture can give rise to a strong and significant marginal effect.

Many LOAD epistasis studies have been done, and, interestingly, many of the significant epistatic interactions involve *APOE* [11] (Table 1) (*see* **Note 3**). Although this may not be terribly surprising given the strong marginal effect of *APOE*, it is key that the many interactions cross a variety of underlying physiological processes. For example, *APOE* interactions include genes affecting cholesterol metabolism and beta amyloid (Aβ) metabolism, both of which are tied to key diagnostic criteria as well as inflammation and oxidative stress (Table 1). It may be interpreted that the strength of the *APOE* marginal effect is in fact a product of the number of genes and pathways with which it interacts to affect LOAD risk. Therefore, although one can detect a clear and reproducible *APOE* marginal effect, *APOE4* is neither necessary nor sufficient to predict AD. This may be because of the many interactions it has such that any effect of *APOE* is always modulated by the presence of specific alleles in other genes. If true, personalization of treatment may be best served by assessing the role of interaction as opposed to this single gene's marginal effect [12].

Similar to *APOE*, there are other genes that have been shown to associate with LOAD, but the findings are not as compelling with respect to marginal effects. Among those best studied are those that are pro-inflammatory, e.g., *IL6*, *TNF*, and *IL1*, as well as those that are anti-inflammatory, e.g., *IL10* and *TGFB1* [13–16] (Table 1). Interestingly, some of these have been shown to interact antagonistically with *APOE*. For example, the well-studied *IL6* - 174G promoter variant has not been shown to consistently associate with LOAD, but it has been shown to decrease risk of *APOE4* carriers [17, 18]. Other cytokines, including monocyte chemoattractant protein 1 (*CCL2*), intercellular adhesion molecule 1 (*ICAM1*), E-selectin (*SELE*), macrophage inflammatory

Table 1
Late onset alzheimer's disease interacting loci

Primary gene	Pathway	Secondary gene	Reference
APOE	Cholesterol metabolism		
		Apolipoprotein a (LPA)	[67]
		ABCA1	[68]
		CETP	[69, 70]
		LRP	[71]
		CYP46	[72]
		OLR1	[73]
		SREBP1a	[74]
	Aβ metabolism		
		ACT	[75, 76]
		BACE1	[77–79]
		BLMH	[80, 81]
		IDE	[82]
		PS2	[83]
		PEN2	[84, 85]
		PRNP	[86]
	Inflammation		
		IL6	[17, 18]
		TNFα	[16]
		TGFB1	[21]
	Oxidative stress		
		MPO	[87]
	Other		
		CTNNA3	[88]
		BCHE	[89–91]
		DNMBP	[92]
		VEGF	[93]
IL6	Inflammation		
		IL10	[13, 14, 94][a]
		MTHFR	[95]
		DBH	[96, 15]

(continued)

Table 1
(continued)

Primary gene	Pathway	Secondary gene	Reference
IL1a/IL1b		DBH	[96, 15]
		PPARa	[97]
		PARP-1	[98, 99]
IL10		CYP19A1	[100]
		PPARa	[97]
	Other		
HFE		TF	[101]
GSTM3		HHEX/IDE/KIF11 cluster	[102]
INS		PPARa	[103]

[a]Ex vivo expression assay

protein 1alpha (*CCL3*), matrix metalloproteinase 3 (*MMP3*) - 1171 5A/6A, and matrix metalloproteinase 9 (*MMP9*), have also been shown to associate with AD both marginally and synergistically with evidence for interaction with *APOE* [19]. The association results for transforming growth factor-beta 1 (*TGFB1*), assessed marginally, have also been less than compelling [20], despite interest in it as a means to direct treatment [20]. However, despite the relatively weak *TGFB1* results when its variants are studied singly, there is evidence that *TGFB1* interacts with *APOE* to modify risk of AD in the absence of marginal effects [21].

It may be argued that given the complex nature of AD, as indicated in the admittedly incomplete summary of AD genetic risk above, it is unlikely that many more genes with strong marginal effects will be identified. In contrast, it is possible that genes that are part of multiple interacting networks may function together to affect AD risk, but marginal effects may be unconvincing. If this is true, as the data so far indicate, then in contrast to designing treatment and preventative measures based on a single gene's associations it would be a better strategy to determine the overall risk model for a given individual, explicitly incorporating not only the major risk alleles, in this case *APOE4*, but the interacting genes that likely provide an increase in the etiological explanatory power. Therefore, even when a strong marginal effect is found, it is probably not descriptive enough of the etiology to help substantially in the design of precise treatments. Such a case is probable for *APOE* as a risk factor as it alone can do little to explain underlying etiology.

3.2 Schizophrenia   As with LOAD, most of the genetic studies of interactions for schizophrenia and schizophrenia-related phenotypes are motivated by candidates known to or suspected to interact biologically (Table 2). However, a key feature of much of the data on epistasis in schizophrenia is that statistical evidence is often supported by functional data either in animal models or from neuroimaging. This latter class of data provides compelling evidence that some of the key interactions significantly change either brain structure or function. Such real-time studies provide data on interaction models not often collected in human disease studies. Finally, as will be shown in the next section, there is evidence that one gene, catechol-O-methyltransferase (*COMT*), not only associates with multiple diseases, but interacts with multiple genes to affect schizophrenia risk. Of note, the most commonly studied *COMT* variant is rs4680, which encodes a Val-Met substitution at amino acid reside 158 and results in a significant change in enzyme activity [22]. In some

Table 2
Schizophrenia-related epistasis interactions

Primary gene	Secondary gene	Phenotype	Reference
ABCB1	*CYP3A5/CYP1A2/CYP2D6*	Schizophrenia	[104]
COMT	*DAOA*	fMRI	[41]
	DAT	Brain activation in executive function in schizophrenia (fMRI)	[43]
	MAOA	Negative results for schizophrenia	[105]
	DTNDP1	Prefrontal function in humans and mice	[42]
	SLCA5A3	Schizophrenia/in vitro biochemical interaction	[52]
	RGS4/G72/GRM3/DISC1	Schizophrenia	[106]
DAOA	*DAO*	Schizophrenia	[107]
DISC1	*NDEL1/NDE1*	Schizophrenia/protein–protein binding	[44]
	RELN/ERBB4/PDE4B	Structural magnetic resonance	[38]
	NKCC1	Schizophrenia/animal model of neurogenesis	[40]
	SLC12A2	fMRI	[39]
DRD3	*SLC18A2*	Schizophrenia/in vitro biochemical interaction	[52]
DTNBP1	*MUTED*	Schizophrenia	[108]
	IL3	Schizophrenia	[109]
ERBB4	*PDE4B*	Structural magnetic resonance	[38]
RELN	*PDE4B/NRG1*	Structural magnetic resonance	[38]
SLC6A3 (DAT)	*SLC18A2*	Schizophrenia/in vitro biochemical interaction	[52]

respects this is not unlike the results for *APOE* in LOAD, except the marginal effects for *COMT* are not quite as compelling and, in many cases, are inconsistent, even in studies of only Europeans [23–35]. This may be explained by the observation that some of the studies that support a significant interaction do not show marginal effects [36]. Adding even more complexity to the *COMT* are data indicating that environmental factors, e.g., marijuana consumption, modulate the role of *COMT* in schizophrenia risk [37] (*see* **Note 4**). The clear-cut role that environmental exposures play in schizophrenia may be, to some extent, responsible for the variation in genetic association even for this gene that has provided the most compelling evidence for epistatic interactions affecting risk.

Given the role of *COMT* in dopamine degradation it remains a compelling candidate as a genetic factor for schizophrenia risk. Data in Table 2 indicate that as part of an interacting network it is often a significant component. One could argue that some of the key interactions have not been explicitly replicated, but as a central player in psychosis its role is difficult to argue against given the results of so many related studies (Table 3). This is especially true since the relationship of *COMT* to other genes is statistically associated not only with the clinical endpoint but also with endophenotypes that are clearly associated with schizophrenia [38–43].

Among the other genes most often found in epistatic models of schizophrenia is the appropriately named "disrupted in schizophrenia 1" (*DISC1*) that has been associated with schizophrenia phenotypes in combination with at least eight other genes (Table 2). There is not only statistical data indicating an association but *DISC1* is known to bind other genes that, together with it, associate with schizophrenia, such as nudE neurodevelopment protein 1 (*NDE1*) and (nudE neurodevelopment protein 1-like 1) *NDEL1*[44]. The inclusion of direct biochemical data, indicating interaction, is key to establishing mechanism and hence plausibility. As with *COMT*, the data tying *DISC1* to schizophrenia-related phenotypes are not purely statistical. First, it was identified as part of a translocation breakpoint region in a large family with behavioral and mental disorders, including but not limited to schizophrenia [45, 46]. *DISC1* also interacts with *PDEB4B* to regulate cAMP in schizophrenia [47], and this is in addition to the interaction these two genes have in their combined effect on brain structure [38] (*see* **Note 5**).

In contrast to LOAD, genetic studies have not been as focused on assessing the role of epistasis in schizophrenia disease risk, and the result has been fewer identified interaction patterns. However, as with LOAD a few genes often play a role in the interactions, such as *COMT* and *DISC1*, indicating that the results represent a true nexus of interactions. The independent effects of these genes have not been as compelling, perhaps due to a series of additional nonlinear effects from strong environmental exposures that are not be as well-documented in LOAD [48–50]. This potential added

Table 3
Catechol-_O_-methyltransferase epistatic interactions in neuropsychiatric phenotypes

Interacting gene(s)	Phenotype	Epistasis evidence	Reference(s)
DTNBP1	Prefrontal cortical function	+	[42]
DRD2	Reward processing	+	[110]
5-HTTLPR	fMRI and facial emotion processing	+	[111]
DRD4	Eating disorders/bulimia	+	[112]
DRD4	Prefrontal cortex response control in ADHD/Working memory	+	[113]
_NR3C1_	DLPFC activation	+	[114]
AKT1/BDNF	MTL structure and memory function/schizophrenia	+	[115]
DRD4 and DAT1	IID	−[a]	[116]
Tryptophan hydroxylase and 5-HTR2A	ASPD	+	[117]
ANKK1	Task-set switching	+	[118]
DAT1/DBH	ADHD	+	[119]
DAOA	fMRI of prefrontal cortex/schizophrenia	+	[41]
DRD2	Social cooperation	+	[120]
DAT1	Stress reactivity-salivary cortisol response	+	[121]
5-HTTLPR	Depression in youth	+	[122]
MAOA	Intellectual ability in ADHD	+	[123]
DAT	Cortical function and schizophrenia	+	[43]
DAT1	Eating behavior	+	[124]
CYP2A6/MAOA/DBH	Smoking	−[b]	[125]
DRD2/ANKK1	Working memory	+	[126]
MAOA	ADHD with comorbid ODD	−[a]	[127]
ALDH3B1	Paranoid schizophrenia	+	[128]
AKT1	Prefrontal cortical structure and function	+	[129]
DRD2	Working memory	+	[130]
DAT	Memory tasks	+	[131]
DAT/SLC6A3	Schizophrenia	+	[52]

(continued)

Table 3
(continued)

Interacting gene(s)	Phenotype	Epistasis evidence	Reference(s)
DRD2	PPI	–	[132]
RGS4	Working memory/DLPFC fMRI	+	[54]
MAOA	Smoking behavior	–	[133]
GAD67	Cortical function and gene expression in schizophrenia	+	[134]
GRM3	Working memory	+	[135]
PRODH	Brain anatomy in schizophrenia	+	[136]
MAOA	Stress in depression	+	[137]
RGS4/DAOA/GRM3/DISC1	Schizophrenia	+	[106]
MAOA	Suicidal behavior	–	[138, 139]
5-HTR2A	Hallucination susceptibility	+	[140]
DRD2	Personality dimensions	+	[141]
MAOA/MAOB	Aggressive behavior	–	[142]
MAOA	Schizophrenia	–	[105]

ADHD attention deficit hyperactivity disorder, *DLFPC* dorsolateral prefrontal cortex, *MTL* medial temporal lobe, *IID* idiopathic intellectual disability, *ASPD* antisocial personality disorder, *ODD* oppositional defiant disorder, *PPI* prepulse inhibition
[a]Epistasis was not found but a marginal effect was present with *COMT*
[b]Although no interaction with COMT was found an interaction between MAOA and CYP2A6 was identified

complexity may explain why strong, consistent marginal effects are harder to discover in schizophrenia. They simply may not exist, but imaging data, with the effects of genetic variants, support a role for epistatic interactions in schizophrenia.

4 *COMT* as a Node in Neuropsychiatric Epistasis

The well-known non-synonymous functional variant in *COMT* (Val158Met, rs4680) has been particularly well interrogated for its role in numerous neuropsychiatric phenotypes through interactions. *COMT* serves as an excellent candidate because of its known effect on dopamine regulation in the brain and the centrality of this process to neuropsychiatric diseases, many directly related to schizophrenia [51]. As such, studies of *COMT* have shown that it interacts with many other genes to affect several neuropsychiatric diseases or related endophenotypes. If, as has been argued, pleiotropy and epistasis are directly related, then it is likely that the many pleiotropic effects of *COMT* provide evidence for epistasis in these phenotypes. The data presented below and in Table 3 are certainly consistent with such an argument.

Although many of the studies reporting *COMT* interactions involve a limited number of subjects and are therefore below the huge numbers generally considered necessary for genetic studies and the studies do replicate findings internally, and the preponderance of the data provides strong suggestive evidence for both pleiotropy and epistasis involving *COMT*. That said, one study did replicate findings between *COMT* and *SLC6A3* (*DAT*) in schizophrenia [52]. These genes and two others (*DRD3* and *SLC18A2*) were selected for interaction analyses following an extensive marginal effect study of 18 genes; these four showed significant marginal association effects. Nonetheless, not only were the interaction terms themselves significant but in testing for interactions among SNPs in these genes the investigators detected a significant excess of interactions relative to what would be expected by chance alone ($p = 0.001–0.008$). Taken together, these data support the conclusion that *COMT* is a significant factor in risk of schizophrenia, and that it probably acts through multiple interactions, especially when one considers the other studies presented in this chapter.

Interaction studies between *COMT* and *RGS4* indicate that the *COMT* genotype affects *RGS4* mRNA levels in dorsolateral prefrontal cortex (DLPFC) activation [53], providing biochemical validation of epistatic interactions detected using purely statistical analyses of fMRI results [54]. The convergence of molecular and functional studies is critical to developing plausible arguments for the role of *COMT* in neuropsychiatric disease, and they abound (*see* **Note 5**). For example, velocardiofacial syndrome (VCFS), a syndrome often involving cleft palate, cardiac anomalies, a typical facies, and learning disabilities, is caused by deletion of 22q11. Of interest to our discussion, it is also associated with high rates of psychotic illness [55, 56]. Interestingly, the chromosomal deletions include *COMT*, which is in this chromosomal region [57], and the Val158Met variant on the intact chromosome in VCFS patients associated with verbal working memory and expressive language performance as well as ADHD [58, 59]. These data are again suggestive of yet other roles for *COMT* in neurological phenotypes. Evidence also indicates that *COMT* may associate with anorexia nervosa in Israelis [60, 61] but not in other populations [62]. There is also data supporting association with obsessive-compulsive disorder [63–65].

Not only does *COMT* associate with multiple neuropsychiatric phenotypes in marginal analyses as described above and noted in the previous section, but it is also well documented that it interacts with multiple genes to putatively affect numerous phenotypes (Table 3). In fact, the same two genes, e.g., *COMT* and *DAT*, can interact in association with multiple albeit related phenotypes, including schizophrenia, ADHD, cortisol response, and eating behavior (Table 3). It is likely these are related to different endpoints of the same pathway involving *COMT*, but without dense clinical data on all phenotypes it is impossible to know. Nonetheless, given the consistency of associations with multiple phenotypes that are due to both the same and different interacting genes provides a wealth of data supporting *COMT* as a key player in many outcomes that vary by the exact nature of the interaction.

5   Conclusions

The brain, where the vast majority of genes are expressed, serves as the ideal location for epistasis to occur. In this brief chapter, data supporting the role of epistasis in two neuropsychiatric diseases, AD and schizophrenia, were presented. A key feature of the presented studies is that the interactions often involve the same "primary" gene, e.g., *APOE* in AD and *COMT* in schizophrenia, in relation to a variable set of "secondary" genes. Importantly, both *APOE* and *COMT* have clear pleiotropic effects, thereby supporting the argument that epistasis and pleiotropy are shadows of each other [10], and ultimately that networks or their disruption defines risk of complex disease (*see* **Note 6**). And even in the presence of multiple epistatic interactions, identifying a strong marginal effect may not be possible depending on their nature. For example, *APOE* independently presents with a strong and significant effects for AD whereas the single gene results for *COMT* with schizophrenia are much weaker.

The nature of the interactions with *APOE* and *COMT* may be critical in understanding and treating disease, as the data presented suggests that the major or primary gene lesions can affect multiple biological processes through complex and multiple interactions. If this is true, then treating on the basis of *COMT* VAL158MET, for example, will run the risk of affecting several of the other related phenotypes, thereby perhaps increasing the risk of adverse events via the other pathways. Hence, tying these seemingly central genes to others that only affect one or two phenotypes may present an alternative strategy to treatment that minimizes the likelihood of competing, negative effects (*see* **Note 7**). This hypothetical argument based on the observed and common patterns of epistasis in neuropsychiatric disease may lead to the conclusion that even knowing a gene with strong marginal effect may be inadequate to understand true genetic models and even worse in extending genetic association data to prevention and treatment in a personalized and safe manner.

6   Notes

1. Epistasis is a hallmark of most complex phenotypes that is well established in model systems. However, epistasis has been significantly understudied in human genetic analyses. This omission is

partially due to the assumption that interacting effects will be smaller than marginal effects and that increasing the number of tests in epistasis analyses will significantly decrease power to identify important genetic signals. This assumption may not be well founded and is explicitly examined here.

2. In the brain, most genes are expressed either at one or more times during development and in one or more compartments. Because of the diversity of genes expressed in the brain, it represents the ideal tissue in which to assess the existence of epistasis. Two adult onset neuropsychiatric diseases were examined for evidence of epistasis, late onset Alzheimer's disease and schizophrenia with the latter representing an earlier onset phenotype.

3. Multiple well-validated interactions were identified for Alzheimer's disease. A majority of the two-way interactions identified included *APOE* as one of the two genes. This is significant because *APOE* is the best validated gene in late onset alzheimer's disease risk. Therefore, it is reasonable to postulate that even though *APOE* has a strong marginal effect, the underlying genetic model that motivates this marginal effect is the cumulative effect of the multiple interactions operating simultaneously. If this is true, then it is likely that designing treatments around *APOE* derived risk or function may not be the best strategy; given its nodal property, any treatment modality aimed at *APOE* function may well increase the chance of generating adverse events in multiple physiological domains.

4. *COMT*, which has been well studied with respect to risk of schizophrenia and is involved in multiple epistatic interactions, does not generate a strong marginal signal. This difference from APOE raises questions about the nature of its interaction and the possibility that extrinsic (environmental) factors may play more important roles in schizophrenia. It has recently been confirmed that schizophrenia risk is at least partially dependent on gene × environment interactions, thereby possibly explaining the smaller marginal effect for *COMT* in schizophrenia than that of *APOE* in Alzheimer's disease. Therefore, it is not only important to understand gene × gene interactions but gene × environment interactions. It is to be expected that the relative importance of each will vary by disease.

5. Statistical association in both AD and schizophrenia has provided important insight into how genes affect each disease. However, these analyses alone are insufficient to provide compelling evidence for biological roles. Functional and anatomical assays in schizophrenia have been successful in tying the statistical evidence to more biological phenomena.

The range of assays extends from direct biochemical interaction studies to structural and functional MRI data. When all three data domains present with evidence consistent with a real effect on phenotype, the conclusions can be stronger than having any single domain even when each piece is relatively weak as compared to standards normally needed to demonstrate a causal relationship. This convergence of evidence was most clear in several gene × gene studies in schizophrenia

6. Pleiotropy can provide semi-independent evidence for interactions. It has been argued that one means to generate pleiotropic effects of a single gene is to have the same gene interact with multiple other genes in the same or unique pathways [66, 10]. If this is true, then simply examining the number of phenotypes that a single gene associates with may be an important factor in assessing the level of epistasis in which it participates. A corollary of this is to examine the number of interactions and the associated phenotypes. *COMT* represents an excellent example of this latter approach, as it has been shown to interact with tens of other genes to generate similar but distinct phenotypes.

7. Studies of epistasis in neuropsychiatric disease have primarily used a candidate gene approach, but despite recent arguments against this approach it has been successful in unraveling the complex genetics of these diseases. These results present compelling evidence that this approach is not outdated.

7   Gene Glossary

5-HTTLPR	Promoter region of the serotonin transporter gene.
5-HTR2A	5-Hydroxytryptmine receptor 2A
ALDH3B1	Aldehyde dehydrogenase 3B1.
AKT1	V-AKT murine thymoma viral oncogene homolog 1
ANKK1	Ankyrin repeat and kinase domain containing 1
BDNF	Brain-derived neurotrophic factor
DAOA	D-Amino acid oxidase activator.
DAT1	Dopamine transporter.
DBH	Dopamine beta-hydroxylase.
DRD2	Dopamine D2 receptor.
DRD4	Dopamine D4 receptor.
DTNBP1	Dystrobrevin-binding protein 1
GAD67	Gamma-aminobutyric acid (GABA) synthetic enzyme glutamic acid decarboxylase 67.
GRM3	Glutamate receptor mgluR3.
MAOA	Monoamine oxidase A.
NR3C1	Nuclear receptor subfamily 3, group C, member 1
PROD	Proline dehydrogenase.
RGS4	Regulator of G-protein signaling 4.

References

1. Manolio TA, Collins FS, Cox NJ, Goldstein DB, Hindorff LA, Hunter DJ, McCarthy MI, Ramos EM, Cardon LR, Chakravarti A, Cho JH, Guttmacher AE, Kong A, Kruglyak L, Mardis E, Rotimi CN, Slatkin M, Valle D, Whittemore AS, Boehnke M, Clark AG, Eichler EE, Gibson G, Haines JL, Mackay TF, McCarroll SA, Visscher PM (2009) Finding the missing heritability of complex diseases. Nature 461(7265):747–753. doi:10.1038/nature08494

2. Eichler EE, Flint J, Gibson G, Kong A, Leal SM, Moore JH, Nadeau JH (2010) Missing heritability and strategies for finding the underlying causes of complex disease. Nat Rev Genet 11(6):446–450. doi:10.1038/nrg2809

3. Siegal ML, Bergman A (2002) Waddington's canalization revisited: developmental stability and evolution. Proc Natl Acad Sci U S A 99(16):10528–10532. doi:10.1073/pnas.102303999

4. Hansen TF (2013) Why epistasis is important for selection and adaptation. Evolution 67(12):3501–3511. doi:10.1111/evo.12214

5. Hemani G, Knott S, Haley C (2013) An evolutionary perspective on epistasis and the missing heritability. PLoS Genet 9(2):e1003295. doi:10.1371/journal.pgen.1003295

6. Greene CS, Penrod NM, Williams SM, Moore JH (2009) Failure to replicate a genetic association may provide important clues about genetic architecture. PLoS One 4(6):e5639. doi:10.1371/journal.pone.0005639

7. Williams SM, Haines JL, Moore JH (2004) The use of animal models in the study of complex disease: all else is never equal or why do so many human studies fail to replicate animal findings? BioEssays 26(2):170–179. doi:10.1002/bies.10401

8. Hawrylycz MJ, Lein ES, Guillozet-Bongaarts AL, Shen EH, Ng L, Miller JA, van de Lagemaat LN, Smith KA, Ebbert A, Riley ZL, Abajian C, Beckmann CF, Bernard A, Bertagnolli D, Boe AF, Cartagena PM, Chakravarty MM, Chapin M, Chong J, Dalley RA, Daly BD, Dang C, Datta S, Dee N, Dolbeare TA, Faber V, Feng D, Fowler DR, Goldy J, Gregor BW, Haradon Z, Haynor DR, Hohmann JG, Horvath S, Howard RE, Jeromin A, Jochim JM, Kinnunen M, Lau C, Lazarz ET, Lee C, Lemon TA, Li L, Li Y, Morris JA, Overly CC, Parker PD, Parry SE, Reding M, Royall JJ, Schulkin J, Sequeira PA, Slaughterbeck CR, Smith SC, Sodt AJ, Sunkin SM, Swanson BE, Vawter MP, Williams D, Wohnoutka P, Zielke HR, Geschwind DH, Hof PR, Smith SM, Koch C, Grant SG, Jones AR (2012) An anatomically comprehensive atlas of the adult human brain transcriptome. Nature 489(7416):391–399. doi:10.1038/nature11405

9. Studies N-NWGoRiA, Chanock SJ, Manolio T, Boehnke M, Boerwinkle E, Hunter DJ, Thomas G, Hirschhorn JN, Abecasis G, Altshuler D, Bailey-Wilson JE, Brooks LD, Cardon LR, Daly M, Donnelly P, Fraumeni JF Jr, Freimer NB, Gerhard DS, Gunter C, Guttmacher AE, Guyer MS, Harris EL, Hoh J, Hoover R, Kong CA, Merikangas KR, Morton CC, Palmer LJ, Phimister EG, Rice JP, Roberts J, Rotimi C, Tucker MA, Vogan KJ, Wacholder S, Wijsman EM, Winn DM, Collins FS (2007) Replicating genotype-phenotype associations. Nature 447(7145):655–660. doi:10.1038/447655a

10. Tyler AL, Asselbergs FW, Williams SM, Moore JH (2009) Shadows of complexity: what biological networks reveal about epistasis and pleiotropy. BioEssays 31(2):220–227. doi:10.1002/bies.200800022

11. Combarros O, Cortina-Borja M, Smith AD, Lehmann DJ (2009) Epistasis in sporadic Alzheimer's disease. Neurobiol Aging 30(9):1333–1349. doi:10.1016/j.neurobiolaging.2007.11.027

12. Moore JH, Williams SM (2009) Epistasis and its implications for personal genetics. Am J Hum Genet 85(3):309–320. doi:10.1016/j.ajhg.2009.08.006

13. Combarros O, van Duijn CM, Hammond N, Belbin O, Arias-Vasquez A, Cortina-Borja M, Lehmann MG, Aulchenko YS, Schuur M, Kolsch H, Heun R, Wilcock GK, Brown K, Kehoe PG, Harrison R, Coto E, Alvarez V, Deloukas P, Mateo I, Gwilliam R, Morgan K, Warden DR, Smith AD, Lehmann DJ (2009) Replication by the Epistasis Project of the interaction between the genes for IL-6 and IL-10 in the risk of Alzheimer's disease. J Neuroinflammation 6:22. doi:10.1186/1742-2094-6-22

14. Infante J, Sanz C, Fernandez-Luna JL, Llorca J, Berciano J, Combarros O (2004) Gene-gene interaction between interleukin-6 and interleukin-10 reduces AD risk. Neurology 63(6):1135–1136

15. Mateo I, Infante J, Rodriguez E, Berciano J, Combarros O, Llorca J (2006) Interaction between dopamine beta-hydroxylase and interleukin genes increases Alzheimer's disease risk. J Neurol Neurosurg Psychiatry 77(2):278–279. doi:10.1136/jnnp.2005.075358

16. McCusker SM, Curran MD, Dynan KB, McCullagh CD, Urquhart DD, Middleton D, Patterson CC, McIlroy SP, Passmore AP

(2001) Association between polymorphism in regulatory region of gene encoding tumour necrosis factor alpha and risk of Alzheimer's disease and vascular dementia: a case-control study. Lancet 357(9254):436–439

17. Bhojak TJ, DeKosky ST, Ganguli M, Kamboh MI (2000) Genetic polymorphisms in the cathespin D and interleukin-6 genes and the risk of Alzheimer's disease. Neurosci Lett 288(1):21–24

18. Koivisto AM, Helisalmi S, Pihlajamaki J, Moilanen L, Kuusisto J, Laakso M, Hiltunen M, Keijo K, Hanninen T, Helkala EL, Kervinen K, Kesaniemi YA, Soininen H (2005) Interleukin-6 promoter polymorphism and late-onset alzheimer's disease in the Finnish population. J Neurogenet 19(3–4):155–161. doi:10.1080/01677060600569721

19. Flex A, Giovannini S, Biscetti F, Liperoti R, Spalletta G, Straface G, Landi F, Angelini F, Caltagirone C, Ghirlanda G, Bernabei R (2014) Effect of proinflammatory gene polymorphisms on the risk of Alzheimer's disease. Neurodegener Dis 13(4):230–236. doi:10.1159/000353395

20. Bosco P, Ferri R, Salluzzo MG, Castellano S, Signorelli M, Nicoletti F, Nuovo SD, Drago F, Caraci F (2013) Role of the Transforming-Growth-Factor-beta1 Gene in Late-Onset alzheimer's Disease: Implications for the Treatment. Curr Genomics 14(2):147–156. doi:10.2174/1389202911314020007

21. Dickson MR, Perry RT, Wiener H, Go RC (2005) Association studies of transforming growth factor-beta 1 and Alzheimer's disease. Am J Med Genet B Neuropsychiatr Genet 139B(1):38–41. doi:10.1002/ajmg.b.30218

22. Lotta T, Vidgren J, Tilgmann C, Ulmanen I, Melen K, Julkunen I, Taskinen J (1995) Kinetics of human soluble and membrane-bound catechol O-methyltransferase: a revised mechanism and description of the thermolabile variant of the enzyme. Biochemistry 34(13):4202–4210

23. Nieratschker V, Frank J, Muhleisen TW, Strohmaier J, Wendland JR, Schumacher J, Treutlein J, Breuer R, Abou Jamra R, Mattheisen M, Herms S, Schmal C, Maier W, Nothen MM, Cichon S, Rietschel M, Schulze TG (2010) The catechol-O-methyl transferase (COMT) gene and its potential association with schizophrenia: findings from a large German case-control and family-based sample. Schizophr Res 122(1–3):24–30. doi:10.1016/j.schres.2010.06.018

24. Pawel K, Hauser J, Skibinska M, Szczepankiewicz A, Dmitrzak-Weglarz M, Gorzkowska K, Pawlak J, Czerski PM (2010) Family based association study of DRD1, DRD2, DRD3, DRD4, DAT, COMT gene polymorphism in schizophrenia. Psychiatr Pol 44(3):405–413

25. Al-Asmary S, Kadasah S, Arfin M, Tariq M, Al-Asmari A (2014) Genetic association of catechol-O-methyltransferase val(158)met polymorphism in Saudi schizophrenia patients. Genet Mol Res 13(2):3079–3088. doi:10.4238/2014.April.17.4

26. Kunugi H, Vallada HP, Sham PC, Hoda F, Arranz MJ, Li T, Nanko S, Murray RM, McGuffin P, Owen M, Gill M, Collier DA (1997) Catechol-O-methyltransferase polymorphisms and schizophrenia: a transmission disequilibrium study in multiply affected families. Psychiatr Genet 7(3):97–101

27. Karayiorgou M, Gogos JA, Galke BL, Wolyniec PS, Nestadt G, Antonarakis SE, Kazazian HH, Housman DE, Pulver AE (1998) Identification of sequence variants and analysis of the role of the catechol-O-methyltransferase gene in schizophrenia susceptibility. Biol Psychiatry 43(6):425–431

28. Rosa A, Peralta V, Cuesta MJ, Zarzuela A, Serrano F, Martinez-Larrea A, Fananas L (2004) New evidence of association between COMT gene and prefrontal neurocognitive function in healthy individuals from sibling pairs discordant for psychosis. Am J Psychiatry 161(6):1110–1112

29. Rosa A, Cuesta MJ, Fatjo-Vilas M, Peralta V, Zarzuela A, Fananas L (2006) The Val66Met polymorphism of the brain-derived neurotrophic factor gene is associated with risk for psychosis: evidence from a family-based association study. Am J Med Genet B Neuropsychiatr Genet 141B(2):135–138. doi:10.1002/ajmg.b.30266

30. Hoenicka J, Garrido E, Martinez I, Ponce G, Aragues M, Rodriguez-Jimenez R, Espana-Serrano L, Alvira-Botero X, Santos JL, Rubio G, Jimenez-Arriero MA, Palomo T, PARGPARG (2010) Gender-specific COMT Val158Met polymorphism association in Spanish schizophrenic patients. Am J Med Genet B Neuropsychiatr Genet 153(1):79–85. doi:10.1002/ajmg.b.30957

31. Costas J, Sanjuan J, Ramos-Rios R, Paz E, Agra S, Ivorra JL, Paramo M, Brenlla J, Arrojo M (2011) Heterozygosity at catechol-O-methyltransferase Val158Met and schizophrenia: new data and meta-analysis. J Psychiatr Res 45(1):7–14. doi:10.1016/j.jpsychires.2010.04.021

32. Glatt SJ, Faraone SV, Tsuang MT (2003) Association between a functional catechol O-methyltransferase gene polymorphism and schizophrenia: meta-analysis of case-control and family-based studies. Am J Psychiatry 160(3):469–476

33. Fan JB, Zhang CS, Gu NF, Li XW, Sun WW, Wang HY, Feng GY, St Clair D, He L (2005)

Catechol-O-methyltransferase gene Val/Met functional polymorphism and risk of schizophrenia: a large-scale association study plus meta-analysis. Biol Psychiatry 57(2):139–144. doi:10.1016/j.biopsych.2004.10.018

34. Munafo MR, Bowes L, Clark TG, Flint J (2005) Lack of association of the COMT (Val158/108 Met) gene and schizophrenia: a meta-analysis of case-control studies. Mol Psychiatry 10(8):765–770. doi:10.1038/sj.mp.4001664

35. Allen NC, Bagade S, McQueen MB, Ioannidis JP, Kavvoura FK, Khoury MJ, Tanzi RE, Bertram L (2008) Systematic meta-analyses and field synopsis of genetic association studies in schizophrenia: the SzGene database. Nat Genet 40(7):827–834. doi:10.1038/ng.171

36. Nicodemus KK, Callicott JH, Higier RG, Luna A, Nixon DC, Lipska BK, Vakkalanka R, Giegling I, Rujescu D, St Clair D, Muglia P, Shugart YY, Weinberger DR (2010) Evidence of statistical epistasis between DISC1, CIT and NDEL1 impacting risk for schizophrenia: biological validation with functional neuroimaging. Hum Genet 127(4):441–452. doi:10.1007/s00439-009-0782-y

37. Costas J, Sanjuan J, Ramos-Rios R, Paz E, Agra S, Tolosa A, Paramo M, Brenlla J, Arrojo M (2011) Interaction between COMT haplotypes and cannabis in schizophrenia: a case-only study in two samples from Spain. Schizophr Res 127(1–3):22–27. doi:10.1016/j.schres.2011.01.014

38. Andreasen NC, Wilcox MA, Ho BC, Epping E, Ziebell S, Zeien E, Weiss B, Wassink T (2012) Statistical epistasis and progressive brain change in schizophrenia: an approach for examining the relationships between multiple genes. Mol Psychiatry 17(11):1093–1102. doi:10.1038/mp.2011.108

39. Callicott JH, Feighery EL, Mattay VS, White MG, Chen Q, Baranger DA, Berman KF, Lu B, Song H, Ming GL, Weinberger DR (2013) DISC1 and SLC12A2 interaction affects human hippocampal function and connectivity. J Clin Invest 123(7):2961–2964. doi:10.1172/jci67510

40. Kim JY, Liu CY, Zhang F, Duan X, Wen Z, Song J, Feighery E, Lu B, Rujescu D, St Clair D, Christian K, Callicott JH, Weinberger DR, Song H, Ming GL (2012) Interplay between DISC1 and GABA signaling regulates neurogenesis in mice and risk for schizophrenia. Cell 148(5):1051–1064. doi:10.1016/j.cell.2011.12.037

41. Nixon DC, Prust MJ, Sambataro F, Tan HY, Mattay VS, Weinberger DR, Callicott JH (2011) Interactive effects of DAOA (G72) and catechol-O-methyltransferase on neurophysiology in prefrontal cortex. Biol Psychiatry

69(10):1006–1008. doi:10.1016/j.biopsych.2010.10.031

42. Papaleo F, Burdick MC, Callicott JH, Weinberger DR (2013) Epistatic interaction between COMT and DTNBP1 modulates prefrontal function in mice and in humans. Mol Psychiatry. doi:10.1038/mp.2013.133

43. Prata DP, Mechelli A, Fu CH, Picchioni M, Toulopoulou T, Bramon E, Walshe M, Murray RM, Collier DA, McGuire P (2009) Epistasis between the DAT 3' UTR VNTR and the COMT Val158Met SNP on cortical function in healthy subjects and patients with schizophrenia. Proc Natl Acad Sci U S A 106(32):13600–13605. doi:10.1073/pnas.0903007106

44. Burdick KE, Kamiya A, Hodgkinson CA, Lencz T, DeRosse P, Ishizuka K, Elashvili S, Arai H, Goldman D, Sawa A, Malhotra AK (2008) Elucidating the relationship between DISC1, NDEL1 and NDE1 and the risk for schizophrenia: evidence of epistasis and competitive binding. Hum Mol Genet 17(16):2462–2473. doi:10.1093/hmg/ddn146

45. Millar JK, Wilson-Annan JC, Anderson S, Christie S, Taylor MS, Semple CA, Devon RS, St Clair DM, Muir WJ, Blackwood DH, Porteous DJ (2000) Disruption of two novel genes by a translocation co-segregating with schizophrenia. Hum Mol Genet 9(9):1415–1423

46. St Clair D, Blackwood D, Muir W, Carothers A, Walker M, Spowart G, Gosden C, Evans HJ (1990) Association within a family of a balanced autosomal translocation with major mental illness. Lancet 336(8706):13–16

47. Millar JK, Pickard BS, Mackie S, James R, Christie S, Buchanan SR, Malloy MP, Chubb JE, Huston E, Baillie GS, Thomson PA, Hill EV, Brandon NJ, Rain JC, Camargo LM, Whiting PJ, Houslay MD, Blackwood DH, Muir WJ, Porteous DJ (2005) DISC1 and PDE4B are interacting genetic factors in schizophrenia that regulate cAMP signaling. Science 310(5751):1187–1191. doi:10.1126/science.1112915

48. van Os J, Rutten BP, Poulton R (2008) Gene-environment interactions in schizophrenia: review of epidemiological findings and future directions. Schizophr Bull 34(6):1066–1082. doi:10.1093/schbul/sbn117

49. van Os J, Murray R (2008) Gene-environment interactions in schizophrenia. Introduction. Schizophr Bull 34(6):1064–1065. doi:10.1093/schbul/sbn116

50. van Os J, Kenis G, Rutten BP (2010) The environment and schizophrenia. Nature 468(7321):203–212. doi:10.1038/nature09563

51. Ira E, Zanoni M, Ruggeri M, Dazzan P, Tosato S (2013) COMT, neuropsychological

function and brain structure in schizophrenia: a systematic review and neurobiological interpretation. J Psychiatry Neurosci 38(6):366–380. doi:10.1503/jpn.120178

52. Talkowski ME, Kirov G, Bamne M, Georgieva L, Torres G, Mansour H, Chowdari KV, Milanova V, Wood J, McClain L, Prasad K, Shirts B, Zhang J, O'Donovan MC, Owen MJ, Devlin B, Nimgaonkar VL (2008) A network of dopaminergic gene variations implicated as risk factors for schizophrenia. Hum Mol Genet 17(5):747–758. doi:10.1093/hmg/ddm347

53. Lipska BK, Mitkus S, Caruso M, Hyde TM, Chen J, Vakkalanka R, Straub RE, Weinberger DR, Kleinman JE (2006) RGS4 mRNA expression in postmortem human cortex is associated with COMT Val158Met genotype and COMT enzyme activity. Hum Mol Genet 15(18):2804–2812. doi:10.1093/hmg/ddl222

54. Buckholtz JW, Sust S, Tan HY, Mattay VS, Straub RE, Meyer-Lindenberg A, Weinberger DR, Callicott JH (2007) FMRI evidence for functional epistasis between COMT and RGS4. Mol Psychiatry 12(10):893–895, 885. doi:10.1038/sj.mp.4002008

55. Shprintzen RJ, Goldberg R, Golding-Kushner KJ, Marion RW (1992) Late-onset psychosis in the velo-cardio-facial syndrome. Am J Med Genet 42(1):141–142. doi:10.1002/ajmg.1320420131

56. Pulver AE, Nestadt G, Goldberg R, Shprintzen RJ, Lamacz M, Wolyniec PS, Morrow B, Karayiorgou M, Antonarakis SE, Housman D et al (1994) Psychotic illness in patients diagnosed with velo-cardio-facial syndrome and their relatives. J Nerv Ment Dis 182(8):476–478

57. Dunham I, Collins J, Wadey R, Scambler P (1992) Possible role for COMT in psychosis associated with velo-cardio-facial syndrome. Lancet 340(8831):1361–1362

58. Gothelf D, Presburger G, Levy D, Nahmani A, Burg M, Berant M, Blieden LC, Finkelstein Y, Frisch A, Apter A, Weizman A (2004) Genetic, developmental, and physical factors associated with attention deficit hyperactivity disorder in patients with velocardiofacial syndrome. Am J Med Genet B Neuropsychiatr Genet 126B(1):116–121. doi:10.1002/ajmg.b.20144

59. Baker K, Baldeweg T, Sivagnanasundaram S, Scambler P, Skuse D (2005) COMT Val108/158 Met modifies mismatch negativity and cognitive function in 22q11 deletion syndrome. Biol Psychiatry 58(1):23–31. doi:10.1016/j.biopsych.2005.03.020

60. Frisch A, Laufer N, Danziger Y, Michaelovsky E, Leor S, Carel C, Stein D, Fenig S, Mimouni M, Apter A, Weizman A (2001) Association of anorexia nervosa with the high activity allele of the COMT gene: a family-based study in Israeli patients. Mol Psychiatry 6(2):243–245. doi:10.1038/sj.mp.4000830

61. Michaelovsky E, Frisch A, Leor S, Stein D, Danziger Y, Carel C, Fennig S, Mimouni M, Klauck SM, Benner A, Poustka A, Apter A, Weizman A (2005) Haplotype analysis of the COMT-ARVCF gene region in Israeli anorexia nervosa family trios. Am J Med Genet B Neuropsychiatr Genet 139B(1):45–50. doi:10.1002/ajmg.b.30230

62. Gabrovsek M, Brecelj-Anderluh M, Bellodi L, Cellini E, Di Bella D, Estivill X, Fernandez-Aranda F, Freeman B, Geller F, Gratacos M, Haigh R, Hebebrand J, Hinney A, Holliday J, Hu X, Karwautz A, Nacmias B, Ribases M, Remschmidt H, Komel R, Sorbi S, Tomori M, Treasure J, Wagner G, Zhao J, Collier DA (2004) Combined family trio and case-control analysis of the COMT Val158Met polymorphism in European patients with anorexia nervosa. Am J Med Genet B Neuropsychiatr Genet 124B(1):68–72. doi:10.1002/ajmg.b.20085

63. Karayiorgou M, Altemus M, Galke BL, Goldman D, Murphy DL, Ott J, Gogos JA (1997) Genotype determining low catechol-O-methyltransferase activity as a risk factor for obsessive-compulsive disorder. Proc Natl Acad Sci U S A 94(9):4572–4575

64. Karayiorgou M, Sobin C, Blundell ML, Galke BL, Malinova L, Goldberg P, Ott J, Gogos JA (1999) Family-based association studies support a sexually dimorphic effect of COMT and MAOA on genetic susceptibility to obsessive-compulsive disorder. Biol Psychiatry 45(9):1178–1189

65. Alsobrook JP 2nd, Zohar AH, Leboyer M, Chabane N, Ebstein RP, Pauls DL (2002) Association between the COMT locus and obsessive-compulsive disorder in females but not males. Am J Med Genet 114(1):116–120

66. Hodgkin J (1998) Seven types of pleiotropy. Int J Dev Biol 42(3):501–505

67. Compton D, Wavrant DeVrieze F, Petersen RC, Tangalos E, Li L, Hardy J (2002) Possible association between genetic variability at the apolipoprotein(a) locus and Alzheimer's disease in apolipoprotein E2 carriers. Neurosci Lett 331(1):60–62

68. Rodriguez-Rodriguez E, Mateo I, Llorca J, Sanchez-Quintana C, Infante J, Garcia-Gorostiaga I, Sanchez-Juan P, Berciano J, Combarros O (2007) Association of genetic variants of ABCA1 with Alzheimer's disease risk. Am J Med Genet B Neuropsychiatr Genet 144B(7):964–968. doi:10.1002/ajmg.b.30552

69. Arias-Vasquez A, Isaacs A, Aulchenko YS, Hofman A, Oostra BA, Breteler M, van Duijn CM (2007) The cholesteryl ester transfer protein (CETP) gene and the risk of Alzheimer's disease. Neurogenetics 8(3):189–193. doi:10.1007/s10048-007-0089-x

70. Rodriguez E, Mateo I, Infante J, Llorca J, Berciano J, Combarros O (2006) Cholesteryl ester transfer protein (CETP) polymorphism modifies the Alzheimer's disease risk associated with APOE epsilon4 allele. J Neurol 253(2):181–185. doi:10.1007/s00415-005-0945-2

71. Hatanaka Y, Kamino K, Fukuo K, Mitsuda N, Nishiwaki-Ueda Y, Sato N, Satoh T, Yamamoto H, Yoneda H, Imagawa M, Miki T, Ohta S, Ogihara T (2000) Low density lipoprotein receptor-related protein gene polymorphisms and risk for late-onset alzheimer's disease in a Japanese population. Clin Genet 58(4):319–323

72. Golanska E, Hulas-Bigoszewska K, Wojcik I, Rieske P, Styczynska M, Peplonska B, Pfeffer A, Luczywek E, Wasiak B, Gabryelewicz T, Religa D, Chodakowska-Zebrowska M, Barcikowska M, Sobow T, Liberski PP (2005) CYP46: a risk factor for Alzheimer's disease or a coincidence? Neurosci Lett 383(1–2):105–108. doi:10.1016/j.neulet.2005.03.049

73. Luedecking-Zimmer E, DeKosky ST, Chen Q, Barmada MM, Kamboh MI (2002) Investigation of oxidized LDL-receptor 1 (OLR1) as the candidate gene for Alzheimer's disease on chromosome 12. Hum Genet 111(4–5):443–451. doi:10.1007/s00439-002-0802-7

74. Spell C, Kolsch H, Lutjohann D, Kerksiek A, Hentschel F, Damian M, von Bergmann K, Rao ML, Maier W, Heun R (2004) SREBP-1a polymorphism influences the risk of Alzheimer's disease in carriers of the ApoE4 allele. Dement Geriatr Cogn Disord 18(3–4):245–249. doi:10.1159/000080023

75. Kamboh MI, Sanghera DK, Ferrell RE, DeKosky ST (1995) APOE*4-associated Alzheimer's disease risk is modified by alpha 1-antichymotrypsin polymorphism. Nat Genet 10(4):486–488. doi:10.1038/ng0895-486

76. Yoshiiwa A, Kamino K, Yamamoto H, Kobayashi T, Imagawa M, Nonomura Y, Yoneda H, Sakai T, Nishiwaki Y, Sato N, Rakugi H, Miki T, Ogihara T (1997) alpha 1-Antichymotrypsin as a risk modifier for late-onset alzheimer's disease in Japanese apolipoprotein E epsilon 4 allele carriers. Ann Neurol 42(1):115–117. doi:10.1002/ana.410420118

77. Cai L, Tang G, Chen L, Zhang B, Jiang S, Ren D (2005) Genetic studies of A2M and BACE1 genes in Chinese Han Alzheimer's disease patients. Neuroreport 16(9):1023–1026

78. Gold G, Blouin JL, Herrmann FR, Michon A, Mulligan R, Duriaux Sail G, Bouras C, Giannakopoulos P, Antonarakis SE (2003) Specific BACE1 genotypes provide additional risk for late-onset alzheimer disease in APOE epsilon 4 carriers. Am J Med Genet B Neuropsychiatr Genet 119B(1):44–47. doi:10.1002/ajmg.b.10010

79. Nowotny P, Kwon JM, Chakraverty S, Nowotny V, Morris JC, Goate AM (2001) Association studies using novel polymorphisms in BACE1 and BACE2. Neuroreport 12(9):1799–1802

80. Montoya SE, Aston CE, DeKosky ST, Kamboh MI, Lazo JS, Ferrell RE (1998) Bleomycin hydrolase is associated with risk of sporadic Alzheimer's disease. Nat Genet 18(3):211–212. doi:10.1038/ng0398-211

81. Papassotiropoulos A, Bagli M, Jessen F, Frahnert C, Rao ML, Maier W, Heun R (2000) Confirmation of the association between bleomycin hydrolase genotype and Alzheimer's disease. Mol Psychiatry 5(2):213–215

82. Edland SD, Wavrant-De Vriese F, Compton D, Smith GE, Ivnik R, Boeve BF, Tangalos EG, Petersen RC (2003) Insulin degrading enzyme (IDE) genetic variants and risk of Alzheimer's disease: evidence of effect modification by apolipoprotein E (APOE). Neurosci Lett 345(1):21–24

83. Riazanskaia N, Lukiw WJ, Grigorenko A, Korovaitseva G, Dvoryanchikov G, Moliaka Y, Nicolaou M, Farrer L, Bazan NG, Rogaev E (2002) Regulatory region variability in the human presenilin-2 (PSEN2) gene: potential contribution to the gene activity and risk for AD. Mol Psychiatry 7(8):891–898. doi:10.1038/sj.mp.4001101

84. Jia L, Ye J, Lv H, Wang W, Zhou C, Zhang X, Xu J, Wang L, Jia J (2007) Genetic association between polymorphisms of Pen2 gene and late onset alzheimer's disease in the North Chinese population. Brain Res 1141:10–14. doi:10.1016/j.brainres.2007.01.005

85. Sala Frigerio C, Piscopo P, Calabrese E, Crestini A, Malvezzi Campeggi L, Civita di Fava R, Fogliarino S, Albani D, Marcon G, Cherchi R, Piras R, Forloni G, Confaloni A (2005) PEN-2 gene mutation in a familial Alzheimer's disease case. J Neurol 252(9):1033–1036. doi:10.1007/s00415-005-0799-7

86. Calero O, Bullido MJ, Clarimon J, Frank-Garcia A, Martinez-Martin P, Lleo A, Rey MJ, Rabano A, Blesa R, Gomez-Isla T, Valdivieso F, de Pedro-Cuesta J, Ferrer I, Calero M (2011) Genetic cross-interaction between APOE and PRNP in sporadic Alzheimer's and Creutzfeldt-Jakob diseases. PLoS One 6(7):e22090. doi:10.1371/journal.pone.0022090

87. Reynolds WF, Hiltunen M, Pirskanen M, Mannermaa A, Helisalmi S, Lehtovirta M, Alafuzoff I, Soininen H (2000) MPO and APOEpsilon4 polymorphisms interact to increase risk for AD in Finnish males. Neurology 55(9):1284–1290

88. Martin ER, Bronson PG, Li YJ, Wall N, Chung RH, Schmechel DE, Small G, Xu PT, Bartlett J, Schnetz-Boutaud N, Haines JL, Gilbert JR, Pericak-Vance MA (2005) Interaction between the alpha-T catenin gene (VR22) and APOE in Alzheimer's disease. J Med Genet 42(10):787–792. doi:10.1136/jmg.2004.029553

89. Lehmann DJ, Johnston C, Smith AD (1997) Synergy between the genes for butyrylcholinesterase K variant and apolipoprotein E4 in late-onset confirmed Alzheimer's disease. Hum Mol Genet 6(11):1933–1936

90. Raygani AV, Zahrai M, Soltanzadeh A, Doosti M, Javadi E, Pourmotabbed T (2004) Analysis of association between butyrylcholinesterase K variant and apolipoprotein E genotypes in Alzheimer's disease. Neurosci Lett 371(2–3):142–146. doi:10.1016/j.neulet.2004.08.057

91. Tilley L, Morgan K, Grainger J, Marsters P, Morgan L, Lowe J, Xuereb J, Wischik C, Harrington C, Kalsheker N (1999) Evaluation of polymorphisms in the presenilin-1 gene and the butyrylcholinesterase gene as risk factors in sporadic Alzheimer's disease. Eur J Hum Genet 7(6):659–663. doi:10.1038/sj.ejhg.5200351

92. Kuwano R, Miyashita A, Arai H, Asada T, Imagawa M, Shoji M, Higuchi S, Urakami K, Kakita A, Takahashi H, Tsukie T, Toyabe S, Akazawa K, Kanazawa I, Ihara Y, Japanese Genetic Study Consortium for Alzheimer's Disease (2006) Dynamin-binding protein gene on chromosome 10q is associated with late-onset alzheimer's disease. Hum Mol Genet 15(13):2170–2182. doi:10.1093/hmg/ddl142

93. Chiappelli M, Borroni B, Archetti S, Calabrese E, Corsi MM, Franceschi M, Padovani A, Licastro F (2006) VEGF gene and phenotype relation with Alzheimer's disease and mild cognitive impairment. Rejuvenation Res 9(4):485–493. doi:10.1089/rej.2006.9.485

94. Remarque EJ, Bollen EL, Weverling-Rijnsburger AW, Laterveer JC, Blauw GJ, Westendorp RG (2001) Patients with Alzheimer's disease display a pro-inflammatory phenotype. Exp Gerontol 36(1):171–176

95. Mansoori N, Tripathi M, Luthra K, Alam R, Lakshmy R, Sharma S, Arulselvi S, Parveen S, Mukhopadhyay AK (2012) MTHFR (677 and 1298) and IL-6-174G/C genes in pathogenesis of Alzheimer's and vascular dementia and

their epistatic interaction. Neurobiol Aging 33(5):1003e1001–1008. doi:10.1016/j.neurobiolaging.2011.09.018

96. Combarros O, Warden DR, Hammond N, Cortina-Borja M, Belbin O, Lehmann MG, Wilcock GK, Brown K, Kehoe PG, Barber R, Coto E, Alvarez V, Deloukas P, Gwilliam R, Heun R, Kolsch H, Mateo I, Oulhaj A, Arias-Vasquez A, Schuur M, Aulchenko YS, Ikram MA, Breteler MM, van Duijn CM, Morgan K, Smith AD, Lehmann DJ (2010) The dopamine beta-hydroxylase -1021C/T polymorphism is associated with the risk of Alzheimer's disease in the Epistasis Project. BMC Med Genet 11:162. doi:10.1186/1471-2350-11-162

97. Heun R, Kolsch H, Ibrahim-Verbaas CA, Combarros O, Aulchenko YS, Breteler M, Schuur M, van Duijn CM, Hammond N, Belbin O, Cortina-Borja M, Wilcock GK, Brown K, Barber R, Kehoe PG, Coto E, Alvarez V, Lehmann MG, Deloukas P, Mateo I, Morgan K, Warden DR, Smith AD, Lehmann DJ (2012) Interactions between PPAR-alpha and inflammation-related cytokine genes on the development of Alzheimer's disease, observed by the Epistasis Project. Int J Mol Epidemiol Genet 3(1):39–47

98. Infante J, Llorca J, Mateo I, Rodriguez-Rodriguez E, Sanchez-Quintana C, Sanchez-Juan P, Fernandez-Viadero C, Pena N, Berciano J, Combarros O (2007) Interaction between poly(ADP-ribose) polymerase 1 and interleukin 1A genes is associated with Alzheimer's disease risk. Dement Geriatr Cogn Disord 23(4):215–218. doi:10.1159/000099471

99. Infante J, Sanchez-Juan P, Mateo I, Rodriguez-Rodriguez E, Sanchez-Quintana C, Llorca J, Fontalba A, Terrazas J, Oterino A, Berciano J, Combarros O (2007) Poly (ADP-ribose) polymerase-1 (PARP-1) genetic variants are protective against Parkinson's disease. J Neurol Sci 256(1–2):68–70. doi:10.1016/j.jns.2007.02.008

100. Medway C, Combarros O, Cortina-Borja M, Butler HT, Ibrahim-Verbaas CA, de Bruijn RF, Koudstaal PJ, van Duijn CM, Ikram MA, Mateo I, Sanchez-Juan P, Lehmann MG, Heun R, Kolsch H, Deloukas P, Hammond N, Coto E, Alvarez V, Kehoe PG, Barber R, Wilcock GK, Brown K, Belbin O, Warden DR, Smith AD, Morgan K, Lehmann DJ (2014) The sex-specific associations of the aromatase gene with Alzheimer's disease and its interaction with IL10 in the Epistasis Project. Eur J Hum Genet 22(2):216–220. doi:10.1038/ejhg.2013.116

101. Lehmann DJ, Schuur M, Warden DR, Hammond N, Belbin O, Kolsch H, Lehmann MG, Wilcock GK, Brown K, Kehoe PG,

Morris CM, Barker R, Coto E, Alvarez V, Deloukas P, Mateo I, Gwilliam R, Combarros O, Arias-Vasquez A, Aulchenko YS, Ikram MA, Breteler MM, van Duijn CM, Oulhaj A, Heun R, Cortina-Borja M, Morgan K, Robson K, Smith AD (2012) Transferrin and HFE genes interact in Alzheimer's disease risk: the Epistasis Project. Neurobiol Aging 33(1):202e201–213. doi:10.1016/j. neurobiolaging.2010.07.018

102. Bullock JM, Medway C, Cortina-Borja M, Turton JC, Prince JA, Ibrahim-Verbaas CA, Schuur M, Breteler MM, van Duijn CM, Kehoe PG, Barber R, Coto E, Alvarez V, Deloukas P, Hammond N, Combarros O, Mateo I, Warden DR, Lehmann MG, Belbin O, Brown K, Wilcock GK, Heun R, Kolsch H, Smith AD, Lehmann DJ, Morgan K (2013) Discovery by the Epistasis Project of an epistatic interaction between the GSTM3 gene and the HHEX/IDE/KIF11 locus in the risk of Alzheimer's disease. Neurobiol Aging 34(4):1309e1301–1307. doi:10.1016/j. neurobiolaging.2012.08.010

103. Kolsch H, Lehmann DJ, Ibrahim-Verbaas CA, Combarros O, van Duijn CM, Hammond N, Belbin O, Cortina-Borja M, Lehmann MG, Aulchenko YS, Schuur M, Breteler M, Wilcock GK, Brown K, Kehoe PG, Barber R, Coto E, Alvarez V, Deloukas P, Mateo I, Maier W, Morgan K, Warden DR, Smith AD, Heun R (2012) Interaction of insulin and PPAR-alpha genes in Alzheimer's disease: the Epistasis Project. J Neural Transm 119(4):473–479. doi:10.1007/s00702-011-0732-4

104. Gasso P, Mas S, Alvarez S, Trias G, Bioque M, Oliveira C, Bernardo M, Lafuente A (2010) Xenobiotic metabolizing and transporter genes: gene-gene interactions in schizophrenia and related disorders. Pharmacogenomics 11(12):1725–1731. doi:10.2217/pgs.10.158

105. Norton N, Kirov G, Zammit S, Jones G, Jones S, Owen R, Krawczak M, Williams NM, O'Donovan MC, Owen MJ (2002) Schizophrenia and functional polymorphisms in the MAOA and COMT genes: no evidence for association or epistasis. Am J Med Genet 114(5):491–496. doi:10.1002/ajmg.10517

106. Nicodemus KK, Kolachana BS, Vakkalanka R, Straub RE, Giegling I, Egan MF, Rujescu D, Weinberger DR (2007) Evidence for statistical epistasis between catechol-O-methyltransferase (COMT) and polymorphisms in RGS4, G72 (DAOA), GRM3, and DISC1: influence on risk of schizophrenia. Hum Genet 120(6):889–906. doi:10.1007/s00439-006-0257-3

107. Corvin A, McGhee KA, Murphy K, Donohoe G, Nangle JM, Schwaiger S, Kenny N, Clarke S, Meagher D, Quinn J, Scully P, Baldwin P, Browne D, Walsh C, Waddington JL, Morris DW, Gill M (2007) Evidence for association and epistasis at the DAOA/G30 and D-amino acid oxidase loci in an Irish schizophrenia sample. Am J Med Genet B Neuropsychiatr Genet 144B(7):949–953. doi:10.1002/ ajmg.b.30452

108. Morris DW, Murphy K, Kenny N, Purcell SM, McGhee KA, Schwaiger S, Nangle JM, Donohoe G, Clarke S, Scully P, Quinn J, Meagher D, Baldwin P, Crumlish N, O'Callaghan E, Waddington JL, Gill M, Corvin AP (2008) Dysbindin (DTNBP1) and the biogenesis of lysosome-related organelles complex 1 (BLOC-1): main and epistatic gene effects are potential contributors to schizophrenia susceptibility. Biol Psychiatry 63(1):24–31. doi:10.1016/j.biopsych.2006.12.025

109. Edwards TL, Wang X, Chen Q, Wormly B, Riley B, O'Neill FA, Walsh D, Ritchie MD, Kendler KS, Chen X (2008) Interaction between interleukin 3 and dystrobrevin-binding protein 1 in schizophrenia. Schizophr Res 106(2–3):208–217. doi:10.1016/j. schres.2008.07.022

110. Balci F, Wiener M, Cavdaroglu B, Branch Coslett H (2013) Epistasis effects of dopamine genes on interval timing and reward magnitude in humans. Neuropsychologia 51(2):293–308. doi:10.1016/j. neuropsychologia.2012.08.002

111. Surguladze SA, Radua J, El-Hage W, Gohier B, Sato JR, Kronhaus DM, Proitsi P, Powell J, Phillips ML (2012) Interaction of catechol O-methyltransferase and serotonin transporter genes modulates effective connectivity in a facial emotion-processing circuitry. Transl Psychiatry 2:e70. doi:10.1038/tp.2011.69

112. Thaler L, Groleau P, Badawi G, Sycz L, Zeramdini N, Too A, Israel M, Joober R, Bruce KR, Steiger H (2012) Epistatic interactions implicating dopaminergic genes in bulimia nervosa (BN): relationships to eating- and personality-related psychopathology. Prog Neuropsychopharmacol Biol Psychiatry 39(1):120–128. doi:10.1016/j. pnpbp.2012.05.019

113. Heinzel S, Dresler T, Baehne CG, Heine M, Boreatti-Hummer A, Jacob CP, Renner TJ, Reif A, Lesch KP, Fallgatter AJ, Ehlis AC (2013) COMT x DRD4 epistasis impacts prefrontal cortex function underlying response control. Cereb Cortex 23(6):1453–1462. doi:10.1093/cercor/bhs132

114. El-Hage W, Phillips ML, Radua J, Gohier B, Zelaya FO, Collier DA, Surguladze SA (2013) Genetic modulation of neural response during working memory in healthy individuals: interaction of glucocorticoid receptor and

dopaminergic genes. Mol Psychiatry 18(2):174–182. doi:10.1038/mp.2011.145

115. Tan HY, Chen AG, Chen Q, Browne LB, Verchinski B, Kolachana B, Zhang F, Apud J, Callicott JH, Mattay VS, Weinberger DR (2012) Epistatic interactions of AKT1 on human medial temporal lobe biology and pharmacogenetic implications. Mol Psychiatry 17(10):1007–1016. doi:10.1038/mp.2011.91

116. Bhowmik AD, Chaudhury S, Dutta S, Shaw J, Chatterjee A, Choudhury A, Saha A, Sadhukhan D, Kar T, Sinha S, Mukhopadhyay K (2011) Role of functional dopaminergic gene polymorphisms in the etiology of idiopathic intellectual disability. Prog Neuropsychopharmacol Biol Psychiatry 35(7):1714–1722. doi:10.1016/j.pnpbp.2011.05.005

117. Cuartas Arias JM, Palacio Acosta CA, Valencia JG, Montoya GJ, Arango Viana JC, Nieto OC, Florez AF, Camarena Medellin BE, Montoya WR, Lopez Jaramillo CA, Achury JG, Fuentes CC, Berrio GB, Ruiz-Linares A (2011) Exploring epistasis in candidate genes for antisocial personality disorder. Psychiatr Genet 21(3):115–124. doi:10.1097/YPG.0b013e3283437175

118. Garcia-Garcia M, Barcelo F, Clemente IC, Escera C (2011) COMT and ANKK1 gene-gene interaction modulates contextual updating of mental representations. Neuroimage 56(3):1641–1647. doi:10.1016/j.neuroimage.2011.02.053

119. Das M, Das Bhowmik A, Bhaduri N, Sarkar K, Ghosh P, Sinha S, Ray A, Chatterjee A, Mukhopadhyay K (2011) Role of gene-gene/gene-environment interaction in the etiology of eastern Indian ADHD probands. Prog Neuropsychopharmacol Biol Psychiatry 35(2):577–587. doi:10.1016/j.pnpbp.2010.12.027

120. Walter NT, Markett SA, Montag C, Reuter M (2011) A genetic contribution to cooperation: dopamine-relevant genes are associated with social facilitation. Soc Neurosci 6(3):289–301. doi:10.1080/17470919.2010.527169

121. Alexander N, Osinsky R, Mueller E, Schmitz A, Guenthert S, Kuepper Y, Hennig J (2011) Genetic variants within the dopaminergic system interact to modulate endocrine stress reactivity and recovery. Behav Brain Res 216(1):53–58. doi:10.1016/j.bbr.2010.07.003

122. Conway CC, Hammen C, Brennan PA, Lind PA, Najman JM (2010) Interaction of chronic stress with serotonin transporter and catechol-O-methyltransferase polymorphisms in predicting youth depression. Depress Anxiety 27(8):737–745. doi:10.1002/da.20715

123. Qian QJ, Yang L, Wang YF, Zhang HB, Guan LL, Chen Y, Ji N, Liu L, Faraone SV (2010)

Gene-gene interaction between COMT and MAOA potentially predicts the intelligence of attention-deficit hyperactivity disorder boys in China. Behav Genet 40(3):357–365. doi:10.1007/s10519-009-9314-8

124. Hersrud SL, Stoltenberg SF (2009) Epistatic interaction between COMT and DAT1 genes on eating behavior: a pilot study. Eat Behav 10(2):131–133. doi:10.1016/j.eatbeh.2009.01.003

125. Tang X, Guo S, Sun H, Song X, Jiang Z, Sheng L, Zhou D, Hu Y, Chen D (2009) Gene-gene interactions of CYP2A6 and MAOA polymorphisms on smoking behavior in Chinese male population. Pharmacogenet Genomics 19(5):345–352

126. Stelzel C, Basten U, Montag C, Reuter M, Fiebach CJ (2009) Effects of dopamine-related gene-gene interactions on working memory component processes. Eur J Neurosci 29(5):1056–1063. doi:10.1111/j.1460-9568.2009.06647.x

127. Qian QJ, Liu J, Wang YF, Yang L, Guan LL, Faraone SV (2009) Attention Deficit Hyperactivity Disorder comorbid oppositional defiant disorder and its predominately inattentive type: evidence for an association with COMT but not MAOA in a Chinese sample. Behav Brain Funct 5:8. doi:10.1186/1744-9081-5-8

128. Wang Y, Hu Y, Fang Y, Zhang K, Yang H, Ma J, Xu Q, Shen Y (2009) Evidence of epistasis between the catechol-O-methyltransferase and aldehyde dehydrogenase 3B1 genes in paranoid schizophrenia. Biol Psychiatry 65(12):1048–1054. doi:10.1016/j.biopsych.2008.11.027

129. Tan HY, Nicodemus KK, Chen Q, Li Z, Brooke JK, Honea R, Kolachana BS, Straub RE, Meyer-Lindenberg A, Sei Y, Mattay VS, Callicott JH, Weinberger DR (2008) Genetic variation in AKT1 is linked to dopamine-associated prefrontal cortical structure and function in humans. J Clin Invest 118(6):2200–2208. doi:10.1172/jci34725

130. Gosso MF, de Geus EJ, Polderman TJ, Boomsma DI, Heutink P, Posthuma D (2008) Catechol O-methyl transferase and dopamine D2 receptor gene polymorphisms: evidence of positive heterosis and gene-gene interaction on working memory functioning. Eur J Hum Genet 16(9):1075–1082. doi:10.1038/ejhg.2008.57

131. Bertolino A, Di Giorgio A, Blasi G, Sambataro F, Caforio G, Sinibaldi L, Latorre V, Rampino A, Taurisano P, Fazio L, Romano R, Douzgou S, Popolizio T, Kolachana B, Nardini M, Weinberger DR, Dallapiccola B (2008) Epistasis between dopamine regulating genes identifies a nonlinear response of the human

hippocampus during memory tasks. Biol Psychiatry 64(3):226–234. doi:10.1016/j.biopsych.2008.02.001

132. Montag C, Hartmann P, Merz M, Burk C, Reuter M (2008) D2 receptor density and prepulse inhibition in humans: negative findings from a molecular genetic approach. Behav Brain Res 187(2):428–432. doi:10.1016/j.bbr.2007.10.006

133. Tochigi M, Suzuki K, Kato C, Otowa T, Hibino H, Umekage T, Kato N, Sasaki T (2007) Association study of monoamine oxidase and catechol-O-methyltransferase genes with smoking behavior. Pharmacogenet Genomics 17(10):867–872. doi:10.1097/FPC.0b013e3282e9a51e

134. Straub RE, Lipska BK, Egan MF, Goldberg TE, Callicott JH, Mayhew MB, Vakkalanka RK, Kolachana BS, Kleinman JE, Weinberger DR (2007) Allelic variation in GAD1 (GAD67) is associated with schizophrenia and influences cortical function and gene expression. Mol Psychiatry 12(9):854–869. doi:10.1038/sj.mp.4001988

135. Tan HY, Chen Q, Sust S, Buckholtz JW, Meyers JD, Egan MF, Mattay VS, Meyer-Lindenberg A, Weinberger DR, Callicott JH (2007) Epistasis between catechol-O-methyltransferase and type II metabotropic glutamate receptor 3 genes on working memory brain function. Proc Natl Acad Sci U S A 104(30):12536–12541. doi:10.1073/pnas.0610125104

136. Zinkstok J, Schmitz N, van Amelsvoort T, Moeton M, Baas F, Linszen D (2008) Genetic variation in COMT and PRODH is associated with brain anatomy in patients with schizophrenia. Genes Brain Behav 7(1):61–69. doi:10.1111/j.1601-183X.2007.00326.x

137. Jabbi M, Korf J, Kema IP, Hartman C, van der Pompe G, Minderaa RB, Ormel J, den Boer JA (2007) Convergent genetic modulation of the endocrine stress response involves polymorphic variations of 5-HTT, COMT and MAOA. Mol Psychiatry 12(5):483–490. doi:10.1038/sj.mp.4001975

138. De Luca V, Tharmalingam S, Muller DJ, Wong G, de Bartolomeis A, Kennedy JL (2006) Gene-gene interaction between MAOA and COMT in suicidal behavior: analysis in schizophrenia. Brain Res 1097(1):26–30. doi:10.1016/j.brainres.2006.04.053

139. De Luca V, Tharmalingam S, Sicard T, Kennedy JL (2005) Gene-gene interaction between MAOA and COMT in suicidal behavior. Neurosci Lett 383(1–2):151–154. doi:10.1016/j.neulet.2005.04.001

140. Ott U, Reuter M, Hennig J, Vaitl D (2005) Evidence for a common biological basis of the Absorption trait, hallucinogen effects, and positive symptoms: epistasis between 5-HT2a and COMT polymorphisms. Am J Med Genet B Neuropsychiatr Genet 137B(1):29–32. doi:10.1002/ajmg.b.30197

141. Reuter M, Schmitz A, Corr P, Hennig J (2006) Molecular genetics support Gray's personality theory: the interaction of COMT and DRD2 polymorphisms predicts the behavioural approach system. Int J Neuropsychopharmacol 9(2):155–166. doi:10.1017/s1461145705005419

142. Zammit S, Jones G, Jones SJ, Norton N, Sanders RD, Milham C, McCarthy GM, Jones LA, Cardno AG, Gray M, Murphy KC, O'Donovan MC, Owen MJ (2004) Polymorphisms in the MAOA, MAOB, and COMT genes and aggressive behavior in schizophrenia. Am J Med Genet B Neuropsychiatr Genet 128B(1):19–20. doi:10.1002/ajmg.b.30021

Chapter 6

On the Partitioning of Genetic Variance with Epistasis

José M. Álvarez-Castro and Arnaud Le Rouzic

Abstract

The decomposition of genetic variance into additive, dominance, and epistatic components is a common procedure in quantitative genetics. Yet, the interpretation of this variance partition is not trivial, especially concerning nonadditive components. In this chapter, we compile various uses of variance partitioning from published analyses, new simulations, and theoretical examples. We show ways in which advanced genetic modeling facilitates the analysis of data through variance partitioning, focusing on the natural and orthogonal interactions (NOIA) model. We also discuss how epistasis and epistatic variance may influence the outcome of selection, a topic that is still a matter of debate among quantitative and evolutionary geneticists.

Key words Genetic effects, Variance decomposition, NOIA, Epistasis

1 Introduction

The decomposition of genetic variance is nuclear in quantitative genetics. Indeed, Fisher [1] developed the analysis of variance (ANOVA) to perform this kind of decomposition in his foundational paper, "The correlation between relatives on the supposition of Mendelian inheritance," almost one century ago. In that work, he bridged the gap between data that could be obtained by direct empirical observation (phenotypes and relatedness of individuals in a population) and indexes of biological interest (heritabilities, breeding values). A cornerstone in Fisher's theory is the additive variance, V_A, which can be estimated in a population as the degree of resemblance among individuals sharing a known proportion of alleles (e.g., groups of half sibs) and enables prediction of selection response (mean phenotype change between generations).

The expression providing a prediction of the one-generation response to directional selection in a population is called the breeder's equation, $R = (V_A / V_P)S$ (see, e.g., [2]). In this simple equation, the selection response, R, is the change in mean trait value between two consecutive generations. V_P is the total

Jason H. Moore and Scott M. Williams (eds.), *Epistasis: Methods and Protocols*, Methods in Molecular Biology, vol. 1253, DOI 10.1007/978-1-4939-2155-3_6, © Springer Science+Business Media New York 2015

phenotypic variance in the initial population, and the selection differential, S, is the departure of the mean of selected individuals from the mean of the initial population. The proportion of additive variance, $h^2 = V_A / V_P$, is the (narrow-sense) heritability of the trait in the initial population.

With his theory, Fisher shed light on the explanation and interpretation of observations the biometricians had been tackling since the nineteenth century, mostly in relation to artificial selection [3]. The pieces sustaining Fisher's bridge between observable data and biologically meaningful indexes are the genetic effects—effects of allele substitutions on the phenotype—as they arise from factors of inheritance in Mendel's theory. Fisher showed, for instance, how additive variance could be expressed in terms of average effects of allele substitutions in populations and, ultimately, in terms of the frequencies of alleles within the genes underlying the trait in the population and the phenotypic expectation of each of the possible genotypes.

It is, however, to be kept in mind that in Fisher's theory the genetic effects lay mostly within a black box at the core of the machinery of decomposition of genetic variance. In point of fact, Fisher could not possibly build a theory based on those effects as observable entities two decades before the first sound evidence that DNA was the genetic material [4], three decades before its structure and replication mechanisms were elucidated [5, 6], or four decades prior to the first step towards a methodology for measuring the effects at chromosomal loci [7]. Up to seven decades were required by quantitative geneticists to setup quantitative trait locus (QTL) analyses as a promising methodological framework for estimating the relative effects of alleles within loci across the genome for any measurable trait [8], an endeavor that some still perceive as disappointing today [9].

The conceptual paradigm shift of QTL analysis made it evident that it was necessary to resume and further improve models of genetic effects. Dealing with epistasis entailed both a manifest motivation and a major challenge in that process. On the one hand, the need for novel implementations to formulate genetic effects and aid the understanding of their influence on the phenotype at the organismal level was argued (e.g., [10–12]). On the other hand, the need for further work along the lines of Fisher's formulation of (average) genetic effects, to analyze changes in mean phenotypes of populations between generations, was also noted (e.g., [13–15]), and updated implementations were developed based on the landmark conceptualization of epistatic effects in the classical formulations by Kempthorne [16] and Cockerham [17]. These two kinds of formulations are often called functional and statistical, respectively.

On the whole, different parameterizations of the models of genetic effects are necessary to investigate different phenomena in different contexts. It is thus convenient to generalize models in order to cover all cases of interest under a common mathematical framework. This way, routine transformations enable researchers to easily compute the values of genetic effects of a given system and with a desired meaning (e.g., providing variance decompositions of a trait in populations with different genetic frequencies) from values with other meanings (e.g., the ones coming from a QTL analysis). This is the broad rationale of the natural and orthogonal interactions (NOIA) model of genetic effects [18], which we use in this chapter to elaborate on decomposition of genetic variance.

This chapter is thus concerned with what can and cannot be inferred from the partition of the genetic variance into additive and interaction. We shall deal with dominance, V_D, and epistasis, V_I, and their relative contributions to the genetic variance ($V_G = V_A + V_D + V_I$). Particularly, we ponder variance decomposition as a tool to investigate the role of epistasis in the outcome of selection. The organization of this chapter is as follows. We first briefly stress some convenient properties of the NOIA model for analyses involving decomposition of genetic variance. Next, we elaborate on some analyses those properties allow us to perform, including new deterministic simulations of selection acting on genetic systems obtained from epistatic QTL analyses of real data. Then, in light of the previous content, we summon applications for analyses of decomposition of genetic variance, particularly where epistasis is concerned. We finalize the chapter with a general discussion and brief concluding remarks.

2   Decomposition of the Genetic Variance with NOIA

The minutia of NOIA for decomposing the genotypic values and the genetic variance have been described elsewhere (e.g., [19–21]). Here, we do not focus on the particular expressions or software to be used to this aim. Instead, we illustrate the potential of NOIA to analyze genetic variance decompositions in the context of selection. For context, we first point out a few convenient properties of NOIA that are crucial for the underlying principles of such analyses.

2.1   Key Properties

First, NOIA has for the first time encompassed under the same umbrella the approaches by Kempthorne [16] and Cockerham [17] towards implementing epistasis in Fisher's [1] theory of decomposition of the genetic variance. Kempthorne [16] worked out models with arbitrary numbers of loci and alleles per locus under the Hardy-Weinberg proportions. Cockerham [17] focused instead on departures from Hardy-Weinberg proportions, although within a more restricted genetic architecture (two biallelic loci). NOIA has recently been extended to generalize these two approaches, thus accounting for arbitrary numbers of loci, alleles per locus, and departures from Hardy-Weinberg proportions [20].

Second, NOIA provides explicit expressions for genetic effects and thus also for decomposition of genetic variance, which were previously given only through implicit expressions (*see*, e.g., [22, 23]). This convenient property holds for the aforementioned case of several loci and several alleles per locus [20]. Although those expressions may be intricate, conveniently, they are explicit, which enables us to easily find minimum values for the additive variance and thus equilibrium points, as expanded below.

Finally, as mentioned above, NOIA enables transformations between statistical genetic effects and functional genetic effects [18, 20, 24]. The former ones are the ones related to the genetic variance decomposition whereas the latest ones can be useful to describe a genetic system using knowledge about interactions between molecules [25]. These transformations make it possible to study decomposition of the genetic variance in such a genetic system, as exemplified in the next section.

2.2   Applied Cases

2.2.1   A Simple Instance of Bateson-Dobzhansky-Muller Incompatibilities

In evolutionary biology, gene-gene interactions (epistasis) are suspected to play significant roles in several major processes, particularly speciation. When a secondary contact occurs between two allopatric populations that have diverged for a significant amount of time, the offspring of interpopulation crosses may suffer from hybrid depression. This low hybrid fitness, generally attributed to interlocus incompatibilities, according to the Bateson-Dobzhansky-Muller (BDM) model, may thus trigger the evolution of additional isolation mechanisms, eventually resulting in two distinct species [26, 27].

Functional and statistical genetic modeling of BDM incompatibilities has been illustrated by Álvarez-Castro [25], who considered a hypothetical enzyme heterodimer formed by two different monomers, encoded by loci A and B. A population with original alleles A_1 and B_1 splits into two populations in which new alleles arise, A_2 and B_2, respectively, with no effects on enzyme function. He then considered two different cases of incompatibilities between the alleles that arose in isolation, which emerge only after the two populations merge. In the first case, the two new monomers bind easily but lead to non-functional dimers, so that concurrence of the new alleles always leads to a disadvantage. In the second case, the two new monomers bind poorly, so the presence of a new allele at one locus together with the absence of the original allele at the other locus leads to a disadvantage (*see* **Note 1**). Both cases fit the BDM model of genetic incompatibilities and differ only in the properties of the double heterozygote, which suffers from the incompatibility between A_2 and B_2 only in the second case (Fig. 1).

The original motivation for illustrating this example was to show that functional formulations of genetic effects can be useful for representing systems that are initially described in terms of interactions between (or among) molecules. Here, we provide

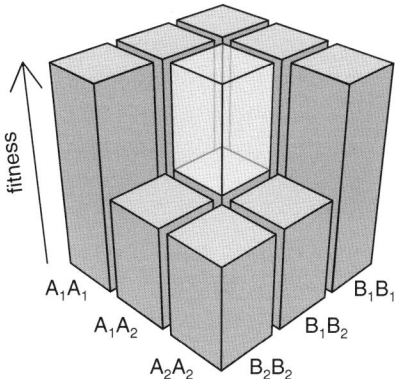

Fig. 1 Genotype-to-fitness (GF) map representing a two-locus genetic system of Bateson-Dobzhansky-Muller (BDM) incompatibilities, as considered in the text. The *transparent bar* indicates the different cases considered (with different incidence of incompatibility in the double heterozygote)

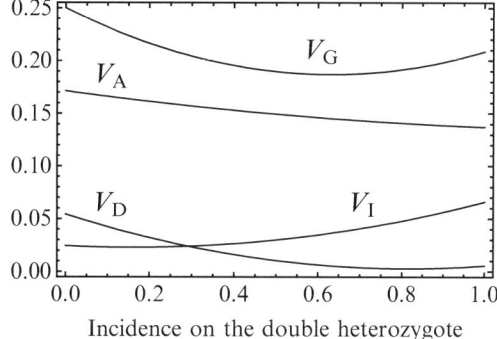

Fig. 2 Variance decomposition of the Bateson-Dobzhansky-Muller (BDM) system considered in the text and Fig. 1, for the different degrees of fitness decline of the double heterozygote (from no decline to the same decline as the other incompatible genotypes). V_G genetic variance, V_A additive variance, V_D dominance variance, V_I epistatic variance

additional information about decomposition of the genetic variance in this genetic system. We consider all cases of populations with similar sizes (up to one of them doubling the other one) and with fitness declines ranging from zero to half the normal fitness. For these cases we obtain that the additive genetic variance always exceeds half of the genetic variance and the proportion of epistatic variance never reaches $1/3$ and may decay to less than 0.1. Some of these cases are shown in (Fig. 2).

With these results, we illustrate a simple fact that is known by many—little evidence of epistatic variance in genetic variance is not necessarily an indication of small epistatic contribution to the outcome of selection. It is indeed clear that epistasis is the responsible mechanism potentially leading to (no less than) speciation under the BDM model. However, the epistatic variance after two populations with BDM incompatibilities merge can be tiny, especially under the broad range of possibilities we considered, whereas the additive variance is always prevalent and can account for almost all the genetic variance.

2.2.2 A Step by a Multiallelic System

Each new allele brings more complexity to the model of genetic effects than the previous one. The k-th allele of a locus brings k new variables (one additive effect and $k-1$ dominance effects) into the system (*see* e.g., [20]). This is along the lines of what happens with epistasis—a new locus not only brings its marginal (additive and dominance) effects into the system, but also effects of interaction with previous loci. Thus, as well as with epistasis, emergent properties arise in the multiallelic case so that the behavior of the system cannot be deduced by analyzing the properties of biallelic subsystems. In the case of the three-allele polymorphism at human acid phosphatase (ACP1) in Europe, the allele C had been claimed to be deleterious by comparing it against (the cluster of) alleles A and B [28]; that is, it was analyzed from a reductionist biallelic perspective. However, Álvarez-Castro and Yang [28] used the expression of additive variance on this multiallelic system to show that the available data are actually consistent with the three alleles having been maintained by balancing selection.

The additive variance of a trait in a population determines the amount of trait variance that selection can use to modify the mean trait value in the next generation (e.g., [2]). It depends on the genotypic frequencies and the genotypic values (the expected trait values of each of the genotypes). Álvarez-Castro and Yang [20] used the aforementioned principle in a reverse manner. Given the genotypic frequencies of a large population sample [29], they minimized the additive variance as a function of the trait values, and they found that the resulting set of trait values actually brought the additive variance to virtually zero. Those values are thus fitness values that explain the sampled genotypic frequencies as a multiallelic equilibrium (at which selection does not affect the mean fitness value due to absence of additive variance for fitness). Reassuringly, those fitness values are in accordance with physiological observations (e.g., the C allele conditions a less efficient use of nutrients but also diminishes the prevalence of fetal macrosomia [30]).

Minimizing the additive variance as a function of some of its variables has been possible because NOIA provides an explicit expression for that function. This advantage enables us to also plot the additive variance for ranges of values of some of its variables. In Fig. 3, we express the ratio V_A/V_G with two of the fitness values as variables (fitness of genotypes AC and AA, ω_{AC} and ω_{AA}, respectively), the other fitness values are the ones that minimize that ratio (and thus also the narrow-sense heritability, $h^2 = V_A/V_P$). The white spot marks the values of ω_{AC} and ω_{AA} that bring V_A/V_G to zero. Note

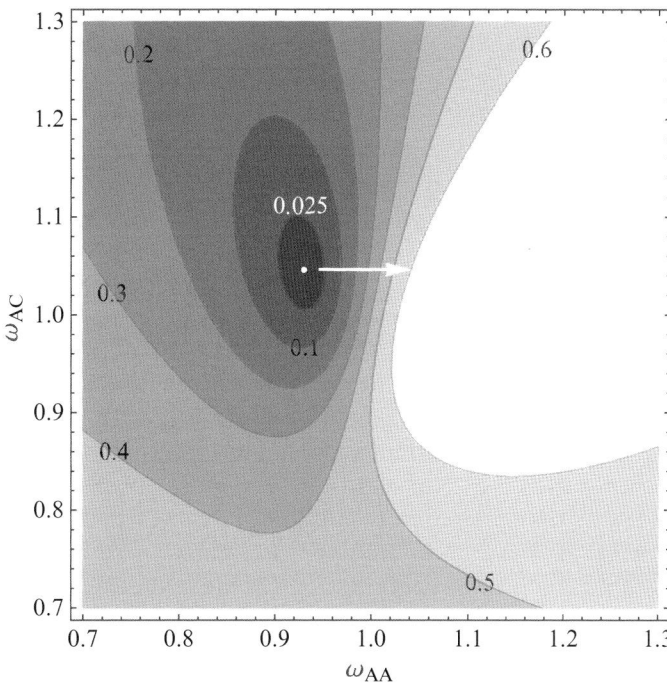

Fig. 3 Contour plot of the ratio V_A/V_G for ACP1 in a German population. The *white spot* marks the minimum value (virtually zero), and the *arrow* indicates the direction of increasing values of ω_{AA}, along which the additive variance rapidly boosts

that small changes in these values may be associated with dramatic increases in the V_A/V_G ratio, which has a clear biological interpretation. In particular, an increase in ω_{AA} (as indicated by the arrow in Fig. 3), would generate a boost of additive variance and thus selection would rapidly change the genotypic frequencies. This could be happening in reality since the polymorphism may have been maintained because genotype AA was affected by fetal macrosomia, which nowadays, with pregnancy monitoring and reasonably safe caesarean sections, is less of a problem than several decades ago.

2.2.3 Mapped Epistasis

The first genetic system we dealt with was built-up (based on a verbal model) and the second one was based on the knowledge of the particular underlying genetics of a trait (one locus with three alleles). In this section, we deal with genetic effects that have been obtained in QTL mapping experiments, leading to more complex (here, four-locus) genotype-to-phenotype (GP) maps.

Epistasis and Hybridization

In recent work, Álvarez-Castro et al. [31] inspected the genetic effects involved in a phenomenon observed in hybrids from two different crosses of chicken. The average growth rate of the F_2 populations was lower than the mid-parental growth rate in two different crosses (hereafter Cross 1 with parental lines being

High × Low [32], two divergent domestic lines selected for high and low growth rate, respectively, and Cross 2 being Wild × Domestic [33]). In both independent crosses, transgressive segregation (F_2 individuals displaying more extreme phenotypes than their parents) were also observed. These observations could not be interpreted in terms of dominance effects alone, and a multi-locus analysis showed that pairwise epistatic effects estimated from QTL scans could explain the empirically observed hybrid effects.

It is noteworthy that the predictive power shown for epistatic effects coming from QTL analyses of a trait has been achieved in two independent crosses, with different characteristics (in particular, the two sets of four loci were not overlapping). The epistatic variance in Cross 1 accounts for about two thirds of the genetic variance explained, whereas in Cross 2 it accounts for about one quarter of it. Thus, this is one more example in which the proportion of epistatic variance, which was very different in the two cases, did not seem to signify much about the epistatic effects (proven to be significant for the two cases). On the other hand, the four-locus networks of Cross 1 and Cross 2 explain about 10 and 20 % of the phenotypic variance, respectively. Thus, relatively small amounts of phenotypic variance do not necessarily preclude considerable predictive power of the detected genetic effects on interesting observed patterns (as also pointed out by Zuk et al. [34]).

Variance Component Dynamics During Selection Response

The way in which mid-term selection response can be affected by dominance and epistasis can be illustrated by simulating the evolution of plausible genetic architectures subject to selection. Therefore, we further inspect the two aforementioned sets of four loci. Note that the loci identified for each cross were different, and thus the two crosses constitute independent genetic architectures. Our general motivation is to illustrate the changes in genetic variance components along generations of selection on genetic systems for which epistasis has been proven to condition observations of evolutionary importance. We have thus run simulations with a range of initial parameters and we show a few of the cases that illustrate examples of evolutionary interest.

Figure 4 shows the simulated selection response (up- and down-directional selection) for Cross 1 with and without epistasis in the GP map. Further additive genetic variation in the up-directional selection line is released as the frequency changes along generations. Epistasis plays a considerable role, as highlighted by the expected dynamics when removing the interactions. A boost in additive variance also occurs with up-directional selection at Cross 2 (Fig. 5a, b). In this case, variation is maintained at one locus at equilibrium due to overdominance (the fitness of heterozygotes being higher than homozygotes), resulting in

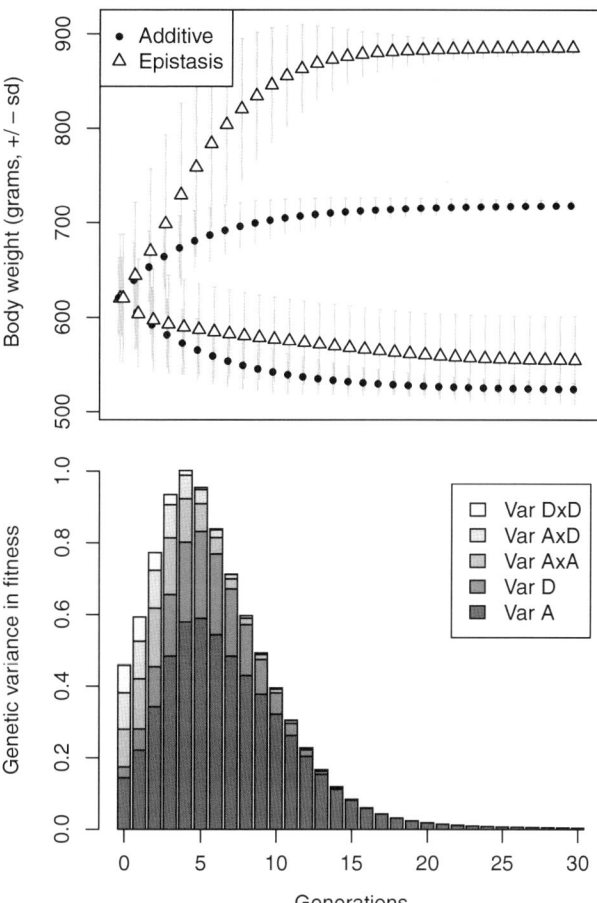

Fig. 4 Simulated evolution of body weight based on the estimated four-locus genotype-to-phenotype (GP) map from Cross 1 (see text for details), starting from an F_2 population. (**a**) Phenotypic evolution (mean and standard deviation) for the epistatic map (all second-order interactions considered between the four loci), compared to the additive map. (**b**) Dynamics of the variance decomposition for the "up" selected line in the epistatic case

significant dominance variance present at equilibrium. Down-directional selection in this cross also leads to maintenance of variation at the same locus (Fig. 5c, d). This is thus a striking case of sign-epistasis on dominance (either over- or under-dominance at one locus occurs on the phenotypic scale, depending on the genetic background).

Figure 6 shows the outcome of an instance of stabilizing selection on Cross 2. After several generations of stasis with very low additive variance (and thus low change in mean phenotype), a boost occurs that brings the system to fixation. Mind that stasis at the level of the trait value does not mean absence of changes in

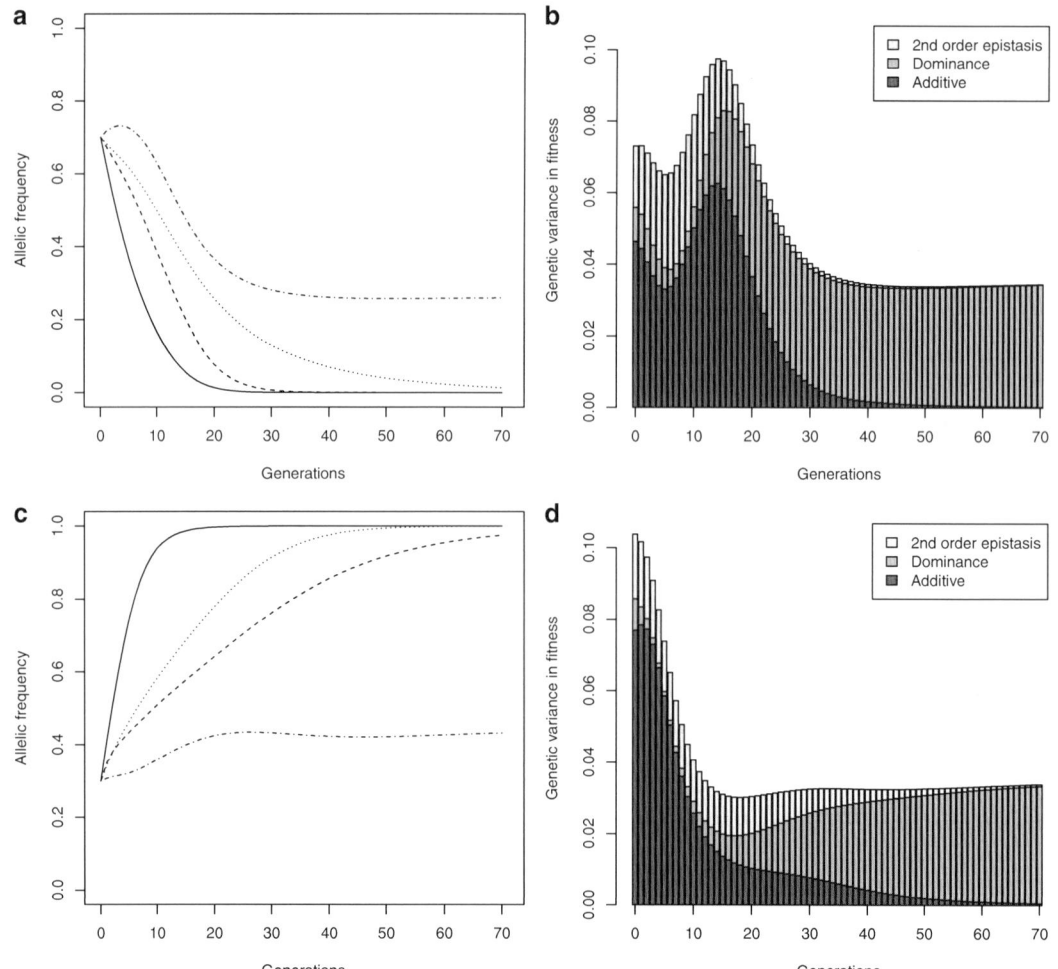

Fig. 5 Simulated evolution for "up" (**a** and **b**) and "down" (**c** and **d**) directional selection of Cross 2 (see text for details). The frequencies of the alleles from the Wild parental line were set to 0.7 in the "up" selection and 0.3 in the "down" selection, to increase the phenotypic space crossed by the population in both cases. (**a**) and (**c**) represent the evolution of allelic frequencies at the four loci; (**b**) and (**d**) display the dynamics of variance components. For both "up" and "down" selection, the best genotype is the heterozygote, and one out of the four loci remains polymorphic at equilibrium as a result of balancing selection. Each allele is associated with the same pattern in (**a**) and (**c**)

allele frequencies (Fig. 6a). In our simulation study, boosts of additive variance were common and several equilibria occurred with various polymorphic loci, which sometimes led to the presence of epistatic variance at equilibrium. Stasis periods followed by boosts may also occur under stabilizing selection on an additive phenotype, and they can last longer with increasing genetic complexity (e.g., with a higher number of epistatic loci involved, results not shown).

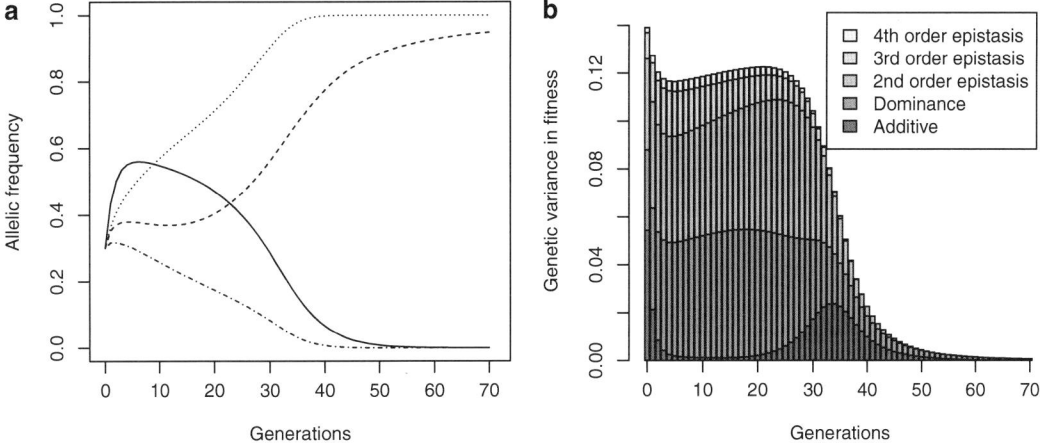

Fig. 6 Simulated evolution under stabilizing selection in Cross 2 (see text for details). The fitness function is Gaussian and centered around a phenotypic optimum of 314 g, the unweighted mean of the GP map. (**a**) Allelic frequencies. (**b**) Variance components. Eventually the four loci fixed. Each allele in (**a**) keeps the same pattern as in Fig. 5

3   Applications and Interpretations of Variance Decomposition

3.1   Variance Decomposition of What?

We first consider the scale in which to measure the trait—the scale that defines what we shall call the phenotype of each individual. Given that scale, the genotypic values (the expected phenotype for each possible genotype) provide the GP map of the trait. The meaning of the decomposition of genetic variance will make sense in the context of that choice, and variance will be expressed in squared phenotypic units. The original works on the decomposition of the genetic variance were carried out with the concept of artificial selection in mind. It is well known that Fisher himself was largely driven by applying his work to eugenics. In this context, the choice of the scale is obvious—it will measure what the breeder considers important for improving production, e.g., growth rate, height, or oil content. The breeder's equation, in which the additive variance plays a central role, then predicts the effect of directional selection on this phenotypic scale [2].

The action of stabilizing selection on the phenotype has traditionally been modeled by modifying the directional selection intensity, so that the selection intensity is zero when the population mean matches the fitness optimum [35]. This way, the decomposition of genetic variance at the optimum will still reflect the amount of genetic variance that is available for directional selection to alter the mean phenotype of the population. In other words, according to this choice, when a population reaches a stabilizing selection optimum, the additive variance for the trait value can be high, but there will be no selection response since selection does not act.

Here, we have chosen instead to consider analyses of the decomposition of genetic variance of the genotype-to-fitness (GF) map. When the fitness function is linear on the phenotype, the difference may not be relevant, since the scale in which the phenotype is measured is directly related to fitness. With stabilizing selection on the phenotype, however, the relationship between phenotype and fitness can be more complex and, in particular, not monotonous. Assuming the resulting GF map, selection will nevertheless be directional in fitness and thus decomposition of the genetic variance in fitness fits to its original meaning. In particular, when a population reaches a stabilizing selection optimum, the additive variance in fitness is zero, regardless of whether genetic variance (either for the trait value or fitness) remains or not. This way, the trait does not change despite the action of selection, although selection is far from negligible; it actually determines the genetic composition of the population at the equilibrium point.

A clear example of the difference between a GP map and its corresponding GF map can be seen in the results of the simulations above. No higher order epistatic variances are present with directional selection on a map with up to pairwise epistasis (Figs. 4 and 5). Stabilizing selection is however nonlinear on the phenotype and thus contributes to the generation of epistasis for fitness. In particular, high-order (third and fourth) interactions occur under stabilizing selection that were absent at the phenotypic level from a phenotype with pairwise epistasis only (Fig. 6). In summary, performing the variance decomposition on fitness thus ensures that selection will always be directional, whether the same holds for the underlying phenotype under study or not.

3.2 The Variance Components

The starting point for interpreting decomposition of genetic variance is the well-known principle that additive variance indicates the genetic variation of a trait that selection can use to modify the population mean in a one-generation step. Conversely, dominance and epistatic variances indicate the amount of variation on the trait in a population that is due to the genetic constitution of the population, but it cannot contribute to selection response in a one-generation step. Although ostensibly simple, some further clarification about these statements often proves valuable.

One of the most important things to keep in mind is that additive variance may be influenced by any type of interaction that occurs at the functional level. Indeed, the distinction between functional and statistical genetic effects helps define how dominance and epistatic interactions contribute to additive genetic variation, depending on allele frequencies in the population. Additive variance is the variance of statistical additive effects that accounts for all trait variation reflecting genotype differences at one generation that allow selection to change the genetic composition in a way that affects the mean trait value of the next generation.

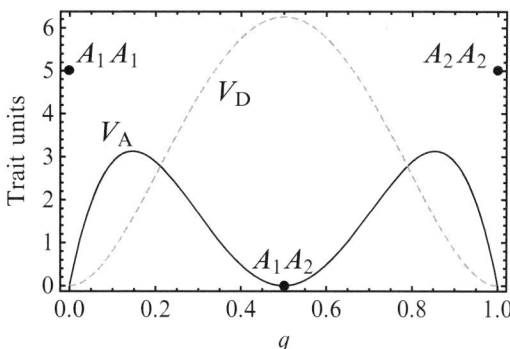

Fig. 7 Variance decomposition of a one-locus, two-allele genetic system with under-dominance, for all possible allele frequencies, q (assuming Hardy-Weinberg proportions). The variances are actually plotted as trait units squared

The original genotype-dependent trait variation may be caused by any kind of functional genetic effects.

Figure 7 shows an extreme, one-locus simple case in which that feature can easily be visualized. Despite the two homozygotes of a biallelic locus having the same trait value and the lack of functional additive effects, the dominance interaction provokes hybrid breakdown and bestows additive variance to the system whenever the frequencies are not at equilibrium (when the two alleles are present and have different frequencies). Analogous outcomes may occur due to epistasis. For instance, synergistic epistasis can contribute to additive variance, and thus speed up selection, whereas antagonistic epistasis can slow down selection by shrinking additive variance—as compared to the additive case.

The definition of the different genetic effects is made in a hierarchical way so that functional interactions may affect lower order variances, and no interaction variance of higher order than the highest functional interaction present in the system will exist. However, absence of significant interaction variances cannot be used to assess the presence of their corresponding functional genetic effects or their importance for evolution. The case in Fig. 7 also illustrates this point, since, close to the extremes, the dominance variance becomes visibly smaller than the additive variance, whereas the functional dominance effect is always higher than the (completely absent) functional additive effect.

Boosts of additive variance during selection responses could be influenced by epistasis since, as a matter of fact, epistasis (or dominance) can shift the allelic frequencies at which the additive variance is maximal, leading to unexpected increases in additive variance when selecting from an F_2 population (the most polymorphic population in an additive model). Yet, the amount of epistatic variance is a poor indicator of the propensity to release selectable additive variation. Incidentally, the popular expectation that

Table 1
Summary of interpretations of variance decomposition

	Additive variance	Dominance and epistatic variances
Meaning	Trait variance available for selection to change the trait mean in a one-generation step	Genetic variance that is not available for selection to change the trait mean in a one-generation step
Link to functional effects	Functional interactions contribute to the additive variance and they can actually enlarge it	Absence of interaction variances may occur in the presence of functional interactions
Mid-term response	Boosts of additive variance can occur whether influenced by interactions or not, whereas interactions are implied in intermediate stasis stages	Interaction variances indicate that the GP map is not linear and that medium-term (directional) selection response can be faster or slower than expected under additivity
Equilibrium	The additive variance is zero at equilibrium points, which can be used to assess genotypic frequencies as equilibrium frequencies and for equilibrium fitness estimation	Dominance variance will always be present at polymorphic equilibria, which does not hold for epistatic variance

variance components could convert into each other (e.g., that an increase in additive variance could be due to a "transfer" from epistatic variance) has been demonstrated to be misleading [36, 37]. In Fig. 6, conversion of interaction variances into additive variance may happen, since the dominance variance decreases during the increase of additive variance. However, boosts of additive variance actually occur regardless of reductions of interaction variances, as is the case in (Fig. 5b).

Table 1 summarizes some major guidelines for reading the decomposition of the genetic variance. Overall, interactions may be important for the outcome of selection even when interaction variances are low, as illustrated above with the BDM case. In other words, statistical genetic effects do not necessarily catch the importance of functional interactions for mid- and long-term evolution. The additive variance (and the decomposition of the genetic variance in general) was built as an index to condense very specific points. Situations that differ in important aspects may coincide for those particular points. In Fig. 8, we provide two simple cases with equal variance decomposition and appreciably different outcomes.

3.3 Variance Decomposition at Equilibrium

By construction, additive variance in fitness at equilibrium is zero. However, if the equilibrium is polymorphic due to overdominance, the genetic variance at equilibrium is non-null. This situation mechanistically generates some dominance variance for fitness, while epistatic variance may (or may not) be present if several loci

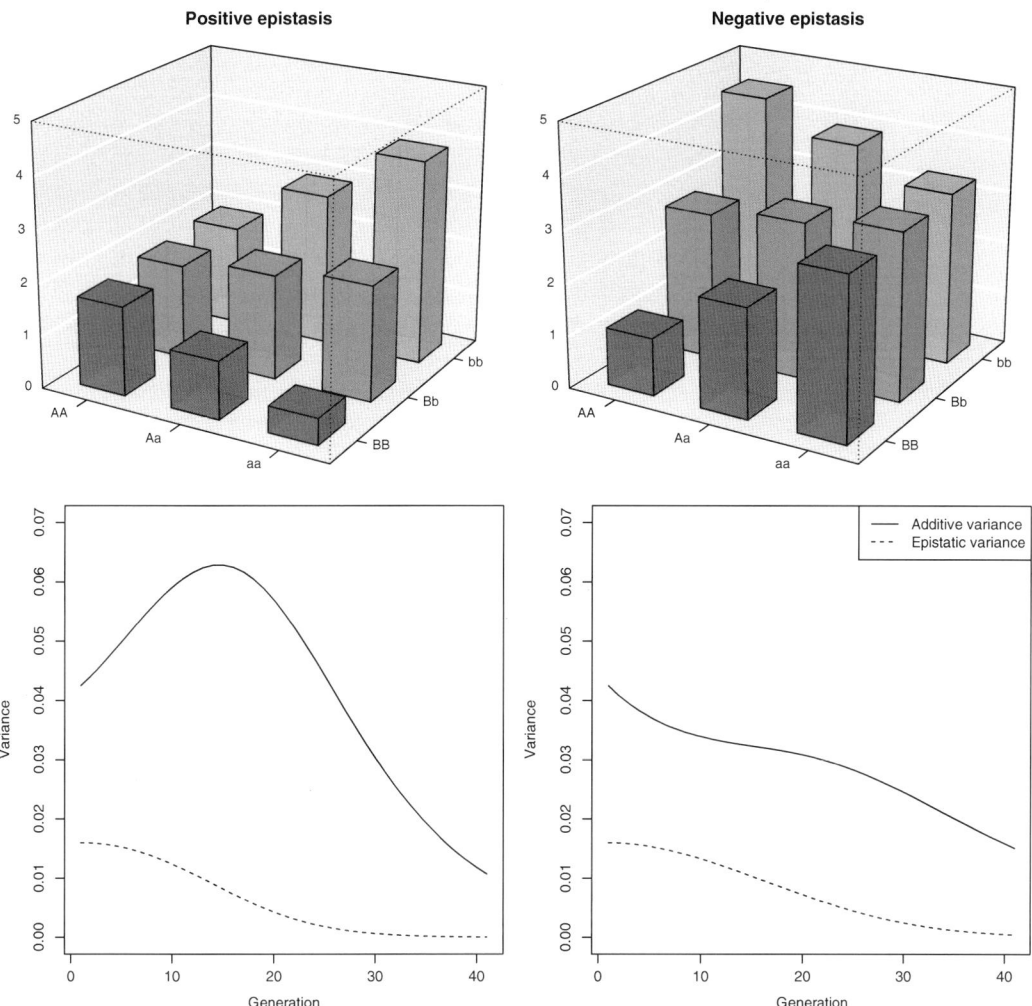

Fig. 8 Two two-locus epistatic maps (**a** and **b**) leading to the same decomposition of the genetic variance in ideal F$_2$ populations. Panels (**c**) and (**d**) show different outcomes of selection through deterministic simulation on the two maps (with fitness proportional to the phenotype), despite the fact that they start from exactly equal variance decompositions

are polymorphic. The meaning of the interaction variances at equilibrium does not change—they account for trait variation maintained at equilibrium that are attributable to genotypic differences.

Since epistasis alone (without any dominance effects) cannot generate stable polymorphic equilibria, statistical dominance (and dominance variance) will be present at stable equilibria. Nevertheless, since statistical dominance can be influenced by functional additive-by-dominance and dominance-by-dominance epistasis, the study of evolutionary equilibria in genetic systems requires information about all kinds of functional genetic interactions in addition to the marginal effects.

In any case, beyond the particular pattern decomposition of genetic variance may take at equilibrium, in this chapter we comment on a different approach based on variance partitioning—the possibility of obtaining a set of parameters that minimize the additive variance in a given situation. We have shown that by minimizing the additive variance for a set of frequencies, it is possible to inspect whether those frequencies are equilibrium frequencies (thus bringing the additive variance to zero) and obtain a putative set of fitness values explaining the polymorphic equilibrium. This is of special importance when either multiple alleles or epistatic interactions are present, since, for such complex cases, general theoretical studies of equilibrium and stability are not available [38].

4   Discussion and Conclusions

4.1   Epistasis and Missing Heritability

The missing heritability problem consists of genetic factors uncovered in thorough mapping studies that account for a non-satisfactory portion of the total narrow-sense heritability of a trait. The claim has been made recently that this issue is mediated by the presence of epistasis. In particular, it has been stressed that the traditional way of computing the decomposition of genetic variance for obtaining total heritability from phenotypic and relatedness data easily generates overestimates (called "phantom heritability") when epistasis is present [34]. Indeed, although epistatic interactions can contribute to the additive variance and revealing epistatic factors may help to increase the explained heritability of a trait in a population, epistasis is by nature known to be difficult to map and estimate properly. Even the kind of epistasis that generates significant phantom heritability may be very difficult to uncover in mapping studies [34].

It has also been shown that slight reductions of linkage disequilibrium (LD) between causative factors and markers may conspicuously erode the observable epistatic variance in genome-wide association studies (GWAS) [39]. On the other hand, even if the missing heritability issue may seem to vanish in mapping populations with large LD blocks, this is but a mirage that comes at the cost of overpassing the real genetic architecture of the trait [40]. Hence, although accurately revealing epistatic networks remains challenging, there seems to be no good reason for considering the detection of missing heritability as a plausible indication of unsatisfactory disclosure of the genetic architecture of a trait; indeed—and paradoxically—sometimes the opposite might actually happen. Consequently, it is reasonable that epistatic networks as revealed in mapping studies may be sufficiently accurate to entail convenient evolutionary predictive power, despite explaining a small proportion of variance, as shown for some empirical cases [31].

4.2 Epistatic Variance and the Role of Epistasis in Selection Response

It is not impossible that the epistatic variance dominates the variance decomposition [34, 41]. However, claims have been made that it is reduced after a period of selection [42]. When considering epistatic genetic architectures with many loci involved, the epistatic variance may dominate the variance components also after a period of selection [39], thus putatively in the vicinity of an equilibrium. A certain portion of epistatic variance may also be present at equilibrium, indicating phenotype variance is maintained at equilibrium due to epistasis.

Interactions are known to maintain variation and genetic variance under directional selection, generating for instance "incomplete selective sweeps" due to over-dominance [43]. In relation to this, we recall that even when epistasis (and thus epistatic variance) is not present at equilibrium, the features of the equilibrium may still be conditioned by epistasis. Indeed, above we have shown a case leading to maintenance of variation in one of the loci through sign epistasis on dominance—generating over-dominance at one locus in both directions. Interestingly, epistasis conditions the maintenance of this equilibrium in the face of directional selection, whereas we have also shown that epistasis can condition fixation under stabilizing selection.

4.3 Estimating the Influence of Epistasis via Its Directionality

For a long time, interaction components of the genetic variance have been mistreated in quantitative genetics, especially when predicting responses to artificial selection [44]. The breeder's equation (shown in the introduction) effectively pools the genetic interaction variances together with environmental variance, hence working as nuisance terms in regards to predicting trait-mean change. However, the breeder's equation is only expected to satisfactorily predict and interpret the short-term dynamics of genetic progress in breeding programs. With time, as selection causes changes in allelic frequencies, additive selectable variance can be revealed or hidden depending on the nature and the intensity of interactions, particularly when sign epistasis is present [45]. Striking cases involve periods of stasis between two periods of phenotypic changes, as shown in this chapter.

For making predictions about selection response in longer time-frames than the breeder's equation, an index explicitly accounting for epistatic interactions has been developed, the directional epistasis index [12, 46]. This index characterizes the nonlinearity of the GP map for an average individual in the population, and contrary to epistatic variance components it provides a measurement of epistasis that has predicting power in terms of evolvability ([47] also *see* Chapter 7). Incidentally, there is a methodological divide between the directional epistasis index and heritability. The second one has been constructed to explain short-term phenomena and consequently it can be measured from observations that take place at the trait level during a short time frame. The directional epistasis index pursues predictions on trait behavior in the long run and consequently it is more difficult to infer it from trait information—it is currently obtained from genetic effect estimates instead [48].

4.4 Concluding Remarks

Overall, indexes are useful because they condense information that is important in particular directions. Here we have reviewed that variance decomposition is informative in several regards. We have focused on selection and thus on the resemblance between parents and offspring. Further, the interaction components of genetic variance have been widely used in classical works to predict the genetic structure within families, since variance components can be directly used to compute expected covariances between relatives in general [1, 22]. But it is crucial to keep in mind that only the additive variance entails a definite predictive power in terms of evolutionary potential.

Quantitative genetics was founded on the pillars of variance decomposition. However, as represented in this chapter, plausible genetic architectures are coming from different approaches nowadays. Thus, beyond variances, genetic effects can currently be inferred, which opens new paths and sheds further light onto our understanding of the influence of epistasis on the genetic determination of traits and the outcomes of selection.

5   Notes

1. We note a mistake in the corresponding sentence of the original paper [25]. Instead of "In this case, a disadvantage arises whenever an allele '2' is present at one locus only—either A2 or B2 but not both of them together—due to ...", the sentence should read, "In this case, a disadvantage arises whenever an allele '2' is present at one locus without an allele '1' being present at the other locus, due to ...".

Acknowledgements

We acknowledge funding from grants BFU2010-20003 and EM2014/024 from the now defunct Spanish Ministry of Science and Innovation and the autonomous administration Xunta de Galicia, respectively, to J.A.C. and by grant ERT 256507 from the European Research Council to A.L.R. We thank Estelle Rünneburger for helpful comments on the manuscript.

References

1. Fisher RA (1918) The correlation between relatives on the supposition of Mendelian inheritance. Trans R Soc Edinburgh 52:339–433

2. Falconer DS, MacKay TFC (1996) Introduction to quantitative genetics, 4th edn. Prentice Hall, Harlow

3. Galton F (1886) Regression towards mediocrity in hereditary stature. J R Anthropol Inst Great Brit Ireland 15:246–263

4. Avery OT, MacLeod CM, McCarty M (1944) Studies on the chemical nature of the substance inducing transformation of pneumococcal types induction of transformation by a desoxyribonucleic acid fraction isolated from pneumococcus type ii. J Exp Med 79:137–158

5. Franklin R, Gosling RG (1953) Molecular configuration in sodium thymonucleate. Nature 171:740–741

6. Watson JD, Crick FHC (1953) A structure for deoxyribose nucleic acid. Nature 171:737–738

7. Thoday JM (1961) Location of polygenes. Nature 191:368–370

8. Lander ES, Botstein D (1989) Mapping mendelian factors underlying quantitative traits using RFLP linkage maps. Genetics 121(1):185–199

9. Travisano M, Shaw RG (2013) Lost in the map. Evolution 67(2):305–314

10. Cheverud JM (2000) Detecting epistasis among quantitative trait loci. In: Wolf JB, Brodie ED, Wade MJ (eds) Epistasis and the evolutionary process. Oxford University Press, Oxford, pp 58–81

11. Cheverud JM, Routman EJ (1995) Epistasis and its contribution to genetic variance components. Genetics 139(3):1455–1461

12. Hansen TF, Wagner GP (2001) Modeling genetic architecture: a multilinear theory of gene interaction. Theor Popul Biol 59(1):61–86

13. Kao CH, Zeng ZB (2002) Modeling epistasis of quantitative trait loci using Cockerham's model. Genetics 160(3):1243–1261

14. Mao Y, London NR, Ma L, Dvorkin D, Da Y (2006) Detection of SNP epistasis effects of quantitative traits using an extended Kempthorne model. Physiol Genomics 28(1):46–52

15. Yang R-C (2004) Epistasis of quantitative trait loci under different gene action models. Genetics 167(3):1493–1505

16. Kempthorne O (1954) The correlation between relatives in a random mating population. Proc R Soc Lond B Biol Sci 143(910):102–113

17. Cockerham CC (1954) An extension of the concept of partitioning hereditary variance for analysis of covariances among relatives when epistasis is present. Genetics 39:859–882

18. Álvarez-Castro JM, Carlborg Ö (2007) A unified model for functional and statistical epistasis and its application in quantitative trait loci analysis. Genetics 176(2):1151–1167

19. Álvarez-Castro JM, Carlborg O, Ronnegard L (2012) Estimation and interpretation of genetic effects with epistasis using the NOIA model. Methods Mol Biol 871:191–204

20. Álvarez-Castro JM, Yang R-C (2011) Multiallelic models of genetic effects and variance decomposition in non-equilibrium populations. Genetica 139(9):1119–1134

21. Le Rouzic A, Álvarez-Castro JM (2008) Estimation of genetic effects and genotype-phenotype maps. Evol Bioinform 4:225–235

22. Kempthorne O (1957) An introduction to genetic statistics. Wiley, New York

23. Lynch M, Walsh B (1998) Genetic analysis of quantitative traits. Sinauer, Sunderland

24. Yang R-C, Álvarez-Castro JM (2008) Functional and statistical genetic effects with miltiple alleles. Curr Top Genet 3:49–62

25. Álvarez-Castro JM (2012) Current applications of models of genetic effects with interactions across the genome. Curr Genomics 13(2):163–175

26. Burke JM, Arnold ML (2001) Genetics and the fitness of hybrids. Annu Rev Genet 35:31–52

27. Dobzhansky T (1936) Studies on hybrid sterility. II. Location of sterility factors in Drosophila pseudoobscura hybrids. Genetics 21:113–135

28. Wilder JA, Hammer MF (2004) European ACP1*C allele has recessive deleterious effects on early life viability. Hum Biol 76(6):817–835

29. Brinkmann B, Hoppe HH, Hennig W, Koops E (1971) Red cell enzyme polymorphisms in a northern German population. Gene frequencies and population genetics of the acid phosphatase (AP), phosphoglucomutase (PGM), adenylate kinase (AK), adenosine deaminase (ADA) and 6-phosphogluconate dehydrogenase (6-PGD). Hum Hered 21(3):278–288

30. Greene LS, Bottini N, Borgiani P, Gloria-Bottini F (2000) Acid phosphatase locus 1 (ACP1): possible relationship of allelic variation to body size and human population adaptation to thermal stress – a theoretical perspective. Am J Hum Biol 12(5):688–701

31. Álvarez-Castro JM, Le Rouzic A, Andersson L, Siegel PB, Carlborg O (2012) Modelling of genetic interactions improves prediction of hybrid patterns – a case study in domestic fowl. Genet Res (Camb) 94(5):255–266

32. Marquez GC, Siegel PB, Lewis RM (2010) Genetic diversity and population structure in lines of chickens divergently selected for high and low 8-week body weight. Poult Sci 89(12):2580–2588

33. Kerje S, Carlborg Ö, Jacobsson L, Schutz K, Hartmann C, Jensen P, Andersson L (2003) The twofold difference in adult size between the red junglefowl and White Leghorn chickens is largely explained by a limited number of QTLs. Anim Genet 34(4):264–274

34. Zuk O, Hechter E, Sunyaev SR, Lander ES (2012) The mystery of missing heritability: genetic interactions create phantom heritability. Proc Natl Acad Sci U S A 109(4):1193–1198

35. Kimura M, Crow JF (1978) Effect of overall phenotypic selection on genetic change at individual loci. Proc Natl Acad Sci U S A 75(12):6168–6171

36. Barton NH, Turelli M (2004) Effects of genetic drift on variance components under a general model of epistasis. Evolution 58(10):2111–2132

37. Turelli M, Barton NH (2006) Will population bottlenecks and multilocus epistasis increase additive genetic variance? Evolution 60(9):1763–1776

38. Templeton AR (2006) Population genetics and microevolutionary theory. Wiley-Liss, Hoboken, NJ

39. Hemani G, Knott S, Haley C (2013) An evolutionary perspective on epistasis and the missing heritability. PLoS Genet 9(2):e1003295

40. Shen X (2013) The curse of the missing heritability. Front Genet 4:225

41. Gjuvsland AB, Vik JO, Woolliams JA, Omholt SW (2011) Order-preserving principles underlying genotype-phenotype maps ensure high additive proportions of genetic variance. J Evol Biol 24(10):2269–2279

42. Hallander J, Waldmann P (2007) The effect of non-additive genetic interactions on selection in multi-locus genetic models. Heredity 98(6):349–359

43. Sellis D, Callahan BJ, Petrov DA, Messer PW (2011) Heterozygote advantage as a natural consequence of adaptation in diploids. Proc Natl Acad Sci U S A 108(51):20666–20671

44. Hansen TF (2013) Why epistasis is important for selection and adaptation. Evolution 67(12):3501–3511

45. Weinreich DM, Watson RA, Chao L (2005) Perspective: sign epistasis and genetic constraint on evolutionary trajectories. Evolution 59(6):1165–1174

46. Carter AJ, Hermisson J, Hansen TF (2005) The role of epistatic gene interactions in the response to selection and the evolution of evolvability. Theor Popul Biol 68:179–196

47. Houle D, Pelabon C, Wagner GP, Hansen TF (2011) Measurement and meaning in biology. Q Rev Biol 86(1):3–34

48. Pavlicev M, Le Rouzic A, Cheverud JM, Wagner GP, Hansen TF (2010) Directionality of epistasis in a murine intercross population. Genetics 185(4):1489–1505

Chapter 7

Measuring Gene Interactions

Thomas F. Hansen

Abstract

Measurement is the assignment of numbers to reality, and valid measurement requires that these numbers represent relevant aspects of reality. I discuss epistatic gene interactions from a measurement-theoretical perspective and argue that the standard measurements of epistasis in classical quantitative genetics have failed to capture aspects of epistasis that are relevant to selection dynamics and adaptation. Instead, the use of statistically motivated measurements such as epistatic variance components has led to the misconception that epistasis is dynamically inert. Here, I review work showing that patterns of epistasis may have profound effects on evolutionary dynamics and discuss how these patterns can be measured.

Key words Epistasis, Gene interaction, Measurement theory, Selection response, Adaptation, Quantitative genetics

1 Introduction

Epistasis is a stepchild of the modern synthesis. The compelling heuristics of considering individual alleles segregating in a population one by one served the purposes of illustrating the power and ubiquity of natural selection, and the simplicity of the resulting models allowed refined representation of the ecological and social contexts in which selection may take place. However, the possibility of epistasis, or gene interaction, is both mathematically and conceptually obvious, and could not be completely ignored. The founders of mathematical population genetics differed sharply on how to treat epistasis. For Sewall Wright, epistasis was a loved child with a central place in his shifting-balance theory of evolution (e.g., Wright [1]). For Ronald Fisher, however, it was a nuisance that clouded the elegance of selection in action. Indeed, Fisher came up with an ingenious way of sidelining epistasis from evolutionary theory. His definitions of average effect and average excess of alleles were essentially ways of averaging out the effects of gene interactions by assuming that alleles segregating in large populations over time will participate in a huge number of interactions

Jason H. Moore and Scott M. Williams (eds.), *Epistasis: Methods and Protocols*, Methods in Molecular Biology, vol. 1253, DOI 10.1007/978-1-4939-2155-3_7, © Springer Science+Business Media New York 2015

with other alleles, so that the only thing that matters for selection is the fitness of the allele averaged across these interactions (Fisher [2]). In this way Fisher was able to capture the dynamically relevant effects of alleles individually without actually having to assume that genes do not interact. The population variance of these average effects summed over loci is the additive genetic variance. As selection acts on variance, the additive variance becomes the determinant of evolutionary potential regardless of whether epistasis is present or not.

The Fisherian notion of an average effect serves as a convenient theoretical underpinning for evolutionary models in which genes/alleles are assigned effects in isolation from the rest of the genome. This single-gene perspective is the basis of most theorizing in fields such as behavioral ecology, life-history theory, evolutionary quantitative genetics, game theory, adaptive dynamics, and also for more popular theorizing based on "selfish" genes and the like.

The widespread success of research strategies based on single-gene and additive models strengthened the impression that evolutionary dynamics could be understood without bothering with gene interaction. Complex epistasis-based theories of evolution, such as Wright's shifting-balance theory, had limited heuristic value, and simple models with constant gene effects became the underpinning of most evolutionary theory. In present-day evolutionary-biology textbooks epistasis is usually defined and then ignored apart from a few brief and often implicit appearances in connection with the evolution of reproductive isolation and coadapted gene complexes. Epistasis is also largely ignored in quantitative genetics. Even if epistatic variance components are often measured, they have little theoretical relevance. Models of phenotypic adaptation are exclusively focused on additive variance.

A consensus has developed that epistasis has little influence on evolutionary dynamics. Within the field of quantitative genetics in particular, it is typically said that epistasis does not have a permanent effect on the response to selection (e.g., Bulmer [3]; Crow [4, 5]; Hill et al. [6]). These claims seem to derive from two closely related theoretical results. The first comes from Griffing [7], who showed that even though some epistatic effects are inherited from parents to offspring, the selection response that this generates is due to a build up of linkage disequilibrium that will decay if selection is relaxed. The second result, from Kimura [8], shows that when the build up of linkage disequilibrium reaches a (quasi) equilibrium between selection and recombination, the ongoing recombinational break down of linkage disequilibrium exactly nullifies the contribution of epistatic variance to the selection response.

I have argued that these two theoretical results have been misinterpreted (Hansen [9]). They strictly apply to the part of the selection response generated by the build-up of linkage disequilibrium and ignore the permanent effects of epistasis that are due to changes in allelic effects. Citing these results as evidence for the unimportance of epistasis in the selection response is flawed. In fact, Carter et al. [10] showed both analytically and by simulation that epistasis could have large effects on the selection response mediated through systematic changes in allelic effects. This should have been obvious from consideration of epistasis in simple two-locus models (e.g., Lewontin [11]; Chapter 6), but the inconsistency between these and the quantitative genetic models has rarely been commented upon. For example, Crow and Kimura [12] dismiss the importance of epistasis in the selection response at the same time as they analyze specific two-locus models where it is clear that epistasis must have dynamical effects. In this chapter I will discuss the conceptual basis of this curious epistasis blindness, and argue that it has to do with how epistasis has been measured. By this, I mean measurement in the conceptual sense of a numerical representation of an empirical phenomenon. I will outline how conceptual advances in the measurement of epistasis that started in the 1990s have set the stage for a better understanding of its evolutionary potential, and on this basis I will review its evolutionary significance and outline some issues and challenges for meaningful measurement of gene interactions.

2   The Conceptual Measurement of Epistasis

2.1   A History of Measuring Epistasis

The term epistasis was introduced by Bateson to describe the situation in which the effect of one gene was blocked by the effect of another gene (*see* Phillips [13]). This foreshadows the use of the term in classical population genetics, where it refers to situations in which the effects of alleles at different loci combine nonadditively. A more recent and precise way of stating this is to say that epistasis occurs whenever the effect of an allele substitution (in an individual) depends on the state of the genotype at other loci (Wagner et al. [14]; Hansen and Wagner [15]). This notion of epistasis has been termed physiological, biological, or functional epistasis (e.g., Cheverud and Routman [16]; Phillips [13]; Hansen and Wagner [15]; Punyani et al. [17]; Moore and Williams [18]; Hansen [19]; Álvarez-Castro and Carlborg [20]), and contrasts with Fisher's [21] statistical representation of epistasis as interaction in a regression model. Fisher's average effects can be thought of as the linear terms in a regression of phenotype on allelic states ("gene content") and statistical epistatic effects are then defined from interaction terms in the same model. The variance components associated with these interaction terms are the standard measures of epistasis in quantitative genetics. These are the familiar additive-by-additive and additive-by-dominance epistatic variance, etc., developed by Cockerham [22] and Kempthorne [23].

The conceptual distinction between biological and statistical epistasis was not clear during the modern synthesis. Although "biological" thinking about epistasis did play a role in some of the theorizing that occurred during the modern synthesis, and proponents for the importance of epistasis would emphasize the context dependency of gene effects (e.g., Lewontin [11]; Templeton [24, 25]; Wright [26]), it was not widely appreciated that these consequences are not detectable from the Fisherian model of statistical epistasis.

Many advances in science happen with advances in measurement. Better measurement is often driven by the development of better instruments or better statistics, but measurement is also conceptual. In formal measurement theory, measurement is defined as a mapping from a system of empirical relations to a number system. Valid measurement requires that the numbers we assign to empirical objects are such that the relations among the numbers represent the (relevant) relations among the objects (e.g., Hand [27]). Any such representation must be theoretically and empirically justified, and many scientific debates have centered on the validity of proposed numerical representations. Well-studied examples from other fields include debates on whether or how IQ or skull measurements represent mental ability (e.g., Michell [28]), whether or how the readings of thermometers represent "temperature" (Chang [29]), or whether or how p-values represent evidence (Vieland and Hodge [30]). Houle et al. [31] recently argued that a measurement-theoretical perspective has been absent from biology, and that this has allowed a conspicuous lack of demand for the justification of proposed methods of measurement (*see* also Schneider [32]). Intuitive statistics are proposed based on statistical convenience and while their statistical properties are often carefully examined, there is little demand to show that they represent anything biologically meaningful. Many biologists seem strangely happy to spend long hours obtaining "measurements" without any justification that these represent anything worth knowing.

The estimation of epistatic variance components is an example. While Fisher's focus on average effects and the additive variance was well justified by their central role in theoretical models describing the effects of natural selection, no theory was used to motivate the epistatic terms. They simply followed a statistical extension of the regression model used to define average effects. Epistatic variance components are statistically well defined and have convenient statistical properties such as orthogonality. They can be estimated with standard techniques, and the properties of these estimates are well understood (e.g., Lynch and Walsh [33]). What was lacking was an argument for why we should care about them in the first place. The main theoretical models ascribing importance to epistasis, such as shifting balance and the Bateson-Dobzhansky-Muller model of reproductive isolation, did not have explicit links to epistatic variance components. Consequently, estimation of these could not help to test or illuminate these models beyond the trivial task of supporting the existence of some form of gene interaction. Models of statistical resemblance between relatives were the only cases in which epistatic variance components appeared as parameters. While this shows that there may be contexts in which they could matter, epistatic variance components have not yet been linked to evolutionary theory in a substantially important manner.

More seriously, Griffing's and Kimura's proofs that the epistatic variance components are not relevant in selection dynamics led to the misinference that epistasis itself is not relevant. This fitted conveniently with the gene-centered view of the modern synthesis, but it was a measurement-theoretical mistake. Without justification, the epistatic variance components were treated as general representations of epistasis, and their dynamical inertia were assumed to be a general property of biological epistasis. This generated inconsistencies in the general neodarwinian theory. On one hand it was claimed that epistasis did not matter for evolution by natural selection and on the other there were arguments about coadapted gene complexes, epistasis generating multiple fitness peaks, epistasis generating genetic incompatibilities, epistasis affecting mutation and inbreeding loads, and epistasis influencing the evolution of sex and recombination, etc. Indeed, I suspect most biologists share the intuition that the evolution of complex organisms requires genes to adapt their interactions to work together as a unified whole.

There seems to have been little explicit effort put toward resolving these inconsistencies. I suppose this was due to the absence of a conceptual framework in which to express a solution. Such a framework started to appear in the 1990s with the realization that epistasis can also influence the additive and dominance variance in the population (e.g., Goodnight [34, 35]; Tachida and Cockerham [36]; Moreno [37]; Cheverud and Routman [16, 38]; Whitlock et al. [39]). The key distinction between "physiological" (i.e. biological) and statistical epistasis was put forward by Cheverud and Routman [16], and this made it entirely clear that biological epistasis has a direct effect on additive variance. Simultaneously, the renewed interest in the evolution of development, embodied in the field of "evodevo," inspired research into the evolution of genetic architecture within a population-genetics framework. So far, most work in population genetics had taken gene effects as given entities with fixed values. In a key paper, Wagner et al. [40] recast the classical Waddingtonian idea of canalization as a question about the evolution of gene and mutational effects, and studied this with population-genetic methods. This required epistasis in the form of genes that acted as modifiers of the effects of other genes. Rice [41] developed a similar approach in an elegant mathematical language that made clear how underlying physiological variables could generate epistasis that allowed the evolution of canalization and generally affected evolutionary dynamics. At the same time, many also started to investigate the links between "physiological" interactions in gene or metabolic networks and showed how these led to biological and statistical epistasis with evolutionary consequences (e.g., Keightley [42, 43]; Szathmary [44]; Gibson [45]; Wagner et al. [14]; Frank [46]; Johnson and Porter [47]; Omholt et al. [48]; Bagheri et al. [49]).

The parameterization of biological epistasis used by Cheverud and Routman [16] was done with reference to a population in which all genotypes appeared at equal frequencies. This is arbitrary, and not a usable framework for general theoretical or methodological work. Hansen and Wagner [15] proposed a more general parameterization of "functional" (i.e., biological) epistasis based on a multilinear representation that is amenable to analytical and statistical work. Their key insight was that biological epistasis and indeed all biological gene effects have to be measured relative to a particular reference genotype, and that the measured effects will change depending on which reference genotype is chosen. This was not explicit in previous models, and I believe this was one of the obstacles to understand the dynamical role of epistasis. Based on this representation, Hansen and Wagner [15] showed how biological epistasis could be consistently measured, and they derived equations for linking these measures to statistical epistasis. This multilinear representation was the basis for the analytical description by Carter et al. [10] of how biological epistasis affects the selection response as discussed further below.

The critical assumption in the multilinear model is that any genotype substitution (including allele substitutions) could change the phenotypic effects of any other genotype substitution, but only as linear functions of its own phenotypic effect. This allows general evolution of the magnitudes of gene effects, but only limited changes in the order (or topology) of the effects. It also allows easy transformation among different reference genotypes, since the multilinear form is invariant to the reference genotype. Still, various formulas for change between reference genotypes under other conditions have been developed by Kirkpatrick et al. [50], Barton and Turelli [51] and Álvarez-Castro and Carlborg [20]. In particular, Jose Álvarez-Castro and collaborators have developed a very general framework for measuring gene and genotype effects that unifies all the previous parameterizations of both biological and statistical epistasis (Álvarez-Castro and Carlborg [20]; Álvarez-Castro et al. [52, 53]; Álvarez-Castro and Yang [54]; reviewed in Le Rouzic and Álvarez-Castro [55]; Álvarez-Castro [56]). This framework, called the NOIA model, can be used to represent any form of epistasis, including the classical statistical measures, and is particularly well suited for estimating and interpreting gene interactions from quantitative trait locus (QTL) data (*see* Chapter 6).

In summary, the conceptual and mathematical representation (i.e., the measurement) of epistasis has advanced from conceptually naive verbal and statistical models to a general mathematical theory with nuanced conceptual distinctions. For example, the fact that any measurement or theoretical result about epistasis depends on a reference genotype is easily overlooked. This dependency arises because the very meaning of epistasis is that gene effects depend on the genetic background, and then the genetic background must be made explicit in the measurement process. It is often the case that epistasis measured in experimental set ups is not congruent with the epistasis defined in theoretical models simply because different reference genotypes are implicitly assumed. In particular, in simulation models, gene effects will often be defined relative to a reference that will soon be far outside the population range when the system starts to evolve. Claims about effects of epistasis or other properties will then refer to entities that have little to do with what can be operationally measured in an experimental setting. Similarly, many studies postulate epistatic modifier genes that can alter the effects of other genes without themselves having effects on the trait. It can be proven that such modifiers will necessarily acquire direct effects as the reference genotype is changing, and observed selection on a modifier can therefore not be taken as evidence that selection is caused by the modification (Hansen [57]). Only in a strictly additive model will gene effects remain invariant to the reference in which they are measured. Below I will outline how the lack of a proper conceptual framework led to unstated implicit assumptions in some of the classical analyses of epistasis in the selection response and to the persistent misconception that epistasis is not important.

2.2 The Relationship Between Biological and Statistical Epistasis

Biological epistasis is a descriptor of the genotype-phenotype map. The definition of biological epistasis as a dependency of the effect of a gene substitution on the genetic background can be graphically represented as a nonlinearity in the genotype-phenotype map as illustrated in Fig. 1a. This representation dates back to Waddington's epigenetic landscape and to Rendel [58], who used it to illustrate the evolution of canalization (see also Rice [41]). Canalization occurs when the population moves into the flat area of the graph, and the phenotypic effects of gene substitutions become smaller or even zero. This makes it immediately obvious how a change in genetic background (position on the x-axis) can also change gene effects and hence the additive genetic variance of the population. If the "molecular" genetic variation on the x-axis is the same, the phenotypic variation on the y-axis will be larger for the population in the steep area (Fig. 1b).

Hansen and Wagner [15] developed a formal description of how statistical measures of genetic effects and variances depend on the properties of the genotype-phenotype map as represented by

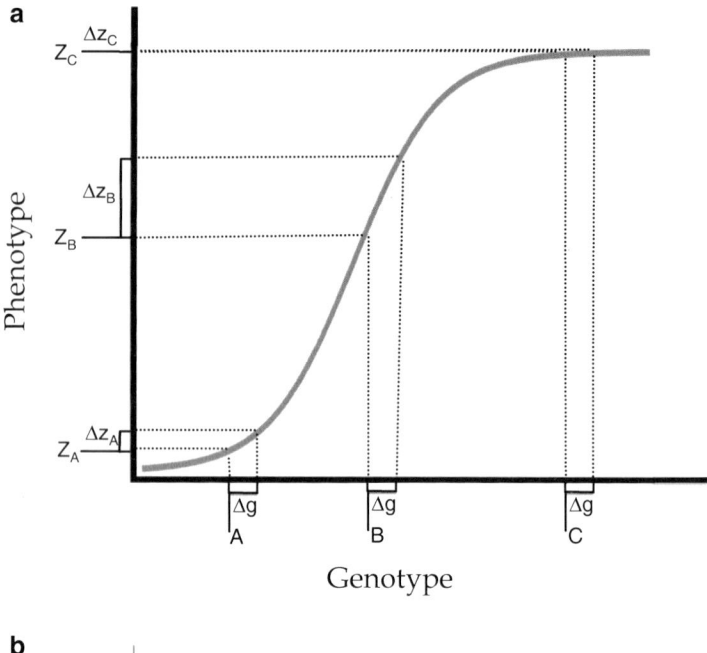

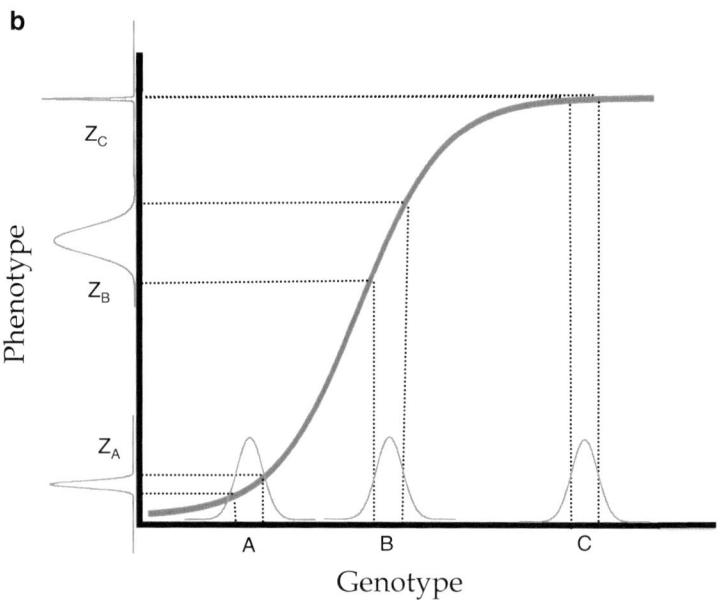

Fig. 1 How systematic epistasis affects evolution. (**a**) The *drawing* shows an idealized genotype-phenotype map in which the phenotypic effects, Δz, of a given molecular gene substitution, Δg, depend on the genetic background (i.e., position in the genotype-phenotype map). In position A, the substitution has effect Δz_A, in position B, it has effect Δz_B, and in position C, it is completely canalized and has the effect $\Delta z_c = 0$. In the range between A and B there is positive directional epistasis for trait z, and in this range selection for increased trait values would increase gene effects, additive variance, and evolvability. In the range between B and C, there is negative directional epistasis with opposite effects. In (**b**) the effects of mapping molecular to phenotypic distributions are shown.

the multilinear model described above. One result is that the average effect of a gene substitution (in Fisher's sense) will change with genetic background as function of the epistatic interactions of the locus with the background. If ^iy is the effect of the substitution at locus i in a given reference genotype, then the effect of the substitution in the background of another genotype will be $f\,^iy$, where the epistasis factor, f, is given as

$$f = 1 + \sum_j{}^{ij}\varepsilon\,^jy + \left[\text{higher order terms}\right],$$

where ^jy is the effect of changing locus j from its state in the reference genotype to its state in the new genotype, the $^{ij}\varepsilon$ are coefficients in epistasis terms, $^{ij}\varepsilon\,^iy\,^jy$, describing the phenotypic effect of the interaction between locus i and j, and the sum is over all loci that differ between the reference genotype and the new background. Higher-order terms describe effects of changes in sets of loci when higher-order epistasis is present. Fisher's average effects are then obtained by taking the average of $f\,^iy$ over all the backgrounds in the population. By choosing a (certain) population mean as the reference genotype in such a way that the average $f = 1$, there is no effect of epistasis, but as soon as this reference is changed (as when the population mean evolves) then epistasis will alter the average effects. The additive variance contributed by this locus will be f^2V_A, where V_A is the additive variance contributed by the locus in the background of the reference genotype. The total additive variance will be the sum of these over all loci. This illustrates how additive variances can be scaled up and down by epistasis as the genetic background changes.

This formalism also yields scaling relationships between the variance components. For example, the additive-by-additive epistatic variance scales with the square of the additive variance as

$$V_{AA} = \frac{1}{2}\overline{\varepsilon^2}V_A^2,$$

where $\overline{\varepsilon^2}$ is a variance-weighted average of the squares of the epistasis coefficients defined above. Higher-order epistatic variances

---

Fig. 1 (continued) If we consider that the genotype-phenotype map in the range between A and B can be approximated with a quadratic function, $z = g^2$, and there is an amount Var[g] of molecular genetic variance, then a population at position A would have additive variance Var$_A$[z] $\approx$ 2AVar[g], and a population at position B would have additive variance Var$_A$[z] $\approx$ 2BVar[g]. This shows how systematic epistasis leads to systematic changes in additive variance as the population is shifted around in the genotype-phenotype map. It does not matter whether the shifts are caused by drift, selection, migration, or other mechanisms. Note also that a population at the completely canalized position C would have additive variance Var$_A$[z] = 0, and not be able to respond to selection

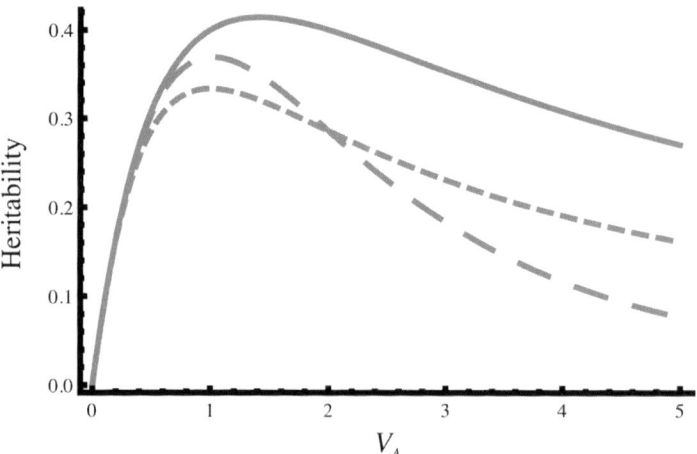

Fig. 2 The relation between additive variance and heritability under three different types of epistasis. The *whole line* shows pairwise epistasis with an average squared epistasis coefficient of $\bar{\varepsilon}^2 = 1$. The *finely-dashed line* shows the same with $\bar{\varepsilon}^2 = 2$, and the *coarsely-dashed line* shows a case where also third- and fourth-order epistasis are included; the second-order, third-order, and fourth-order squared coefficients are all 1. The additive variance on the *x*-axis is measured in units of environmental variance. Based on scaling relationships derived in Hansen and Wagner [15]. See also Hansen et al. [59]

scale with the corresponding powers of the additive and dominance variances. These relationships can be understood by thinking of the genetic model as a Taylor approximation to the genotype-phenotype map. When there is little "molecular" variation, a linear (additive) approximation fits well and most of the genetic variance is additive. As the molecular variance increases, the nonlinearities become more important and the epistatic terms describe relatively more of the variance; the additive variance will increase, but the epistatic variances will increase faster. This is one reason why heritabilities should not be used to measure levels of additive genetic variance (Hansen et al. [59]). As illustrated in Fig. 2, the heritability, which is a ratio between additive and total variance, may correlate positively or negatively with additive variance depending on the level of genetic variance and strength of epistasis.

In general, this illustrates how the statistical variance components are not biologically independent and do not properly delineate biological additivity and epistasis. If there is relatively little genetic variation in the population, the additive model will fit well, but this does not rule out strong effects of epistasis if the population is shifted to new positions on the genotype-phenotype map. Hence, epistasis becomes essential to understand evolutionary dynamics on all but the smallest scales.

Griffing [7] noted that one quarter of the additive-by-additive epistatic component of variance, and smaller fractions of higher-order epistatic variance, are transferred from parent to offspring. Hence, selection on epistatic variance components will generate a response in the trait mean, but he then showed that this response is mediated entirely through changes in linkage disequilibrium and does not produce any change in individual allele frequencies. This follows from the Fisherian model, because the epistatic interactions are by definition orthogonal to the population and allelic averages, and cannot change these averages. Because linkage disequilibrium is continuously broken down by recombination, Griffing then concluded that the response to selection on epistatic variance was not permanent, and would decay as soon as selection ceased. In his analysis, Griffing assumed that the additive variance stayed constant, which may seem reasonable when the goal is to understand the effects of selection on the epistatic variance, but it carries the implicit assumption that there are no effects of epistasis on the additive variance. In the statistical sense this is true by definition, but as I demonstrated above it is far from true for the effects of biological epistasis.

Kimura [8] derived a related result. He showed that a two-locus, two-allele system under directional selection will reach a state in which linkage disequilibrium stays constant in a balance between selection and recombination. In this state of "quasi linkage equilibrium" the response to selection is determined by the additive variance with no contribution from epistatic variance components (but see Karlin [60]). Although cited as a basis for the irrelevance of epistasis in selection (e.g., Crow [5]), this result does not imply that the response is unaffected by epistasis. In fact, numerical simulations in Kimura's paper show huge changes in the additive variance, and my reanalysis of a version of his model in Fig. 3 shows that these changes and hence the response to selection are strongly influenced by biological epistasis (see Hansen [9] for details).

Interestingly, some numerical simulation studies dating back to the 1960s show effects of epistasis on the response to selection. For example, Young [61] concluded that heritabilities are poor predictors of responses to selection in the presence of epistasis, but he lacked the conceptual tools to discuss why this happens. With hindsight we can see that this was because a simulation model cannot avoid a representation of biological epistasis, and Young's assumptions implied systematic directional epistatic effects, which will affect the response as discussed below and illustrated in Fig. 3. See Hallander and Waldmann [62] for a review of such studies and a new numerical study showing the impact of biological epistasis.

I conjecture that the reason why results such as those of Griffing and Kimura were, and still are, misinterpreted is that the statistical independence (orthogonality) of the variance components in the

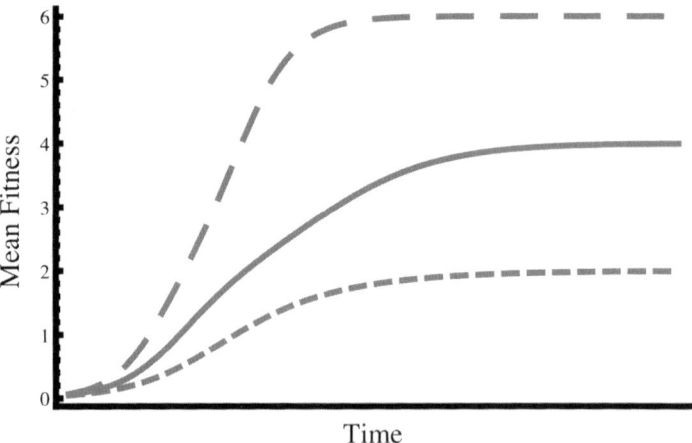

Fig. 3 Evolutionary dynamics of mean (Malthusian) fitness for a two-allele, two-locus epistatic model with different types of epistasis. The whole line gives the dynamics of an additive genotype-phenotype map, the *coarsely-dashed line* shows the dynamics of a positive epistatic map, and the *finely-dashed line* shows the dynamics of the negative epistatic map. If we refer to the alleles at the two loci as a and A, and b and B, respectively, then the positive epistasis is set such that the effect of the substitution b→B is enhanced by 50 % if the substitution a→A has already happened, and the negative epistasis is set such that that the effect of the substitution b→B is diminished by 50 % if the substitution a→A has already happened. In the additive map the effect of any allele substitution is independent of the state of the other locus. The shown dynamics are with free recombination, but even strong linkage cause only small differences to the dynamics. The starting haplotype frequencies were 0.98 for ab, 0.01 for Ab, and 0.005 for aB and AB. Random mating is assumed. Based on numerical integration of dynamical equations given in Hansen [9]. Figure reproduced from Hansen [9] with permission from John Wiley and Sons Limited

Fisherian model are thought to represent the properties of epistasis in general. This is a measurement-theoretical mistake. The properties of the measurements are taken to be the properties of the natural objects in the absence of a theoretical link between the two. The development of such a theory is now taking place on the foundation of a conceptual and mathematical distinction between biological and statistical epistasis.

3   Epistasis and Evolutionary Dynamics

3.1   Epistasis and Genetic Drift

The rather obvious fact that genes have to work together to make a complex organism suggests that their interactions are an important factor in evolutionary dynamics. But having convinced themselves that epistasis did not have permanent effects on the selection response, evolutionary quantitative geneticists were forced into a search for the keys under the lamp post, which in this case was provided by the idea that genetic drift during bottlenecks could convert epistatic variance into additive variance, which subsequently could affect adaptive evolution (e.g., Goodnight [34, 35]; Cockerham and Tachida [63]; Cheverud and Routman [38]). This resulted in a substantial research industry with theory, simulations, and experimental studies. Indeed, the juxtaposition of so many studies on interactions between epistasis and drift with the near absence of studies on the interaction between epistasis and selection illustrates better than anything how severely research on epistasis has been emasculated by the classical quantitative genetics theory.

Moreover, the conception of epistatic variance being converted into additive variance is misleading. Although genetic drift, like any other change in gene frequencies, can alter gene effects and additive variances in the presence of epistasis, there is no meaningful conversion of variance in the sense that additive variance is expected to increase in proportion to changes of epistatic variances in a way that suggests a conversion law. Instead, random changes in allele frequencies that are caused by drift lead to random changes in the average effects of other epistatically interacting alleles. This may cause random changes in all variance components (beyond the changes directly caused by drift). Although the expectation of these changes depends on the patterns of gene interactions and distributions of genotype frequencies, and are thus not quantitatively predictable from the observation of the variance components alone, there is a general bias towards increases of variance, such that the presence of pairwise interactions will tend to increase additive variances, and the presence of three-way interactions will tend to increase pairwise epistatic variance components, etc. (Hansen and Wagner [15]; Barton and Turelli [51]; Hermisson and Wagner [64]; Hill et al. [65]; Zhang [66]). These effects are, however, not dramatic, and require very strong epistasis and very small population sizes to matter in evolutionary dynamics (Turelli and Barton [67]).

3.2 Epistasis and Directional Selection

If biological epistasis is present, any change in allele frequencies, not just those caused by drift, may lead to changes in the additive genetic variance. Hence, epistatic systems will typically behave differently from additive systems under selection, because epistasis causes changes in the additive variance. Carter et al. [10] presented analytical equations to describe these changes under linear selection on a multilinear epistatic system (ignoring effects of linkage disequilibrium). The first terms of the equations for the per generation changes in the trait mean and the additive variance, $\bar{z}$ and, V_A, respectively, are

$$\Delta \bar{z} = \beta V_{A} + \frac{1}{2} \varepsilon \left(\beta V_{A} \right)^{2} + \frac{1}{6} \tau \left(\beta V_{A} \right)^{3} + \dots$$

$$\Delta V_{A} = 2\beta \sum_{i} {}^{i}C_{3} - 2\beta^{2} \sum_{i} {}^{i}C_{2}^{2} + 2\beta \varepsilon V_{A}^{2} + \beta^{2} \left(\tau + \omega^{2} \right) V_{A}^{3} + 4\beta^{2} V_{A} \sum_{i} {}^{i}\varepsilon \, {}^{i}C_{3} + \dots,$$

where β is the selection gradient, ${}^{i}C_{2}$ and ${}^{i}C_{3}$ are the second and third cumulants of allelic (reference) effects at locus i, summations are over all loci, and ε, τ, and ω^{2} are composite parameters describing patterns of epistasis. By inspecting these equations, we can see that even the change in the trait mean is affected by epistasis, but this effect is generally small unless selection and epistasis are very strong, and it also tends to be counteracted by the dynamics of linkage disequilibrium making the additive prediction very good. The important effects of epistasis are mediated through changes in the additive variance (and higher moments, which we will not consider here). The two first terms in the dynamics of the variance are the standard terms we also find in an additive model, which describes how changes in additive variance depend on the skew and squared variance of the allelic effect distributions at individual loci. The subsequent terms describe effects of epistasis. Of particular importance is the parameter ε, which is a weighted average of the epistasis coefficients defined above. The ε parameter thus measures the average tendency for positive genetic change to elevate the effects of other positive genetic changes. This is called positive directional epistasis, because there is a systematic tendency for positive changes to reinforce each other. A negative value of ε means that there is an average tendency for positive changes to reduce the effects of other positive changes. This is called negative directional epistasis. The equation shows that positive directional epistasis tends to elevate additive genetic variance under positive directional selection while negative directional epistasis has the opposite effect, reducing additive variance. Over time this also affects the dynamics of the trait mean, which depends on the additive genetic variance. Simulations in Carter et al. [10] show that directional epistasis can have dramatic effects on the response of the trait mean to directional selection over more than a few generations (see also Hansen et al. [68]; Hallander and Waldmann [62] and Fig. 3).

As hinted by the equations above, the selection dynamics of epistatic systems are complicated. The two other epistasis parameters given above describe third-order directional epistasis (τ), and "locus-directional epistasis" (ω^{2}), which is the average tendency for individual loci to have consistent directional interactions with other loci. Locus-directional epistasis could be present even when there is no overall directionality in the system as a whole, because the specific directionality of loci could differ. These patterns have smaller, but not negligible effects. The second-most important effect, however, comes from the effects of directional epistasis on the skew (the third cumulant), which to some extent counteracts the effects of directional epistasis on the variance, since positive directional epistasis tends to generate negative skew in the locus distributions (see Carter et al. [10] and Hansen et al. [68] for details).

Even if the mathematical description of the effects of epistasis on the selection response is complex, the underlying biology is easy to understand. Directional selection changes allele frequencies, generally increasing the frequencies of alleles that increase the trait in the direction favored by selection. With epistasis these changes will change the effects of other potential allelic substitutions. If the epistasis is systematic (i.e., directional) then there will be systematic change in the effects of potential allele substitutions, which manifest in systematic change in the additive variance, which will again lead to systematic change in the response of the trait mean to selection. As illustrated in Fig. 3, epistasis is even important in simple two-locus, two-allele systems.

A geometric perspective on this process is to think of biological epistasis as a result of nonlinearities in the genotype-phenotype map, as shown in Fig. 1. As described mathematically by Rice [41, 69, 70] such nonlinearities can lead to systematic changes in the phenotypic effects of underlying physiological or genetic changes as the population moves about in the genotype space. That this can have strong effects on evolutionary dynamics is obvious. As made clear already by Rendel [58], systematic nonlinearities allow the evolution of canalization and decanalization. In population genetic terms, systematic nonlinearity in the genotype-phenotype map will manifest as directional epistasis, and the changes in evolvability described by the equations above can be described as canalization or decanalization under directional selection.

Perhaps due to the lack of a theoretical expectation, there are few direct empirical investigations of the role of epistasis in the selection response. The most detailed example is provided by Carlborg et al. [71] and Le Rouzic et al. [72], who analyzed populations of chicken that had been selected up and down for body size over 42 generations, and showed that the influence of what was originally thought to be a major locus was in fact a system of four epistatically interacting loci (see also Álvarez-Castro et al. [73]). They showed that these epistatic interactions mediated a larger selection response. Similarly, Pavlicev et al. [74] quantified the influence of directional epistasis on the response to selection on body size in mice. Le Rouzic et al. [75] looked for an influence of epistasis in fruitflies selected for wing traits, but the effects were small and open to alternative interpretations. Ungerer et al. [76] looked for effects of genetic background on selection responses in *Arabidopsis* and found some small effects. This was only over three generations of selection, however, and was also based on averaging over several lines, so it could be consistent with quite strong epistasis. More generally there is evidence for directional epistasis from many line-cross analyses of selected lines, which is strong, indirect evidence that epistasis has played a role in the response.

3.3 Epistasis, Evolvability, and the Maintenance of Additive Variance

If a prolonged response to directional selection over hundreds or thousands of generations is possible, almost any pattern of epistasis will eventually have dramatic and different effects, rendering the response unpredictable in the absence of genetic details (Hansen et al. [68]; Yukilevich et al. [77]). A sustained long-term directional response is not biologically realistic, however. It will be stopped by changes in the selection regime (ever more extreme trait values are rarely beneficial) or by genetic constraints (which may be manifest as negative directional epistasis). It is therefore more realistic to consider the effects of epistasis on the consequences of stabilizing and fluctuating forms of selection, and in particular when these are in mutation equilibrium. With my collaborators, I have attempted a systematic investigation of the dynamics of multilinear epistatic systems under some simple selection regimes, which I will briefly review here.

First, multilinear epistasis has general effects on the amount and type of genetic variance maintained in balance between mutation and stabilizing selection (Hermisson et al. [78]; Álvarez-Castro et al. [79]; Fierst and Hansen [80]; Le Rouzic et al. [81]). To understand this situation, it is important to consider evolution on two levels. The first is the establishment of a mutation-selection equilibrium given a genetic architecture (i.e., a pattern of mutational effects, epistasis), and the second is the slower evolution of the genetic architecture made possible by epistasis. Hermisson et al. [78] showed that the standing additive genetic variance, and in most cases the total genetic variance, will always be lower in mutation-selection balance with multilinear epistasis (and no overdominance) as compared to an additive genotype-phenotype map with the same mutational effect sizes. In addition, the epistasis also allows the evolution of reduced mutational effect sizes (i.e., canalization), which will further reduce the equilibrium variances. Complete canalization, where all additive genetic variance disappears, is a theoretical possibility, but requires fine-scale adjustment of the genetic interaction effects and is not likely to happen in practice. Le Rouzic et al. [81] analyzed this situation further and extended to fluctuating selection. They showed that the canalizing selection induced by the negative curvature of stabilizing fitness functions also exists under most forms of stationary fluctuating fitness functions, but that the potential for canalization is still lower under fluctuating selection simply because the back and forth movement of the trait mean induce "random" changes of the genetic architecture that makes it harder to organize strong canalization. A result is that mutational effects and the additive genetic variance maintained in mutation-selection balance will be elevated under symmetric stationary fluctuating selection relative to stabilizing selection. With asymmetries in the fitness landscape the results become more dependent on the directionality of epistasis as discussed for directional selection above.

These results are based on the assumption that there is no overdominance. With multilinear epistasis, it is not possible to evolve overdominance on a locus if it was not initially present, and conversely, not possible to remove it if it was initially present. With some exceptions, multilinearity allows only the evolution of what has been called the "magnitude" (Weinreich et al. [82]) or "order-preserving" (Gjuvsland et al. [83, 84]) aspects of epistasis. With more general forms of epistasis, the evolutionary dynamics become more complicated, because the possibility of changes in the ordering of gene effects may then allow the evolution of heterozygote superiority, which will have strong effects on the maintenance of genetic variance. See Gimelfarb [85], Gavrilets [86], Liberman and Feldman [87], or Weinreich et al. [82] for some relevant results and discussion. Although it is obvious that order-breaking epistasis may also have dramatic effects on the short-term selection response, this has not been studied systematically.

In addition to its influence on the evolution of allele effects and additive variance, including pleiotropic effects and genetic covariances (e.g., Rice [70]; Pavlicev et al. [88, 89]; Griswold and Henry [90]), epistasis also allows the evolution of dominance and epistatic effects (Hansen and Wagner [15]; Bagheri and Wagner [91]; Bagheri [92]), which can be canalized or decanalized by higher-order epistasis in the same way as allele effects are canalized or decanalized by second-order epistasis (Hansen et al. [68]; Le Rouzic et al. [81]).

---

4 Epistasis in Adaptation and Macroevolution

4.1 Epistasis, Adaptation, and Constraint

The fact that epistasis becomes progressively more important with larger genetic change makes it essential for understanding macroevolution. The additive model predicts few genetic constraints on longer time scales. If gene effects remain the same regardless of how much the phenotype is changing, then it is always possible that new mutations can keep evolution going. In contrast, negative epistasis predicts a slowdown of evolution and raises the possibility of epistatic constraints that could stop adaptive evolution. Epistatic constraint is thus a possible solution to the "paradox of stasis," which refers to the prevalence of macroevolutionary stasis in morphological traits with abundant additive genetic variation (Hansen and Houle [93]; Estes and Arnold [94]; Uyeda et al. [95]). In the analysis of the multilinear model discussed above, we have seen that both stabilizing and directional selection may lead to complete canalization in theory, but that this is unlikely in practice (Hermisson et al. [78]; Hansen et al. [68]; Le Rouzic et al. [81]). This is partially because canalization is not a structurally stable state in the multilinear model, and it is therefore an open question whether more robust epistatic constraints are likely to evolve in more general models.

Constraints may be more likely to evolve when epistasis is order breaking in the sense that phenotypic ordering of genotypes change with changes in genetic background. In such cases the direction of selection on individual loci may change, and there is potential for creating polymorphic fitness maxima or saddle points that make evolutionary dynamics qualitatively different from the additive model. Weinreich et al. [82] proved an important result that motivated their definition of sign epistasis as cases in which an allele substitution (= "mutation") has positive fitness effects in some genetic backgrounds and negative fitness effects in other genetic backgrounds. They showed that the shortest path from any genotype to a fitness peak is selectively accessible if and only if there is no sign epistasis in the set of loci that must change to produce the optimal genotype. Hence, sign epistasis always implies some constraints on paths to optimality. More research is needed to understand systematic relationships between patterns of epistasis and both quantitative and absolute constraints. The hypothesis that macroevolutionary stasis is due to epistatic constraints has not been rejected.

Even if epistatic constraints are not absolute, epistasis may still be extremely important in affecting the possible paths that evolution may take (e.g., Weinreich [96]). The most obvious example is that epistasis may create multiple peaks in the mapping from genotype to fitness (Kauffman [97]; Whitlock et al. [39]). Sewall Wright formulated his shifting-balance theory of macroevolution based on shifts between such alternative states. While the original formulation based on drift-induced jumps between peaks may not be particularly realistic (debated in Coyne et al. [98, 99]; Peck et al. [100]; Wade and Goodnight [101], Goodnight and Wade [102]), ideas of macroevolution being influenced by complex epistatic genotype-phenotype maps have seen a renaissance (e.g., Kauffman [97]; Gavrilets and Gravner [103]; Wagner [104]; Haag [105]).

Although quantitative genetics has kept the door closed for epistasis in adaptive evolution, some epistasis has inevitably squeezed in through studies of the genetic architecture of local adaptation (e.g., many entries in Wolf et al. [106]). When linecrosses are made between phenotypically different natural populations, the epistatic signal is often strong. Well-studied examples include the adaptation of photoperiod to latitude in pitcher-plant mosquitoes (Bradshaw and Holzapfel [107]; Bradshaw et al. [108]), life-history traits in the legume *Chamaecrista fasciculata* (Fenster and Galloway [109, 110]), and beak morphology in soapberry bugs (Carroll et al. [111]; Carroll [112]). In such work the causal role of epistasis is usually attributed to some form of founder events, but we have seen that this is not necessary or likely. Instead, epistatic effect detected from line-cross means are necessarily directional and must have influenced and interacted with selection on the traits.

Evidence for an epistatic influence on adaptation can also come in the form of coadapted gene complexes in which the presence of specific unique alleles at different loci is necessary to produce the required phenotype. Some classical examples involve the cosegregation of alternative complexes, as in the Batesian mimicry morphs in *Papillio* butterflies (Clarke and Sheppard [113]). The importance of cosegregating complexes for adaptation has been debated, as it requires strong linkage to protect against recombinational breakdown. There is little doubt, however, that "coadapted" alleles may fix within local populations (e.g., Takahasi and Tajima [114]), and that this is an important component of adaptive diversification and speciation.

4.2 Epistasis, Speciation, and Reproductive Isolation

Although quantitative geneticists did not find much use for epistasis within populations, it has always been recognized that epistatic interactions are important in crosses between genetically distinct populations. As I have argued throughout, epistasis becomes more and more important the larger the genetic differences there are and the more genetic variation there is. At the extreme end we find crosses between populations or incipient species that have been genetically isolated for a long time or have experienced divergent selection to make them different. Indeed, epistatic incompatibilities are generally thought to be the main mechanism behind postzygotic reproductive isolation (Coyne and Orr [115]; Gavrilets [116]).

The Bateson-Dobzhansky-Muller model is the major account of how this happens. Any population will experience genetic changes, and different isolated populations will not experience the same gene substitutions. Since combinations of substitutions that have not happened in the same population are not seen by selection, incompatibilities will accumulate in a stochastically regular manner until any intercross will be inviable due to deleterious epistatic interactions (Orr [117]; Orr and Turelli [118]; Gavrilets [116]; Fierst and Hansen [80]; Bank et al. [119]; Wang et al. [120]). Although this process can be sped up by divergent selection (e.g., Porter and Johnson [121]), it will also happen in isolated populations under identical stabilizing selection regimes due to quasi-neutral evolution among equivalent genotypes (Gavrilets [122, 116]; Fierst and Hansen [80]). There is abundant empirical evidence for such epistatic incompatibilities (critically reviewed in Coyne and Orr [115]; Welch [123]).

Epistasis is essential to a suite of models for speciation variously known as genetic revolution, founder-flush, transilience, or holey-landscape models (e.g., Templeton [24, 124]; Gavrilets [116]). All these models center on peak shifts or quasi-neutral systems drift in epistatic genotype-phenotype maps. They have links to the shifting-balance theory of evolution and have also been a motivation for the study of changes in additive variance during population bottlenecks. The main idea is that genetic rearrangements due to drift in small populations could cause novel adaptations and incompatibilities relative to the mother species, which then provided the starting point for a speciation event. These models are controversial (e.g., Coyne and Orr [115]). In light of the above discussion, we can note that such theories do not need to rely on drift and small population size. Any change of allele frequency can cause changes of gene and genotype effects with epistasis, and episodes of directional selection or gene flow (e.g., hybridization) could have similar but much stronger and faster effects. For example, Porter and Johnson [121] and Johnson and Porter [125] have shown how genetic incompatibilities can rapidly evolve under directional selection on epistatic systems.

4.3 Epistasis, Sex, and Recombination

It has been understood for quite some time that epistasis may be central to the evolution of sex and recombination. The costs and benefits of breaking up old and creating new gene combinations depend on how genes interact (e.g., Burt [126]). For example, if fitness improvements depend on substitutions at several loci, then a sexually recombining population can evolve much faster than an asexual population (e.g., Maynard Smith [127]). Directional forms of epistasis also have strong effects on the standing mutation load and on the operation of Muller's ratchet, rendering an advantage to sexually reproducing populations when deleterious mutations show "synergistic" epistasis (references below). The evolution of recombination rates also depends on the patterns of epistasis for fitness (Barton [128]; Otto and Barton [129]).

5   How to Study Epistasis

5.1 Meaningful Measurement of Epistasis

"Epistasis is everything else" (Rice [130]). A term covering any combination of gene action that is not strictly additive hides a huge variety of interactions. To make progress we must stop treating epistasis as a patternless residual. We have seen that different patterns of biological epistasis such as directionality may have important and varied effects on selection dynamics, and this must be the tip of an iceberg of interesting biology. From the measurement-theoretical perspective the important thing is to identify aspects of nature that are theoretically relevant and then find appropriate numerical representations for them. The epistatic interaction terms in the Fisher model were statistical constructs not motivated by dynamical relevance. Hence, it is not surprising that they turned out to be dynamically inert. We need to identify types and patterns of epistasis that are biologically relevant. If these can be turned into measurable parameters, we are in a position to test hypotheses and learn how nature is really working.

A good example of this procedure is the work on the deterministic-mutation hypothesis for the maintenance of sex (e.g., Kondrashov [131]). Kimura and Maruyama [132] showed that the mutation load in sexually recombining populations is reduced if there are synergistic interactions among deleterious mutations (=negative directional epistasis for fitness), and the effects can be quantified and assessed in relation to the twofold cost of sex. This theory identified the total deleterious mutation rate and the strength of "synergistic" epistasis as the relevant parameters, and motivated work to estimate them. In this case, the empirical study of epistasis was not obscured by focus on irrelevant statistical measurements, but was explicitly directed at the key directional patterns (e.g., Clark and Wang [133]; de Visser et al. [134, 135]; Elena and Lenski [136]; Charlesworth [137]; Peters and Keightley [138]; Whitlock and Borguet [139]; Kelly [140]; Sanjuan and Elena [141]; Dickinson [142]). The results were mixed, and also in some cases hard to interpret because the epistatic effects were not necessarily measured with reference to the optimal state, which is the reference used in most of the theoretical predictions, but this work illustrates the path we have to take to improve our understanding. The theory was also rather simplistic in that it postulated highly specific regular changes in mutational effects with number of substitutions, and as shown by Butcher [143] and Otto and Feldman [144] variation in epistatic effects may be important in understanding both the mutation load and the operation of Muller's ratchet. For some further theoretical results see also Charlesworth [145], Hansen and Wagner [146], and Otto and Barton [147].

The importance of directional epistasis goes far beyond its influence on the mutation load, and needs to be studied in phenotypic traits in general and not just as a property of unconditionally deleterious fitness mutations. In principle, there are many straightforward ways of detecting directional epistasis. The simplest method is to use line-cross means. Traditional line-cross methods are not well suited for this as they are focused on sequential testing and on separating AxA, AxD, and DxD epistasis (e.g., Lynch and Walsh [33]), but reparameterizations can be done to focus on directionality (Hansen and Wagner [15]). The same principles can also be applied to QTL analyses. Pavlicev et al. [74] developed this to estimate directional epistasis from QTL data in crosses between mouse lines selected up and down for size (*see* also Pavlicev et al. [88, 89] on multivariate epistasis in this data). Jannink et al. [148] and Slatkin and Kirkpatrick [149] have developed model-based approaches for the detection of some forms of directional epistasis from QTL data. Methods for detecting directional epistasis from the analysis of time series of selection responses have also been developed, but require high-quality data and careful interpretation due to the many factors that may influence the dynamics (Le Rouzic et al. [150, 75]). Demuth and Wade [151] showed how to connect line crosses to parameters relevant for the Bateson-Dobzhansky-Muller model (see also Alvarez-Castro et al. [73]).

Unfortunately, standard marker-based studies are poorly designed for studying epistasis of any form. The need to pass through steps where individual genes have to be detected by the use of significance thresholds before inferences about effect sizes can be made means that individually small but cumulatively important effects will go undetected. This is particularly important for epistasis, since the number of potential interaction effects is much larger. The common use of sequential significance testing also raises the threshold for "detecting" epistasis. In practice, only the extreme tail of the effect distribution is studied, and these extremes may be both atypical and evolutionarily irrelevant (e.g., Rockman [152]). Although including effects of all possible (pairwise) interactions among detected genes can improve inferences about epistasis (Malmberg and Mauricio [153]), fundamental progress awaits the development of parametric models of effect distributions and gene occurrences that bypass the estimation of individual effects and gene positions. Such models would focus on estimating a few parameters of relevance, such as the average and variance of interaction effects (e.g., the ε and ε^2 parameters above). Otto and Jones [154] provide an example of such an approach to the estimation of gene number.

Awaiting this, the most efficient ways of studying directional epistasis are to rely on large effects in line crosses and chromosome-substitution studies. The heuristics of this can be read off Fig. 1. Interactions of individual QTLs amount to small shifts in the map, and the deviation from a linear approximation is not expected to be large. When the genetic effects are compared across very different backgrounds however, epistatic effects are more apparent (e.g., Moreno [37]). For example, it has been observed that the sum of the effects of individual chromosomes substituted into the genetic background of another strain is often very different from the total difference between the strains (see Shao et al. [155] and references therein). This is evidence for directional epistasis.

Directional epistasis describes the systematic aspects of local curvature in the genotype-phenotype map, and is the obvious first step towards dynamical understanding beyond the additive model. A better understanding of macroevolutionary potential requires a more general description of systematic features of the genotype-phenotype map. This can be achieved through model-based estimation, and the development of suitable parameterized models of the genotype-phenotype map is an important area for further work. There are many models of genotype-phenotype maps for specific systems in the evodevo and systems-biology literature (reviewed in Polly [156] and Omholt [157], respectively), but these need to be connected to dynamically relevant parameters such as additive variance and directional epistasis.

Detailed system-specific or metabolic and gene-network models are still very useful to build intuition about variational potential in the systems under study. Important insights on both the manifestation and dynamical potential of epistasis have been derived from models of metabolic control (Keightley [42, 43]; Szathmary [44]; Bagheri et al. [49]; Bagheri and Wagner [91]), gene regulation and gene networks (Gibson [45]; Frank [46]; Johnson and Porter [125]; Omholt et al. [48]; Gjuvsland et al. [83, 84]; Gertz et al. [158]), and physiological or developmental systems (Omholt [157]).

5.2 Scaling and Transformation

Scaling and transformation are important in both mathematical and statistical modeling of epistasis (Mani et al. [159]; Pavlicev et al. [74]). Any nonlinear transformation of a trait with an additive architecture will create epistasis. This observation alone should be enough to put to rest claims that epistasis is not common. Bone lengths, bone areas, and bone volumes cannot simultaneously be additive, and even if one of them is, one would then have to argue that this is the only relevant scale. Transformation is impossible whenever there is any biological, economical, or medical context that defines the trait. The square root of milk yield or the log of blood pressure may simply not be of any interest. Epistasis is always defined in relation to a given trait scale, and this scale is given by the theoretical or practical context (Houle et al. [31]). Fitness, for example, plays a defined theoretical role and cannot be transformed in any way and still be fitness (Wagner [160]; Houle et al. [31]). Wagner [160] shows how epistasis for Wrightian fitness needs be defined on a multiplicative scale, and for Malthusian fitness it needs be defined on an additive scale (*see* also Punyani et al. [17]).

In Fig. 2, we can see how the scaling relationships that exist between genetic variance components have strong implications for how to standardize and measure any genetic effect. The standard practice of measuring traits on a scale determined by population variation can totally obscure the relationship between effects. If genetic variance components are measured on the scale of the total population variance (e.g., using the heritability to measure additive variance), they are measured on a rubber scale that is stretched when measuring something large (Houle et al. [31]). This effect is so extreme that heritability is empirically uncorrelated with additive genetic variance (Hansen et al. [59]).

6 Conclusions

The problem of how to incorporate gene interaction was an important in the formation of the modern synthesis, and a central bone of contention in the debates between Fisher and Wright (e.g., Frank [161]). Fisher's concept of the average allelic effect was an elegant and powerful methodological device that sidelined epistasis for a long time. But an average effect is not the same as a constant effect, and the dynamical importance of epistasis derives precisely from its ability to change the biological and hence the average effects of alleles. Refocusing the study of gene interaction around this crucial role may help make epistasis a central element of an "extended synthesis" (Pigliucci [162]).

Acknowledgements

I thank the editors for the invitation to contribute, and Jose Álvarez-Castro for discussions and comments on the manuscript.

References

1. Wright S (1932) The roles of mutation, inbreeding, crossbreeding and selection in evolution. Proc 6th Int Cong Genet 1:356–366

2. Fisher RA (1941) Average excess and average effect of a gene substitution. Ann Eugen 11:53–63

3. Bulmer MG (1980) The mathematical theory of quantitative genetics. Claredon, Oxford

4. Crow JF (2008) Maintaining evolvability. J Genet 87:349–353

5. Crow JF (2010) On epistasis: why it is unimportant in polygenic directional selection. Phil Trans R Soc B 365:1241–1244

6. Hill WG, Goddard ME, Visscher PM (2008) Data and theory point to mainly additive genetic variance for complex traits. PLoS Genet 4:e100008

7. Griffing B (1960) Theoretical consequences of truncation selection based on the individual phenotype. Aust J Biol Sci 13:307–343

8. Kimura M (1965) Attainment of quasi linkage equilibrium when gene frequencies are changing by natural selection. Genetics 52:875–890

9. Hansen TF (2013) Why epistasis is important for selection and adaptation. Evolution 67:3501–3511

10. Carter AJR, Hermisson J, Hansen TF (2005) The role of epistatic gene interactions in the response to selection and the evolution of evolvability. Theor Popul Biol 68:179–196

11. Lewontin RC (1974) The genetic basis of evolutionary change. Columbia University Press, New York

12. Crow JF, Kimura M (1970) An introduction to population genetics theory. Harper & Row, New York

13. Phillips PC (1998) The language of gene interaction. Genetics 149:1167–1171

14. Wagner GP, Laubichler MD, Bagheri-Chaichian H (1998) Genetic measurement theory of epistatic effects. Genetica 102(103): 569–580

15. Hansen TF, Wagner GP (2001) Modeling genetic architecture: a multilinear theory of gene interaction. Theor Popul Biol 59:61–86

16. Cheverud JM, Routman EJ (1995) Epistasis and its contribution to genetic variance components. Genetics 139:1455–1461

17. Puniyani A, Liberman U, Feldman MW (2004) On the meaning of non-epistatic selection. Theor Pop Biol 66:317–321

18. Moore JH, Williams SM (2005) Biological vs. statistical epistasis. Bioessays 12:637–646

19. Hansen TF (2006) The evolution of genetic architecture. Annu Rev Ecol Evol Syst 37:123–157

20. Álvarez-Castro JM, Carlborg Ö (2007) A unified model for functional and statistical

epistasis and its application in quantitative trait loci analysis. Genetics 176:1151–1167

21. Fisher RA (1918) The correlation between relatives on the supposition of Mendelian inheritance. Trans R Soc Edinburgh 3:399–433

22. Cockerham CC (1954) An extension of the concept of partitioning hereditary variance for analysis of covariances among relatives when epistasis is present. Genetics 39:859–882

23. Kempthorne O (1954) The correlation between relatives in a random mating population. Proc R Soc Lond B 143:103–113

24. Templeton AR (1980) The theory of speciation via the founder principle. Genetics 92:1265–1282

25. Templeton AR (2000) Epistasis and complex traits. In: Wolf JB, Broodie ED, Wade MJ (eds) Epistasis and the evolutionary process. Oxford University Press, Oxford, pp 41–57

26. Wright S (1977) Evolution and the genetics of populations, Vol 3. Experimental results and evolutionary deductions. Chicago University Press, Chicago, IL

27. Hand DJ (2004) Measurement theory and practice: the world through quantification. Arnold, London

28. Michell J (1999) Measurement in psychology: a critical history of a methodological concept. Cambridge University Press, Cambridge

29. Chang H (2004) Inventing temperature: measurement and scientific progress. Oxford University Press, Oxford

30. Vieland VJ, Hodge SE (2011) Measurement of evidence and evidence of measurement. Stat Appl Genet Mol Biol 10:35

31. Houle D, Pélabon C, Wagner GP et al (2011) Measurement and meaning in biology. Quart Rev Biol 86:3–34

32. Schneider DC (2009) Quantitative ecology: measurement, models, and scaling, 2nd edn. Elsevier, New York

33. Lynch M, Walsh B (1998) Genetics and analysis of quantitative characters. Sinauer, Sunderland

34. Goodnight C (1987) On the effect of founder events on the epistatic genetic variance. Evolution 41:80–91

35. Goodnight C (1988) Epistasis and the effect of founder events on the additive genetic variance. Evolution 42:441–454

36. Tachida H, Cockerham CC (1989) A building block model for quantitative genetics. Genetics 121:839–844

37. Moreno G (1994) Genetic architecture, genetic behavior, and character evolution. Ann Rev Ecol Syst 25:31–45

38. Cheverud JM, Routman EJ (1996) Epistasis as a source of increased additive genetic variance at population bottlenecks. Evolution 50:1042–1051

39. Whitlock MC, Phillips PC, Moore FB-G et al (1995) Multiple fitness peaks and epistasis. Ann Rev Ecol Syst 26:601–629

40. Wagner GP, Booth G, Bagheri-Chaichian H (1997) A population genetic theory of canalization. Evolution 51:329–347

41. Rice SH (1998) The evolution of canalization and the breaking of von Baer's laws: modeling the evolution of development with epistasis. Evolution 52:647–656

42. Keightley PD (1989) Models of quantitative variation of flux in metabolic pathways. Genetics 121:869–876

43. Keightley PD (1996) Metabolic models in selection response. J Theor Biol 182:311–316

44. Szathmáry E (1993) Do deleterious mutations act synergistically? Metabolic control theory provides a partial answer. Genetics 133:127–132

45. Gibson G (1996) Epistasis and pleiotropy as natural properties of transcriptional regulation. Theor Pop Biol 49:58–89

46. Frank SA (1999) Population and quantitative genetics for regulatory networks. J Theor Biol 197:281–294

47. Johnson NA, Porter AH (2000) Rapid speciation via parallel, directional selection on regulatory genetic pathways. J Theor Biol 205:527–542

48. Omholt SW, Plahte E, Øyehaug L et al (2000) Gene regulatory networks generating the phenomena of additivity, dominance and epistasis. Genetics 155:969–980

49. Bagheri HC, Hermisson J, Vaisnys JR et al (2003) Effects of epistasis on phenotypic robustness in metabolic pathways. Math Biosci 184:27–51

50. Kirkpatrick M, Johnson T, Barton NH (2002) General models of multilocus evolution. Genetics 161:1727–1750

51. Barton NH, Turelli M (2004) Effects of genetic drift on variance components under a general model of epistasis. Evolution 58:2111–2132

52. Álvarez-Castro JM, Le Rouzic A, Carlborg Ö (2008) How to perform meaningful estimates of genetic effects. PLoS Genet 4:e1000062

53. Álvarez-Castro JM, Carlborg Ö, Rönnegård L (2012) Estimation and interpretation of genetic effects with epistasis using the NOIA model. Methods Mol Biol 871:191–204

54. Álvarez-Castro JM, Yang R-C (2011) Multiallelic models of genetic effects and variance decomposition in non-equilibrium populations. Genetica 139:119–1134

55. Le Rouzic A, Álvarez-Castro JM (2008) Estimation of genetic effects and genotype-phenotype maps. Evolutionary bioinformatics 4:225–235

56. Álvarez-Castro JM (2012) Current applications of models of genetic effects with interactions across the genome. Curr Genomics 13:163–175

57. Hansen TF (2011) Epigenetics: Adaptation or contingency? In: Hallgrimsson B, Hall BK (eds) Epigenetics: linking genotype and phenotype in development and evolution. University of California press, Berkeley, CA, pp 357–376

58. Rendel R (1967) Canalization and gene control. Logos Press, London

59. Hansen TF, Pélabon C, Houle D (2011) Heritability is not evolvability. Evolutionary Biology 38:258–277

60. Karlin S (1975) General two-locus selection models: some objectives, results, and interpretations. Theor Pop Biol 7:364–398

61. Young SSY (1967) Computer simulation of directional selection in large populations II. The additive x additive and mixed model. Genetics 56:73–87

62. Hallander J, Waldmann P (2007) The effect of non-additive genetic interactions on selection in multi-locus genetic models. Heredity 98:349–359

63. Cockerham CC, Tachida H (1987) Evolution and maintainance of quantitative genetic variation by mutation. Proc Natl Acad Sci 84:6205–6209

64. Hermisson J, Wagner GP (2004) The population genetic theory of hidden variation and genetic robustness. Genetics 168:2271–2284

65. Hill WG, Barton NH, Turelli M (2006) Predictions of effects of genetic drift on variance components under a general model of epistasis. Theor Pop Biol 70:56–62

66. Zhang XS (2008) Increase in quantitative variation after exposure to environmental stresses and/or introduction of a major mutation: G x E interaction and epistasis or canalization. Genetics 180:687–695

67. Turelli M, Barton NH (2006) Will population bottlenecks and multilocus epistasis increase additive genetic variance? Evolution 60:1763–1776

68. Hansen TF, Álvarez-Castro JM, Carter AJR et al (2006) Evolution of genetic architecture under directional selection. Evolution 60:1523–1536

69. Rice SH (2002) A general population genetic theory for the evolution of developmental interactions. Proc Natl Acad Sci U S A 99:15518–15523

70. Rice SH (2004) Developmental associations between traits: covariance and beyond. Genetics 166:513–526

71. Carlborg Ö, Jacobson L, Åhgren P et al (2006) Epistasis and the release of genetic variation during long-term selection. Nat Genet 38:418–420

72. Le Rouzic A, Álvarez-Castro JM, Carlborg Ö (2008) Dissection of the genetic architecture of body weight in chicken reveals the impact of epistasis on domestication traits. Genetics 179:1591–1599

73. Álvarez-Castro JM, Le Rouzic A, Andersson L et al (2012) Modeling of genetic interactions improves predicton of hybrid patterns – a case study in domestic fowl. Genet Res Camb 94:255–266

74. Pavlicev M, Le Rouzic A, Cheverud JM et al (2010) Directionality of epistasis in a murine intercross population. Genetics 185:1489–1505

75. Le Rouzic A, Houle D, Hansen TF (2011) A modeling framework for the anlysis of artificial-selection time series. Genet Res 93:155–173

76. Ungerer MC, Linder CR, Rieseberg LH (2003) effects of genetic background on response to selection in experimental populations of *Arabidopsis thaliana*. Genetics 163:277–286

77. Yukilevich R, Lachance J, Aoki F et al (2008) Long-term adaptation of epistatic genetic networks. Evolution 62:2215–2235

78. Hermisson J, Hansen TF, Wagner GP (2003) Epistasis in polygenic traits and the evolution of genetic architecture under stabilizing selection. American Naturalist 161:708–734

79. Álvarez-Castro JM, Kopp M, Hermisson J (2009) Effects of epistasis and the evolution of genetic architecture: exact results for a 2-locus model. Theor Popul Biol 75:109–122

80. Fierst JL, Hansen TF (2010) Genetic architecture and postzygotic reproductive isolation: evolution of Bateson-Dobzhansky-Muller incompatibilities in a polygenic model. Evolution 64:675–693

81. Le Rouzic A, Álvarez-Castro JM, Hansen TF (2013) The evolution of canalization and evolvability in stable and fluctuating environments. Evolutionary Biology 40:317–340

82. Weinreich DM, Watson RA, Chao L (2005) Sign epistasis and genetic constraint on evolutionary trajectories. Evolution 59:1165–1174

83. Gjuvsland AB, Hayes BJ, Omholt SW et al (2007) Statistical epistasis is a generic feature of gene regulatory networks. Genetics 175:411–420

84. Gjuvsland AB, Vik JO, Wooliams JA et al (2011) Order-preserving principles underlying genotype-phenotype maps ensure high additive proportions of genetic variance. J Evol Biol 24:2269–2279

85. Gimelfarb A (1989) Genotypic variation for a quantitative character maintained under stabilizing selection without mutations: epistasis. Genetics 123:217–227

86. Gavrilets S (1993) Equilibria in an epistatic viability model under arbitrary strength of selection. J Math Biol 31:397–410

87. Liberman U, Feldman MW (2005) On the evolution of epistasis I: diploids under selection. Theor Pop Biol 67:141–160

88. Pavlicev M, Kenney-Hunt JP, Norgard EA et al (2008) Genetic variation in pleiotropy: differential epistasis as a source of variation in the allometric relationship between long bone lengths and body weight. Evolution 62:199–213

89. Pavlicev M, Norgard EA, Fawcett GI et al (2011) Evolution of pleiotropy: epistatic interaction pattern supports a mechanistic model underlying variation in genotype-phenotype map. J Exp Zool (Mol Dev Evol) 316B:371–385

90. Griswold CK, Henry TA (2012) Epistasis can increase multivariate trait diversity in haploid non-recombining populations. Theor Pop Biol 82:209–221

91. Bagheri HC, Wagner GP (2004) The evolution of dominance in metabolic pathways. Genetics 168:1713–1735

92. Bagheri HC (2006) Unresolved boundaries of evolutionary theory and the question of how inheritance systems evolve: 75 years of debate on the evolution of dominance. J Exper Zool 306B:329–359

93. Hansen TF, Houle D (2004) Evolvability, stabilizing selection, and the problem of stasis. In: Pigliucci M, Preston K (eds) Phenotypic integration: studying the ecology and evolution of complex phenotypes. Oxford University Press, Oxford, pp 130–150

94. Estes S, Arnold SJ (2007) Resolving the paradox of stasis: models with stabilizing selection explain evolutionary divergence on all timescales. Am Nat 169:227–244

95. Uyeda JC, Hansen TF, Arnold SJ et al (2011) The million-year wait for macroevolutionary bursts. Proc Natl Acad Sci U S A 108: 15908–15913

96. Weinreich DM (2005) The rank ordering of genotypic fitness values predicts genetic constraint on natural selection on landscapes lacking sign epistasis. Genetics 171:1397–1405

97. Kauffman SA (1993) The origins of order. Self-organization and selection in evolution. Oxford University Press, Oxford

98. Coyne JA, Barton NH, Turelli M (1997) A critique of Sewall Wright's shifting balance theory of evolution. Evolution 51:643–671

99. Coyne JA, Barton NH, Turelli M (2000) Is Wright's shifting balance process important in evolution? Evolution 54:306–317

100. Peck SL, Ellner SP, Gould F (1998) A spatially explicit stochastic model demonstrates the feasibility of Wright's shifting balance theory. Evolution 52:1834–1839

101. Wade MJ, Goodnight CJ (1998) The theories of Fisher and Wright in the context of metapopulations: when nature does many small experiments. Evolution 52:1537–1553

102. Goodnight CJ, Wade MJ (2000) The ongoing synthesis: a reply to Coyne, Barton, and Turelli. Evolution 54:317–324

103. Gavrilets S, Gravner J (1997) Percolation on the fitness hypercube and the evolution of reproductive isolation. J Theor Biol 184:51–64

104. Wagner A (2005) Robustness and evolvability in living systems. Princeton University Press, Princeton, NJ

105. Haag ES (2007) Compensatory vs. pseudocompensatory evolution in molecular and developmental interactions. Genetica 129:45–55

106. Wolf JB, Broodie ED III, Wade MJ (eds) (2000) Epistasis and the evolutionary process. Oxford University Press, Oxford

107. Bradshaw WE, Holzapfel CM (2000) The evolution of genetic architectures and the divergence of natural populations. In: Wolf JB, Broodie ED, Wade MJ (eds) Epistasis and the evolutionary process. Oxford University Press, Oxford, pp 245–263

108. Bradshaw WE, Haggerty BP, Holzapfel CM (2005) Epistasis underlying a fitness trait within a natural population of the pitcher-plant mosquito, *Wyeomyia smithii*. Genetics 169:485–488

109. Fenster CB, Galloway LF (2000) The contribution of epistasis to the evolution of natural populations: A case study of an annual plant. In: Wolf JB, Broodie ED, Wade MJ (eds) Epistasis and the evolutionary process. Oxford University Press, Oxford, pp 232–244

110. Fenster CB, Galloway LF (2000) Population differentiation in an annual legume: genetic architecture. Evolution 54:1157–1172

111. Carroll SP, Dingle H, Famula TR (2003) Rapid appearance of epistasis during adaptive divergence following colonization. Proc R Soc Lond B 270:S80–S83

112. Carroll SP (2007) Brave new world: the epistatic foundations of natives adapting to invaders. Genetica 129:193–204

113. Clarke CA, Sheppard PM (1960) Super-genes and mimicry. Heredity 14:175–185

114. Takahasi KR, Tajima F (2005) Evolution of coadaptation in a two-locus epistatic system. Evolution 59:2324–2332

115. Coyne JA, Orr HA (2004) Speciation. Sinauer, Sunderland

116. Gavrilets S (2004) Fitness landscapes and the origin of species. Princeton University Press, Princeton, NJ

117. Orr HA (1995) The population genetics of speciation: the evolution of hybrid incompatibilities. Genetics 139:1805–1813

118. Orr HA, Turelli M (2001) The evolution of postzygotic isolation: accumulating Dobzhansky-Muller incompatibilities. Evolution 55:1085–1094

119. Bank C, Bürger R, Hermisson J (2012) The limits to parapatric speciation: Dobzhansky-Muller incompatibilities in a continent-island model. Genetics 191:845–863

120. Wang RJ, Ane C, Payseur BA (2013) The evolution of hybrid incompatibilities along a phylogeny. Evolution 67:2905–2922

121. Porter AH, Johnson NA (2002) Speciation despite gene flow when developmental pathways evolve. Evolution 56:2103–2111

122. Gavrilets S (1999) A dynamical theory of speciation on holey adaptive landscapes. Am Nat 154:1–22

123. Welch JJ (2004) Accumulating Dobzhansky-Muller incompatibilities: reconciling theory and data. Evolution 58:1145–1156

124. Templeton AR (1981) Mechanisms of speciation—a population genetics approach. Ann Rev Ecol Syst 12:23–48

125. Johnson NA, Porter AH (2007) Evolution of branched regulatory genetic pathways: directional selection on pleiotropic loci accelerates developmental system drift. Genetica 129:57–70

126. Burt A (2000) Sex, recombination, and the efficacy of selection – was Weismann right? Evolution 54:337–351

127. Maynard Smith J (1978) The evolution of sex. Cambridge University Press, Cambridge

128. Barton NH (1995) A general model for the evolution of recombination. Genet Res 65:123–144

129. Barton NH, Otto SP (2005) Evolution of recombination due to random drift. Genetics 169:2353–2370

130. Rice SH (2000) The evolution of developmental interactions: Epistasis, canalization, and integration. In: Wolf JB, Broodie ED, Wade MJ (eds) Epistasis and the evolutionary process. Oxford University Press, Oxford, pp 82–98

131. Kondrashov AS (1988) Deleterious mutations and the evolution of sexual reproduction. Nature 336:435–440

132. Kimura M, Maruyama T (1966) The mutational load with epistatic gene interactions in fitness. Genetics 54:1337–1351

133. Clark AG, Wang L (1997) Epistasis in measured genotypes: *Drosophila* P-element insertions. Genetics 147:157–163

134. de Visser JAGM, Hoekstra RF, van den Ende H (1997) An experimental test for synergistic epistasis in *Chlamydomonas*. Genetics 145:815–819

135. de Visser JAGM, Hoekstra RF, van den Ende H (1997) Test of interaction between genetic markers that affect fitness in *Aspergillus niger*. Evolution 51:1499–1505

136. Elena SF, Lenski RE (1997) Test of synergistic interactions among deleterious mutations in bacteria. Nature 390:395–398

137. Charlesworth B (1998) The effect of synergistic epistasis on the inbreeding load. Genet Res 71:85–89

138. Peters AD, Keightley PD (2000) A test for epistasis among induced mutations in *Caenorhabditis elegans*. Genetics 156:1635–1647

139. Whitlock MC, Bourguet DB (2000) Factors affecting the genetic load in Drosophila: synergistic epistasis and correlations among fitness components. Evolution 54:1654–1660

140. Kelly JK (2005) Epistasis in monkey flowers. Genetics 171:1917–1931

141. Sanjuan R, Elena SF (2006) Epistasis correlates to genomic complexity. Proc Natl Acad Sci U S A 103:14402–14405

142. Dickinson WJ (2008) Synergistic fitness interactions and a high frequency of beneficial changes among mutations accumulated under relaxed selection in *Saccharomyces cerevisiae*. Genetics 178:1571–1578

143. Butcher DL (1995) Muller's ratchet, epistasis and mutation effects. Genetics 141:431–437

144. Otto SP, Feldman MW (1997) Deleterious mutations, variable epistatic interactions, and

the evolution of recombination. Theor Pop Biol 51:134–147

145. Charlesworth B (1990) Mutation-selection balance and the evolutionary advantage of sex and recombination. Genet Res 55:199–221

146. Hansen TF, Wagner GP (2001) Epistasis and the mutation load: a measurement-theoretical approach. Genetics 158:477–485

147. Otto SP, Barton NH (2001) Selection for recombination in small populations. Evolution 55:1921–1931

148. Jannink JL, Moreau L, Charmet G et al (2009) Overview of QTL detection in plants and tests for synergistic epistatic interactions. Genetica 136:225–236

149. Slatkin M, Kirkpatrick M (2012) Using known QTLs to detect directional epistatic interactions. Genet Res 94:39–48

150. Le Rouzic A, Skaug HJ, Hansen TF (2010) Estimating genetic architectures from artificial-selection responses: a random-effect framework. Theor Popul Biol 77:119–130

151. Demuth JP, Wade MJ (2005) On the theoretical and empricial framework for studying genetic interactions within and among species. Am Nat 165:524–536

152. Rockman MV (2012) The QTN program and the alleles that matter for evolution: all that's gold does not glitter. Evolution 66:1–17

153. Malmberg RL, Mauricio R (2005) QTL-based evidence for the role of epistasis in evolution. Genet Res 86:89–95

154. Otto SP, Jones CD (2000) Detecting the undetected: estimating the total number of loci underlying a quantitative trait. Genetics 156:2093–2107

155. Shao H, Burrage LC, Sinasac DS et al (2008) Genetic architecture of complex traits: large phenotypic effects and pervasive epistasis. Proc Natl Acad Sci U S A 105:19910–19914

156. Polly PD (2008) Developmental dynamics and G-matrices: can morphometric spaces be used to model phenotypic evolution? Evolutionary Biology 35:83–96

157. Omholt SW (2012) From sequence to consequence and back. Prog Biophys Mol Biol 111:75–82

158. Gertz J, Gerke JP, Cohen B (2010) Epistasis in quantitative trait captured by a molecular model of transcription factor interactions. Theor Pop Biol 77:1–5

159. Mani R, Onge RPS, Hartman JL et al (2008) Defining genetic interaction. Proc Natl Acad Sci U S A 105:3461–3466

160. Wagner GP (2010) The measurement theory of fitness. Evolution 64:1358–1376

161. Frank SA (2012) Wright's adaptive landscape versus Fisher's fundamental theorem. In: Svensson E, Calsbeek R (eds) The adaptive landscape in evolutionary biology. Oxford University Press, Oxford, pp 41–57

162. Pigliucci M (2007) Do we need an extended evolutionary synthesis? Evolution 61:2743–2749

Chapter 8

Two Rules for the Detection and Quantification of Epistasis and Other Interaction Effects

Günter P. Wagner

Abstract

Traditionally statistical interactions and epistasis are defined with respect to the ANOVA model of additive effects; that is, epistasis is defined as a deviation from the additive mode of combining main effects of gene substitutions. Furthermore, the definition is relative to a particular scale and epistasis can potentially be eliminated by a non-linear transformation of the underlying phenotype variables. The latter fact raises questions of the scientific validity of the concept if interaction, given its presumed arbitrariness. Here I am arguing that the arbitrariness in the definition and detection of epistasis, and any other interaction, can be eliminated if we observe measurement theoretical constraints on the treatment of quantitative data. I propose two principles for determining the appropriate reference model for the detection of epistasis. The first is the *principle of effect propagation* stating that the scale type of the effect measure determines the reference model for defining epistasis. For instance, if effects are measured as differences, then the reference model has to be additive. If the reference effects are fold differences, then the reference model has to be multiplicative. A mathematical justification for this rule is provided. The second principle is called *irrelevant effects* and derives from the principle of meaningfulness in measurement theory. In short, the rule says that the reference model is determined by the allowable scale transformations of the variables measured. The justification for this rule is that any mathematical model in which these variables figure have to be invariant to allowable scale transformations. These two rules can effectively eliminate the arbitrariness in the definition, detection, and quantification of epistasis or any other interaction effect.

Key words Epistasis, Interaction, Measurement theory

1 Introduction

The detection and measurement of interactions plays a fundamental role in the sciences, however, interactions come in different flavors. There are physical interactions where material entities (molecules, organisms) directly engage with each other, like protein-protein interactions or predator prey interactions. Then, there are statistical interactions that are defined within the Analysis of Variance framework, as deviations from a statistical model [1]. Finally there are interactions among effects of certain experimental manipulations, which can be called "functional interactions."

Jason H. Moore and Scott M. Williams (eds.), *Epistasis: Methods and Protocols*, Methods in Molecular Biology, vol. 1253, DOI 10.1007/978-1-4939-2155-3_8, © Springer Science+Business Media New York 2015

145

Statistical and functional interactions are closely related, but I want to distinguish between them to emphasize that the problem of interaction detection is not limited to a particular statistical model, like ANOVA. Functional and statistical interactions have to do with the combination of individual effects of experimental interventions or naturally occurring changes on a quantitative response variable. The distinction between statistical and functional interactions has, for instance, been introduced in the context of gene interactions (aka epistasis), where statistical interaction is defined as variance components in the ANOVA model that depend on gene frequencies in a population. In contrast functional interactions are defined on the basis of the deviations of individual and combined gene effects on a particular phenotype [2], and thus ignore, for instance, gene frequencies in a specific population.

Gene interaction is conventionally defined as a deviation of the combined effects of two gene substitutions on some quantitative character from the sum of the two individual effects, i.e., a deviation from additivity [3]. However, this definition depends on the scale used to measure the phenotypic trait. A nonlinear transformation of the scale type could lead to the elimination of the "interaction effect." Furthermore, statistical and physical interactions do not need to be congruent, so that not every statistical interaction actually is indicative of a physical interaction and vice versa. This as well as the mentioned scale dependence of statistical and functional interaction raises the question of how meaningful the notions of statistical and functional interactions are. If a scale transformation can eliminate them, then the notion of interactions seems to be arbitrary and may not reflect a scientifically meaningful fact.

In this short contribution I want to propose two rules for the detection and quantification of interaction effects that are intended to remove the arbitrariness from interaction measures. One is the *principle of effect propagation*, which says that the definition of interaction effects depends on the way the primary effects have been measured. The second rule uses an argument about *allowable scale transformations* to define interaction effects. In both cases it is argued that the proper definition of interaction effects depends on the properties of measurement scales and thus on the meaningful representation of empirical facts in the form of quantitative relationships between numbers or numerical variables [4]. We will see that the two rules apply to two slightly different situations. Rule 1 applies to situations where effects of interventions of the same kind are combined, for instance, two mutations in one gene. In this case we have effects defined for each individual intervention (substitution) and the question is how they combine, assuming they make independent contributions. Rule 2 applies when we consider two factors, each with different realizations, like in a factorial ANOVA design, and where the effects of one factor (say genotype) is modeled by a mathematical model, as in population genetic models of natural selection. Before we discuss these rules I want to trace the additivity definition of interaction to its historical roots, as far as it is represented in the teaching of science today. That is to say, I am not making a historical claim, based on original historical research, but refer to the way these ideas are transmitted during the teaching of science. My conclusion will be that we are living with the additivity definition of interaction largely because of mathematical convenience rather than scientific principle.

2   The Root of the Additivity Definition of Epistasis/Interaction

The notion of interaction effects arises primarily in the context of the Analysis of Variance (ANOVA) [1], particularly in factorial experimental designs. Note: It is more accurate to say that ANOVA is the most influential source of the concept of statistical interaction. In fact, the fundamental concepts of quantitative genetics can be seen as an application of ANOVA concepts to genetics. The additive effect of a gene substitution is equal to the main effect of an ANOVA, and epistasis corresponds to the interaction effect in a factorial ANOVA.

A factorial experimental design aims to quantify the combined effects of two kinds of experimental variables. For instance, an experiment that simultaneously aims at measuring the effects of soil pH and temperature on plant growth is a factorial design, as sketched in Fig. 1. Each sample represents plants grown in one of the possible combinations of pH and temperature. Treatment and

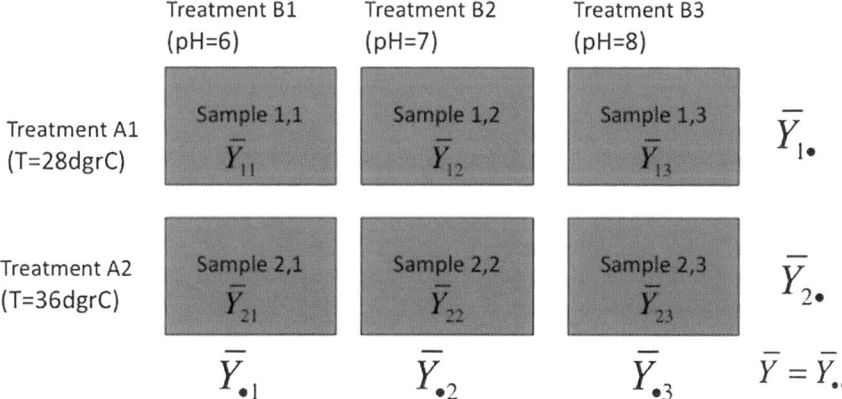

Fig. 1 Schematic example of a factorial experimental design in which two types of treatments, called factors *A* and *B*, are simultaneously varied and measured. In this example one factor is pH the other temperature. Each cell of the matrix represents a sample of observations made under the corresponding regime of pH and temperature. The objective is to find the effect of each treatment regime and test for the interaction between the two factors. For more details refer to the text

interaction effects are then calculated as additive deviations of sample means, $\bar{Y}_{\mathrm{pH}=i,T=j}$, from the overall mean over all samples $\bar{Y}$. A sample mean is the mean over the samples of plants grown in soil of a particular pH and at a certain temperature. The main effect of the treatment with $T = 36$ °C, $A_{T=36\,°C}$, is the deviation from the overall mean of the marginal mean, $\bar{Y}_{\bullet,T=36°C}$: $A_{T=36°C} = \bar{Y}_{\bullet,T=36°C} - \bar{Y}$. The marginal mean, $\bar{Y}_{\bullet,T=36°C}$, is the average over all samples that where treated at that temperature. Similarly, the interaction effect among two treatments factors $AB_{\mathrm{pH}=i,T=j}$ is defined as the deviation of the sample mean $\bar{Y}_{\mathrm{pH}=i,T=j}$ from the sum of the main effects $A_{T=j}$ and $B_{\mathrm{pH}=i}$: $AB_{\mathrm{pH}=i,T=j} = \bar{Y}_{\mathrm{pH}=i,T=j} - \left(\bar{Y} + A_{T=j} + B_{\mathrm{pH}=i} \right)$.

The total model of this ANOVA design then is $Y_{\mathrm{pH}=i,T=j,k} = \bar{Y} + A_{T=j} + B_{\mathrm{pH}=i} + AB_{\mathrm{pH}=i,T=j} + \varepsilon_{\mathrm{pH}=i,T=j,k}$, where $Y_{\mathrm{pH}=i,T=j,k}$ is the measured value of specimen k from the sample with treatments pH $= i$ and $T = j$, and $\varepsilon_{\mathrm{pH}=i,T=j,k}$ is the individual deviation from the main and interaction effects. The latter term is also called the "error term" even though it does not mean that anyone had committed an error. It is just the remaining individual variation that is not explained by the main effects of the treatments or by the interaction effect.

The advantage of this formulation is that the measure of variation caused by the main and the interaction effects, measured in sum of squares, SS (for details see [1]), are additive if the effects are defined as additive:

$$SS_{Total} = SS_A + SS_B + SS_{AB} + SS_{within}$$

This fact has major advantages in testing for the presence of statistically significant main and interaction effects. The point here is that the definition of interaction as the additive deviation from additive main effects is dictated by a mathematical property facilitating the decomposition of variance into its contributions from treatments and interactions [1]. What is completely missing in this approach, though, is a consideration of the empirical meaning of the response variable Y. The latter, i.e., considerations of the empirical meaning of quantitative measures, are the realm of measurement theory [5].

The strained relationship between statistics and measurement theory over the proper treatment of quantitative data is too complicated to be explored here [6], but it has been argued recently that the lack of measurement theoretical sensibilities among biologists is the reason for some of the more problematic aspects of quantitative biology [4]. Measurement theory is a branch of applied mathematics that deals with the question how empirical observations can be represented in numerical mathematical structures [7]. The main objective of measurement theory is to ensure that the mathematical structures faithfully reflect the empirical facts they are supposed to represent. Here I am arguing that the ambiguity of how to measure statistical and functional interactions is in its core a measurement theoretical problem. Below I propose two measurement theoretical rules to remove the arbitrariness of measures of interaction and to ensure empirical meaningfulness.

2.1 Rule I: Effect Propagation

I want to introduce the rule on effect propagation with the example that led me to recognize this rule. It occurred in a study on the evolution of the uterine promoter of the prolactin gene (PRL) performed in my lab. This is a recently published study by Dr. Deena Emera and myself on the evolution of gene regulation in endometrial stromal cells [8].

PRL is mostly known as a pituitary hormone involved in the regulation of lactation in mammals, but is also expressed in the uterus as well as other non-pituitary tissues [9]. In humans, PRL is among the most strongly induced genes upon differentiation of endometrial stromal cells, aka decidualization [10, 11]. Deena Emera has shown that fast and strong decidual expression is a character of decidual PRL promoters from great apes and that this derived gene expression is due to cis-regulatory changes in the promoter/enhancer region of the PRL locus. Dr Emera also identified the nucleotide substitutions that caused this expression difference and identified seven relevant nucleotide differences that account for a considerable amount of the differences between great apes and Old World monkeys. The question then was to determine to what degree these nucleotide changes interact in their effects on gene expression. A small number of these results are shown in Fig. 2.

In this figure the y-axis is the fold difference in reporter gene activity (measured as relative fluorescence values) relative to the wild type promoter; for that reason the wild-type fold change is $=1$, per definition. Mutation 1 reduces reporter gene activity to 66 % of the wild type; that is, the fold change is 0.66. Similarly, mutation 4,5 has an effect of 0.80, i.e. the activity is 80 % of the wild type. How then should we calculate the predicted combined effect of mutations 1 and 4,5? Invoking the additive model one could say that mutation 1 is decreasing the activity by $[1-0.66=0.34]$ or 34 %, and the mutation 4,5 decreases the activity by 20 %, $1-0.8=0.2$. Then the predicted combined effect would be $0.34+0.2=0.54$ and the relative reporter gene activity of the promoter with the combined mutations should be $1-0.54=0.46$. The actual observed activity of the combined mutations is 0.37.

Another approach is to say that the first mutation has the multiplicative effect of 0.66, and the mutation 4,5 has the multiplicative effect 0.8 and thus the predicted combined effect should be $0.66 \times 0.8 = 0.528$. Both ways to predicting the combined effect, assuming no interaction, differ from the actual observed effect and thus indicate epistasis or an interaction between mutations 1 and 4,5. But, the predictions do differ in the degree of deviation from the actual observed value. The additive prediction is actually closer

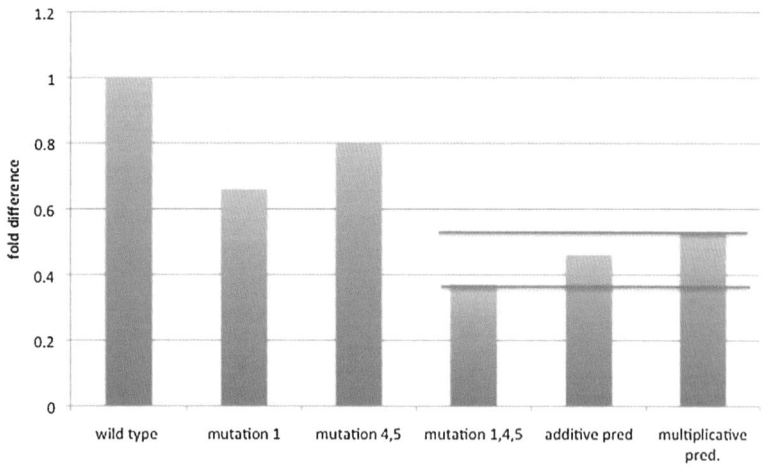

Fig. 2 Some results from the study on the evolution of the decidual PRL promoter by Emera and Wagner [8]. The *y*-axis is relative reporter gene expression in human cells, measured in ratios of fluorescence intensity between the reporter gene and a control gene. The latter is to account for variation between experiments. The values are also relative to the expression of the reporter gene driven by the wild-type orang promoter. Here we compare the effects of three mutations, mutation 1 and mutations 4 and 5, as well as the combination of all three mutations. In all cases, the mutations reduce reporter gene activity. The effect of all three mutations 1,4,5 is compared with the predictions of the additive and the multiplicative models. All of them show differences between the activity under three mutations and the models indicating that mutation 1 interacts with mutations 4,5. The rational why the multiplicative model is the correct one to compare to the triple mutation result is explained in the text. For details about the experiments *see* Emera and Wagner [8]

to the measured combined effect than the multiplicative prediction (*see* Fig. 2). Thus, these two approaches lead to different conclusions: one, the additive, suggests weak interaction, whereas the other, the multiplicative method, suggests strong interaction. Is there a rational way to determine which non-epistatic model, the additive or the multiplicative, is the correct one? Below I show that in this specific example the multiplicative model is the correct one to use, then I will generalized the principle. The result will be the *rule of effect propagation*, which can either lead to an additive or multiplicative model or another model, depending on how the original effects of the individual interventions (or mutations) are measured.

In the example discussed here, the actual experimentally measured variable is the intensity of fluorescence, which indicates the expression level of the reporter gene, normalized with the fluorescence of another protein (expressed from a standard promoter) to correct for differences in transfection efficiency. Let us call this variable v, with v(wild type) related to the v(mutation 1) by fold difference defined by

$$f(\text{mut } 1) = \frac{\nu(\text{mutation } 1)}{\nu(\text{wild type})}$$

which implies that the measured fluorescence caused by the enhancer with mutation 1 can be calculated from that of the wild type by

$$\nu(\text{mutation } 1) = \nu(\text{wild type}) \times f(\text{mut } 1)$$

and analogously for mutation 4,5:

$$\nu(\text{mutation } 4,5) = \nu(\text{wild type}) \times f(\text{mut } 4,5)$$

If the mutation effect of mutation 4,5, as measured by fold change, is independent of the genotype to which mutation 4,5 is introduced, then the combined effect of mutation should be

$$\nu(\text{mutation } 1,4,5) = \nu(\text{mutation } 1) \times f(\text{mut } 4,5)$$

which

$$\nu(\text{mutation } 1,4,5) = \left[\nu(\text{wild type}) \times f(\text{mut } 1)\right] \times f(\text{mut } 4,5)$$
$$\nu(\text{mutation } 1,4,5) = \nu(\text{wild type}) \times \left[f(\text{mut } 1) \times f(\text{mut } 4,5)\right]$$
$$f(\text{mut } 1,4,5) = f(\text{mut } 1) \times f(\text{mut } 4,5)$$

Consequently, the multiplicative model is the correct model to predict how independent mutations should affect the gene activity. The reason is that the effects of the individual mutations are measured as multiplicative factors, i.e., fold changes. In other words the correct model to predict the outcome of combinations of non-epistatic mutations is determined by the scale of the measures of the mutation effects themselves. In this case the effect measure is fold change, which is multiplicative. This result can be generalized in the form of the following principle:

Rule I, Effect Propagation: *Let X and Y be two interventions (e.g., two mutations or experimental treatments), and C is a control case. V is a measured variable (size, gene activity, etc.), and let $m(X)$ and $m(Y)$ be measures of the effects of the interventions X and Y (compatible with the scale type of V) on the variable V such that $V(X) = V(C) \circ m(X)$ and $V(Y) = V(C) \circ m(Y)$, respectively. If the operation "$\circ$" is invertible, associative and commutative, then the predicted joint effect of X&Y in the absence of interaction is $m_{\text{no interaction}}(X \& Y) = m(X) \circ m(Y)$. Note: an operation is associative if $[a \circ b] \circ c = a \circ [b \circ c]$ and commutative if $a \circ b = b \circ a$.*

Justification: As per the definition we have $V(X) = V(C) \circ m(X)$, and in the absence of interaction we also have $V_{\text{no interaction}}(X \& Y) = V(X) \circ m(Y) = [V(C) \circ m(X)] \circ m(Y)$. Because of associativity, we can rewrite this as

$$V_{\text{no interaction}}(X \& \Upsilon) = V(C) \circ \big[m(X) \circ m(\Upsilon)\big]$$

and thus obtain $m_{\text{no interaction}}(X \& \Upsilon) = m(X) \circ m(\Upsilon)$.

In this justification we only explicitly used the associativity of the binary operation "$\circ$." Commutativity is important if the effect of the joint interventions is independent of the order in which they occur. For instance in the case of two mutations, the effect on the phenotype only depends on the composition of the final genotype and not on the order in which the mutations were introduced. Since we have

$$V(X \& \Upsilon) = V(\Upsilon \& X),$$

we have to require that

$$m_{\text{no interaction}}(X \& \Upsilon) = m(X) \circ m(\Upsilon) = m(\Upsilon) \circ m(X),$$

which is commutativity.

Rule 1 is entirely abstract and can be applied to any mathematical operation able to define an intervention effect. It clearly covers the additive model with $\circ \equiv +$ and the multiplicative model with $\circ \equiv *$. In other words, the appropriate model for defining the standard for detecting interaction effects is determined by the nature of the effect measure. If effects on the measured variables V are defined by addition/subtraction, then the model for predicting the joint effect of interventions is additivity. If the original effect measure is multiplicative, e.g., fold changes, then the prediction is based on the multiplicative model. Moreover quantification of the degree of interaction can implicitly be defined by

$$V(X \& \Upsilon) = V(C) \circ m(X) \circ m(\Upsilon) \circ m(X \& \Upsilon),$$

where $m(X \& \Upsilon)$ is the measure of the interaction effect.

While Rule I is open to any mathematical operation to define effect measures as long as it is invertible, associative, and commutative, the scale properties of the variable V pose additional constraints on the admissible effect measures. Quantitative variables are classified according to the admissible scale transformations (for an overview *see* [5, 12]). For instance, ratio scale variables like distance and weight are defined up to a multiplicative constant. Length measures, like inch, cm, and Å differ by a "conversion constant." For ratio scale measures, both differences as well as ratios are meaningful. However, conventional temperature measures (degrees Celsius, Fahrenheit) are defined up to both a multiplicative and an additive constant. For this reason, ratios of temperatures, and thus multiplicative effect measures, are inadmissible for measuring effects on temperature. For a deeper discussion of scale types and their implications *see* ref. 5. Table 1 provides a list of scale types and the admissible ways to define effect measures.

Table 1
The relationship between scale type and meaningful effect measures and measures of interaction for the most commonly used scale types

Scale type	Allowable transformation	Effect measures	Interaction measure
Nominal	Any 1–1 transformation	N/A	N/A
Ordinal	Any monotone increasing transformation	N/A	N/A
Difference	$y = x + \delta$	Additive	Additive
Interval	$y = ax + \delta,\ a > 0$	Additive	Additive
Ratio	$y = ax,\ a > 0$	Additive or multiplicative	Additive or multiplicative
Log interval	$y = ax^b,\ a,\ b > 0$	Multiplicative	Multiplicative

Note that nominal and ordinal scales cannot be used to define effect measures of any kind, because on these scales differences or ratios are not meaningful. In the case of ratio scales, both additivity and multiplicativity are meaningful effect measures, but the interaction measure needs to be corresponding. If the effect measure is additive the interaction measure has to be additive as well, and if the effect measure is multiplicative then the interaction measure also has to be multiplicative, per Rule I. For a discussion of scale types *see* ref. 12

2.2 Rule II: Irrelevant Effects

The second rule for determining the presence of interaction effects arose in the context of measuring genotype by environment effects with respect to fitness [13]. Hence, I want to introduce this idea with reference to this problem. On the most general level one speaks of a genotype-environment interaction if the effect of a mutation differs between environments. For instance the fitness effect of certain Drosophila alleles can be beneficial in low population densities and deleterious in high-density populations [14]. However, certain kinds of environmentally caused fitness changes will have no influence on natural selection among the alleles. This can easily be seen by assuming that the two genotypes C and M (standing for control and mutation genotype) have fitness values $w(C)$ and $w(M)$ in environment 1 and fitness values $w(C)'$ and $w(M)'$ in environment 2. Now let us assume that the fitness values in environment 2 differ from that in environment 1 by a constant factor c:

$$w'(C) = c \times w(C)$$
$$w'(M) = c \times w(M)$$

How would this difference between environment 1 and 2 affect the natural selection acting on strains C and M. After all, fitness values are predictors for the outcome of natural selection. Assuming no recombination and ignoring mutations and a variety of other complicating factors (for a review of mathematical selection theory *see* [15]), the frequency of the strain M will be

$$p' = \frac{w(M)p}{\overline{w}}$$

where p is the frequency of M in the current generation and $\overline{w}$ is the mean fitness of the population. But what will happen in environment 2? In that environment, the fitness values are different and thus potentially the dynamic and/or the outcome of natural selection could be different. So let us put the environment specific fitness values into our selection equation and see what happens:

$$p' = \frac{w'(M)p}{\overline{w}'}$$

To compare whether the environment has an influence on the dynamic of natural selection, let us compare the two equations:

$$p' = \frac{w'(M)p}{\overline{w}'} = \frac{cw(M)p}{c\overline{w}} = \frac{w(M)p}{\overline{w}}$$

because:

$$\overline{w}' = w'(M)p + w'(C)(1-p)$$
$$\overline{w}' = cw(M)p + cw(C)(1-p)$$
$$\overline{w}' = c\left[w(M)p + w(C)(1-p)\right] = c\overline{w}$$

In words, the dynamics of natural selection are the same in environment 1 and 2. Hence, the effect of the environmental change on fitness is in fact irrelevant for the consequences of natural selection. This phenomenon is called invariance of the selection equation with respect to a scale transformation and is related to one of the most important insights of measurement theory.

As mentioned above, the numerical values assigned to some natural quantities, like weight and length and many more, are constrained by empirical facts up to a multiplicative constant [12]. Other quantities like temperature are determined up to one multiplicative and an additive constant. These conversion constants do not affect the meaning of the measurement, only the numerical values we use to express the state of a quantity. An important consequence of this fact is that any mathematical model that uses these quantities have to be independent of the scale in which the various quantities are expressed [7]. Technically any law needs to be invariant to an allowable scale transformation of its variables. This is exactly what we saw above with respect to Wright's selection equation. Wright's equation is invariant to ratio scale transformations of the fitness values, and it can be shown that Wrightian fitness values are measured on a ratio scale [13]. From that it also follows that any environmental change that affects the fitness values in the same way as an allowable scale transformation is inconsequential. Only when the environmental effects change the fitness values in another way, does the environmental change and the genetic differences interact; that is, the environment affects the dynamics of natural selection. As a consequence, we can speak of genotype-environment interactions if the effects of an environmental factor on the fitness of various genotypes deviate from a multiplicative model. This leads to the following rule:

Rule II—irrelevant effects: *Let us consider a variable V with a family of allowable scale transformations T, and let us consider the effects of two factors A and B on V. The two factors do not interact if a change in one factor affects the values of V across the realizations of the other factor in the same way as an allowable scale transformation T.*

The justification for this rule follows from the fact that any mathematical model in which we want to use the measured quantities *V* has to be invariant with respect to allowable scale transformations. A corollary of this rule is that the reference model for the detection of interaction effects is determined by the scale type of the response variable *V*. We are dealing with an interaction between the two factors if the combined effects of the two factors are not equivalent to an allowable scale transformation of the variable *V*.

If we have a ratio scale variable, which is the majority of physical quantities, the reference model for judging whether two treatments or factors interact in their influence on the variable *V* is the multiplicative model. This implies that if we are dealing with log transformed ratio scale variables, the reference model is additivity. This is also the reason why log transformations are so frequently used in ANOVA applications, since many quantities are measured on a ratio scale and additivity of the log-transformed response variable are the appropriate model for detecting interaction, as in the classical ANOVA model discussed above.

Considering Rule II, the question arises whether the relationship described in the rule is symmetrical: let us assume that factor *A* affects the variable values across factor *B* by a multiplicative constant. Are the values across factor *A* affected by factor *B* also by a multiplicative constant? In fact it can be easily shown in the following proposition.

Proposition
If factor B affects variable V across factor A by a multiplicative constant c, then there exists a factor k such that the variable values across factor B are affected by factor A by this multiplicative constant.

Proof
For simplicity let us assume that each factor has two realizations, A_1 and A_2 as well as B_1 and B_2. The values of the variable *V* then are

$$
\begin{array}{c c c}
 & B_1 & B_2 \\
A_1 & V_{11} & V_{12} \\
A_2 & V_{21} & V_{22}
\end{array}
$$

As per the assumption we have $V_{12} = cV_{11}$ and $V_{22} = cV_{21}$. Without loss of generality, we can define $k = \dfrac{V_{21}}{V_{11}}$ and thus have $V_{21} = kV_{11}$. As per assumption we have $V_{22} = cV_{21}$; if we substitute in this equation by $V_{21} = kV_{11}$, we obtain $V_{22} = ckV_{11} = kcV_{11} = kV_{12}$. Hence, if the table is multiplicative across B, it is also multiplicative across A and vice versa.

3   Conclusions and Discussion

In this short contribution, I have argued that the detection and measurement of interactions is determined by considerations of scale type. Furthermore, I maintain, in accordance with representative measurement theory [5, 7, 12], that scale types are not a matter of mathematical or formal convenience or convention, but rather are a direct reflection of the empirical facts they are meant to reflect. Thus, I venture to suggest that interaction effects defined on the basis of scale type considerations have empirical meaning. This is most obvious with respect to the principle of effect propagation. If, for instance, a certain mutation or intervention affects gene expression by a factor X in genotype 1 and by a different factor Y in genotype 2, then the differences between genotype 1 and genotype 2 must be mechanistically relevant for the way the mutation affects the phenotype. No such inference is justified if interactions are considered as relative to an arbitrary choice or transformations of scale types, e.g., by log transforming the original variables. The reason is that if a transformation can eliminate a statistical interaction, then we have no reason to think that it reflects a mechanistic reality. To make statistical and functional interactions mechanistically meaningful, it is essential to respect the formal rules for the treatment of quantitative variables, which reflect the empirical meaning of the variables.

4   Notes

Here I am arguing that the inherent arbitrariness of the classic definition of epistasis and other forms of interaction can be eliminated by observing measurement theoretical constraints on the use of quantitative variables. The most salient points are:

- The scale type of the phenotypic variable measured determines the allowable effect measures (e.g., differences, fold differences, and others more).

- The scale type of the effect measure then determines the reference model for the definition of interaction:

- Epistasis is a deviation from the additive model if the main effects of gene substitutions are measured as a difference.

- Epistasis is a deviation from the multiplicative model if the main effects of gene substitutions are measured as a fold difference.

- Other reference models are possible, given other effect measures.

• These simple rules are straight forward to apply and guarantee that the so defined measures of epistasis/interaction are biologically meaningful.

References

1. Sokal RR, Rohlf FJ (1981) Biometry. The principles and practice of statistics in biological research, 2nd edn. W. H. Freeman and Company, New York

2. Cheverud J, Routman E (1995) Epistasis and its contribution to genetic variance components. Genetics 130:1455–1461

3. Lynch M, Walsh B (1998) Genetics and analysis of quantitative traits. Sinauer Associates, Inc., Sunderland, MA

4. Houle D, Pelabon C, Wagner GP, Hansen TF (2011) Measurement and meaning in biology. Q Rev Biol 86(1):3–34

5. Hand DL (2004) Measurement theory and practice: the world through quantification. Arnold, London

6. Hand DJ (1996) Statistics and the theory of measurement. J R Stat Soc A 159:445–492

7. Krantz DH, Luce RD, Suppes P, Tversky A (1971) Foundations of measurement, vol I. Academic Press, San Diego

8. Emera D, Wagner GP (2012) Transformation of a transposon into a derived prolactin promoter with function during human pregnancy. Proc Natl Acad Sci U S A 109(28):11246–11251. doi:10.1073/pnas.1118566109

9. Ben-Jonathan N, Mershon JL, Allen DL, Steinmetz RW (1996) Extrapituitary prolactin: distribution, regulation, functions, and clinical aspects. Endocr Rev 17(6):639–669

10. Riddick DH, Luciano AA, Kusmik WF, Maslar IA (1978) De novo synthesis of prolactin by human decidua. Life Sci 23(19):1913–1921

11. Golander A, Hurley T, Barrett J, Hizi A, Handwerger S (1978) Prolactin synthesis by human chorion-decidual tissue: a possible source of prolactin in the amniotic fluid. Science 202(4365):311–313

12. Roberts FS (1979) Measurement theory with applications to decisionmaking, utility, and the social sciences. Addison-Wesley Publ. Co., Reading, MA

13. Wagner GP (2010) The measurement theory of fitness. Evolution 64(5):1358–1376. doi:10.1111/j.1558-.00909.x

14. Mackay TFC, Lyman RF, Jackson MS (1992) Effects of *P* element insertions on quantitative traits in *Drosophila melanogaster*. Genetics 130:315–332

15. Bürger R (2000) The mathematical theory of selection, recombination, and mutation. Wiley, New York

Chapter 9

Direct Approach to Modeling Epistasis

Rong-Cai Yang

Abstract

Genome-wide association studies have recently been conducted in humans and domesticated animals and plants to locate and identify chromosomal regions or genes (quantitative trait loci or QTLs) to select individuals with superior performance and qualities. QTL or genetic effects, including epistatic effects, can be defined at the genotypic (functional) and gene (statistical) levels. In the past, the functional or statistical genetic effects have been defined indirectly, and genotypic values were expressed as linear functions of additive, dominance, and epistatic genetic effects. In this chapter, we propose to reverse the thinking and define genetic effects as linear functions of genotypic values. The direct definition of functional genetic effects is straightforward for well-known gene action models [e.g., unweighted (UW), F_2, and F_∞ models]. However, the direct definition of statistical genetic effects is based on Fisher's concept of average excess, which is closely related to the well-known concept of the average effect of a gene substitution. These definitions can be easily extended to cases of two or more loci as long as the loci are independent of each other. Two numerical examples are used to illustrate the properties of the direct approach.

Key words Average excess, Average effect, Nonepistatic and epistatic effects, Functional genetic effects, Statistical genetic effects, Reference populations

1 Introduction

Genome-wide association studies have recently been conducted in humans and domesticated animals and plants to locate and identify chromosomal regions or genes (quantitative trait loci or QTLs) to select individuals with superior qualities. These studies are possible because single-nucleotide polymorphisms (SNPs) are abundant and assays for them are readily available and affordable for many species. The increased marker density increases linkage disequilibrium (LD) between SNP markers and QTLs so that the marker effects may serve as reliable surrogates for QTL effects [1]. Consequently, while the effort to locate and estimate the effects of particular QTLs remains important, a growing focus is now on modeling actions and interactions (epistasis) of detected genome-wide QTL effects that contribute collectively to the overall genetic merits of animals [2] and plants [3].

Jason H. Moore and Scott M. Williams (eds.), *Epistasis: Methods and Protocols*, Methods in Molecular Biology, vol. 1253, DOI 10.1007/978-1-4939-2155-3_9, © Springer Science+Business Media New York 2015

QTL effects, including epistatic effects, are often defined using the two classic quantitative genetic models of Fisher [4], one for defining genotypic values (often known as functional models) and the other for defining the average effect of a single gene (statistical models). For a locus in a diploid population, Fisher [4] defined a as the functional additive effect equal to half the difference between the values of the two homozygotes and d as the functional dominance effect equal to the deviation of the heterozygote value from the mean (m) of the two homozygotes. This model is subsequently known as the F_∞ model [5–7] and can be transformed into other equivalent models, including an F_2 model [8, 9], an unweighted (UW) model [10], or a model of arbitrary reference [11]. This simple model has been extended to include the presence of multiple alleles or multiple loci [10, 12–14].

However, a diploid individual passes on its genes, not its genotype, to offspring, and measures of such genic effects are statistical because they depend on both the frequencies and values of genotypes in a given population. Thus, Fisher [4] defined the average effect of a gene as a partial regression coefficient from the linear regression of the genotypic values on the number of copies of that gene. This regression analysis is a weighted analysis with weights being genotypic frequencies in the population. The sum of the average effects of two alleles in a diploid genotype are added to the population mean to give the "predicted" genotypic value (or breeding value) of the individual and deviations from the predicted value are due to the interactions of the two alleles (dominance) at a given locus [7]. Thus, Fisher's concept of "average effect" allows a genotypic value at a locus to be partitioned into two components due to statistical additive and dominance effects. This model of partitioning has been extended to include statistical epistatic effects at two or more loci with two or more alleles per locus [15–18].

In the past, Cockerham [15] and others [6, 9, 11, 12, 18] developed a two-step indirect approach (regression analysis) to define the genetic effects. First, a linear model is used to construct a set of indicator variables that serve as elements of the genetic-effect design matrix for the genotypic values to be expressed as functions of genetic (additive, dominance, and epistatic) effects. Second, the genetic effects are then obtained by pre-multiplying the vector of genotypic values by the inverse of the genetic-effect design matrix. However, as there is no clear guideline for how to construct the indicator variables, it is not always easy to convey their biological meaning. Thus, it is desirable to develop a direct approach to defining the genetic effects based on clearly defined rules. In this chapter, we describe such a direct approach. We first provide an overview of the conventional indirect approach. We then describe the new direct approach to defining the functional and statistical genetic effects. Finally, we illustrate the new approach with empirical examples.

1.1 Reference Populations

Regardless of whether functional or statistical genetic effects are used to decompose the genotypic value into components due to allelic (additive and dominance) effects at individual loci and non-allelic interactions (epistatic effects) between loci, they are meaningful only with reference to a particular population (the so-called reference population). Therefore, choosing which population serves as the reference population is critical. If the actual population and the reference population are to be the same, then the components in the decomposition will have special properties including orthogonality in the absence of linkage disequilibrium (LD) between loci. However, because the number of actual populations that one could study is indefinitely large, the possibility that the actual and reference populations coincide is small.

For convenience, the populations considered are of two types: (1) those derived from inbred-line crosses with known gene and genotypic frequencies, and (2) those from natural outbreeding populations with arbitrary (unknown) gene and genotypic frequencies. Of all type (1) populations, the F_2 and F_∞ populations (e.g., doubled haploid lines or recombinant inbred lines) are commonly used as the reference populations to define the genetic effects and the genetic models known as the F_2 model and the F_∞ model, respectively [8]. In contrast, of all possible populations in type (2), the two populations with arbitrary gene frequencies and Hardy-Weinberg equilibrium (HWE) or Hardy-Weinberg disequilibrium (HWD) and with independence between loci (i.e., no LD) are commonly used as reference populations, and the genetic model is known as the Cockerham model [15]. The genetic effects defined for the inbred populations represent fixed differences *between* parental inbred lines or even species, whereas those defined for outbred populations represent genetic variation *within* a population. Thus, the scale and magnitude of the genetic effects in inbred populations are expected to be larger than those in outbred populations.

Let us consider a general non-equilibrium population in which all specific reference populations are special cases. For simplicity and relevance to the present study, we will only consider cases of two alleles per locus. Let A and a be the two alleles at a locus A with allele frequencies of p_A and p_a, and AA, Aa, and aa are three genotypes with frequencies of P_A^A, P_a^A, and P_a^a, respectively (Table 1). Clearly, the frequency of allele A is,

$$p_A = P_A^A + \frac{1}{2} P_a^A$$

and that of allele a is $p_a = 1 - p_A$. HWD at this locus is measured by

$$D_A = P_A^A - p_A^2 = P_a^a - p_a^2 = -\left(\frac{1}{2} P_a^A - p_A p_a\right).$$

Table 1
One- and two-locus genotypic frequency distributions for commonly used reference populations

Genotype	Genotypic value	Genotypic frequencies					
		General	F_∞	Haploid	F_2	HWE	HWD
One-locus							
AA	G_A^A	P_A^A	$\frac{1}{2}$	p_A	$\frac{1}{4}$	p_A^2	P_A^A
Aa	G_a^A	P_a^A	0	0	$\frac{1}{2}$	$2p_Ap_a$	P_a^A
aa	G_a^a	P_a^a	$\frac{1}{2}$	p_a	$\frac{1}{4}$	p_a^2	P_a^a
Total		1	1	1	1	1	1
Two-locus							
$AABB$	G_{AB}^{AB}	P_{AB}^{AB}	$\frac{1}{4}$	p_Ap_B	$\frac{1}{16}$	$p_A^2p_B^2$	$P_{A\cdot}^A P_{\cdot B}^B$
$AABb$	G_{Ab}^{AB}	P_{Ab}^{AB}	0	0	$\frac{1}{8}$	$2p_A^2p_Bp_b$	$P_{A\cdot}^A P_{\cdot b}^B$
$AAbb$	G_{Ab}^{Ab}	P_{Ab}^{Ab}	$\frac{1}{4}$	p_Ap_b	$\frac{1}{16}$	$p_A^2p_b^2$	$P_{A\cdot}^A P_{\cdot b}^b$
$AaBB$	G_{aB}^{AB}	P_{aB}^{AB}	0	0	$\frac{1}{8}$	$2p_Ap_ap_B^2$	$P_{a\cdot}^A P_{\cdot B}^B$
$AaBb$	G_{ab}^{AB}	P_{ab}^{AB}	0	0	$\frac{1}{4}$	$4p_Ap_ap_Bp_b$	$P_{a\cdot}^A P_{\cdot b}^B$
$Aabb$	G_{ab}^{Ab}	P_{ab}^{Ab}	0	0	$\frac{1}{8}$	$2p_Ap_ap_b^2$	$P_{a\cdot}^A P_{\cdot b}^b$
$aaBB$	G_{aB}^{aB}	P_{aB}^{aB}	$\frac{1}{4}$	p_ap_B	$\frac{1}{16}$	$p_a^2p_B^2$	$P_{a\cdot}^a P_{\cdot B}^B$
$aaBb$	G_{ab}^{aB}	P_{ab}^{aB}	0	0	$\frac{1}{8}$	$2p_a^2p_Bp_b$	$P_{a\cdot}^a P_{\cdot b}^B$
$aabb$	G_{ab}^{ab}	P_{ab}^{ab}	$\frac{1}{4}$	p_ap_b	$\frac{1}{16}$	$p_a^2p_b^2$	$P_{a\cdot}^a P_{\cdot b}^b$
Total		1	1	1	1	1	1

One- and two-locus genotypic frequencies for general non-equilibrium, F_∞, haploid, F_2, Hardy-Weinberg equilibrium (HWE), and Hardy-Weinberg disequilibrium (HWD) populations, respectively

Thus, the genotypic frequencies can be written as sums of HWE frequencies and deviations due to HWD. For convenience of subsequent discussion, we write these relationships in matrix form:

$$\mathbf{P}_A = \mathbf{P}_{HWE\cdot A} + \mathbf{D}_A \qquad (1)$$

where $\mathbf{P}_A = \mathrm{diag}\{P_A^A\ P_a^A\ P_a^a\}$, $\mathbf{P}_{HWE\cdot A} = \mathrm{diag}\{p_A^2\ 2p_Ap_a\ p_a^2\}$, and $\mathbf{D}_A = \mathrm{diag}\{D_A\ -2D_A\ D_A\}$ with $\mathrm{diag}\{\}$ indicating a diagonal matrix. For example,

$$\mathbf{P}_A = \mathrm{diag}\left\{P_A^A\ \ P_a^A\ \ P_a^a\right\} = \begin{bmatrix} P_A^A & 0 & 0 \\ 0 & P_a^A & 0 \\ 0 & 0 & P_a^a \end{bmatrix}.$$

Thus, the genotypic frequencies in an F_2 population are $\mathbf{P}_{F_2\cdot A} = \mathrm{diag}\left\{\frac{1}{4}\ \frac{1}{2}\ \frac{1}{4}\right\}$, and those in an F_∞ population are $\mathbf{P}_{F_\infty\cdot A} = \mathrm{diag}\left\{\frac{1}{2}\ 0\ \frac{1}{2}\right\}$. Cheverud and Routman [10] and

Cheverud [12] used a hypothetical population with unweighted (UW) or equal genotypic frequencies. The genotypic frequencies in a UW population are $\mathbf{P}_{UW \cdot A} = \mathrm{diag}\left\{\frac{1}{3} \quad \frac{1}{3} \quad \frac{1}{3}\right\}$.

Now consider two loci, each with two alleles, A and a at locus A and B and b at locus B. In this case, there are nine possible genotypes assuming that the coupling and repulsion double heterozygotes are indistinguishable. Following Yang [19], we write frequencies of these genotypes as, $P_{vz}^{uy} = P_{uy}^{vz}$, which result from the union of gametes uy and vz with u, $v = A$ or a, and y, $z = B$ or b (Table 1). The genotypic frequencies at individual loci are the marginal totals of the appropriate two-locus genotypic frequencies. Thus, the frequency of the genotype uv at locus A is

$$P_{v\cdot}^{u\cdot} = P_{vB}^{uB} + P_{vb}^{uB} + P_{vb}^{ub},$$

and the frequency of genotype yz at locus B is

$$P_{\cdot z}^{\cdot y} = P_{Az}^{Ay} + P_{az}^{Ay} + P_{az}^{ay}.$$

A deviation of observed two-locus genotypic frequency (P_{vz}^{uy}) from the product of single-locus genotypic frequencies ($P_{v\cdot}^{u\cdot} \times P_{\cdot z}^{\cdot y}$) under interlocus independence is called zygotic association or disequilibrium [19], $\omega_{vz}^{uy} = P_{vz}^{uy} - P_{v\cdot}^{u\cdot} P_{\cdot z}^{\cdot y}$. In this case, each two-locus genotypic frequency is simply the sum of its expected value under the zygotic equilibrium ($\omega_{vz}^{uy} = 0$ and $P_{vz}^{uy} = P_{v\cdot}^{u\cdot} \ P_{\cdot z}^{\cdot y}$) and zygotic association. Writing all two-locus genotypic frequencies in matrix form, we have

$$\mathbf{P}_{AB} = \Psi_{AB} + \Omega_{AB}, \tag{2}$$

where

$$\mathbf{P}_{AB} = \mathrm{diag}\{P_{AB}^{AB} \quad P_{Ab}^{AB} \quad P_{Ab}^{Ab} \quad P_{aB}^{AB} \quad P_{ab}^{AB} + P_{Ab}^{aB} \quad P_{ab}^{Ab} \quad P_{aB}^{aB} \quad P_{ab}^{Ab} \quad P_{ab}^{ab}\},$$

$$\Psi_{AB} = \mathrm{diag}\{P_{A\cdot}^{A}\cdot P_{\cdot B}^{B} \quad P_{A\cdot}^{A}\cdot P_{\cdot b}^{B} \quad P_{A\cdot}^{A}\cdot P_{\cdot b}^{b} \quad P_{a\cdot}^{A}\cdot P_{\cdot B}^{B} \quad P_{a\cdot}^{A}\cdot P_{\cdot b}^{b} \quad P_{a\cdot}^{A}\cdot P_{\cdot b}^{b} \quad P_{a\cdot}^{a}\cdot P_{\cdot B}^{B} \quad P_{a\cdot}^{a}\cdot P_{\cdot B}^{B} \quad P_{a\cdot}^{a}\cdot P_{\cdot b}^{b}\}$$

and

$$\Omega_{AB} = \mathrm{diag}\{\omega_{AB}^{AB} \quad \omega_{Ab}^{AB} \quad \omega_{Ab}^{Ab} \quad \omega_{aB}^{AB} \quad \omega_{ab}^{AB} \quad \omega_{ab}^{Ab} \quad \omega_{aB}^{aB} \quad \omega_{ab}^{Ab} \quad \omega_{ab}^{ab}\}$$

with diag{} indicating a diagonal matrix. Clearly the matrix of two-locus expected genotypic frequencies (Ψ_{AB}) can be written as a Kronecker product of the two matrices of genotypic frequencies at loci A and B,

$$\Psi_{AB} = \mathbf{P}_A \otimes \mathbf{P}_B. \tag{3}$$

Most gene action models for quantitative genetic analyses and QTL mapping assume the interlocus independence ($\Omega = 0$ and $\mathbf{P} = \Psi$).

The two-locus results above can be readily generalized to cases of three or more loci. For example, with three loci, A, B, and C, the three-locus genotypic frequencies can be written as the sum of the expected values under complete interlocus independence and three-locus zygotic disequilibria, i.e.,

$$\mathbf{P}_{ABC} = \mathbf{\Psi}_{ABC} + \mathbf{\Omega}_{ABC}. \tag{4}$$

The matrix of three-locus genotypic frequencies under complete interlocus independence ($\mathbf{\Psi}_{ABC}$) can be obtained from the reverse-order Kronecker products of the three matrices of genotypic frequencies at loci A, B, and C:

$$\mathbf{\Psi}_{ABC} = \mathbf{P}_C \otimes \mathbf{P}_B \otimes \mathbf{P}_A. \tag{5}$$

The three-locus zygotic disequilibria ($\mathbf{\Omega}_{ABC}$) can be further subdivided to account for partial dependence between pairs of loci, but this matter is beyond the scope of the present study.

2   Materials

Two computer programs, DIRECT_EPISTASIS.SAS and DIRECT_EPISTASIS.SAS, written in SAS PROC IML [20], are available at our website (http://www.statgen.ualberta.ca/index.html?open=software.html). These programs implement our new direct approach and the traditional indirect two-step approach to modeling epistasis that will be described in detail below. The implementation is done through the analysis of an empirical data set taken from Table 8 of Doebley et al. [21].

3   Methods

3.1   Indirect Approach

The conventional (indirect) approach is based on the premise that for a vector of genotypic value $\mathbf{G}$, we can always find a full-rank genetic-effect design matrix $\mathbf{W}$ and its inverse $\mathbf{W}^{-1}$ to transform the vector of genotypic values into the vector of genetic effects $\mathbf{\beta}$,

$$\mathbf{G} = \mathbf{W}\mathbf{W}^{-1}\mathbf{G} = \mathbf{W}\mathbf{\beta} \quad \text{and} \quad \mathbf{\beta} = \mathbf{W}^{-1}\mathbf{G}.$$

Clearly, it is the different constructs of $\mathbf{W}$ or $\mathbf{W}^{-1}$ that lead to different gene action models (i.e., different solutions for $\mathbf{\beta}$). Thus, the indirect approach consists of two steps. First, a genetic-effect design matrix is constructed so that the genotypic values are expressed as functions of genetic (additive, dominance, and epistatic) effects. Second, the genetic effects are obtained by pre-multiplying the vector of genotypic values by the inverse of the genetic-effect design matrix. This approach has been widely described and used in the literature [6, 11–13]. We now look at specific cases of one or more loci.

With two alleles, A and a, at locus A, three parameters (β_0, β_1, and β_2) are required for a complete specification of three genotypes (AA, Aa, and aa), and individual genotypic values are expressed as a linear function of the three parameters using the following regression equation:

$$\mathbf{G}_A = \mathbf{1}\beta_0 + \mathbf{w}_1\beta_1 + \mathbf{w}_2\beta_2 \tag{6}$$

where $\mathbf{G}_A = \begin{bmatrix} G_A^A & G_a^A & G_a^a \end{bmatrix}'$ is the vector of genotypic values with the prime ($'$) denoting matrix or vector transposition, $\mathbf{1}$ is a vector of ones, $\beta_0 = \mathbf{G}_A'\mathbf{P}_A\mathbf{1}$ is the mean of a given reference population, $\mathbf{w}_1$ and $\mathbf{w}_2$ are the vectors of values for the two design variables, and β_1 and β_2 are the coefficients of partial regression of G_A on $\mathbf{w}_1$ and $\mathbf{w}_2$, respectively,

$$\beta_i = \frac{\mathbf{G}_A'\mathbf{P}_A\mathbf{w}_i - \left(\mathbf{G}_A'\mathbf{P}_A\mathbf{1}\right)\left(\mathbf{w}_i'\mathbf{P}_A\mathbf{1}\right)}{\mathbf{w}_i'\mathbf{P}_A\mathbf{w}_i - \left(\mathbf{w}_i'\mathbf{P}_A\mathbf{1}\right)^2}. \tag{7}$$

for $i = 1, 2$. Cockerham [15] and subsequent workers have constructed vectors $\mathbf{w}_1$ and $\mathbf{w}_2$ with the special requirements of orthogonality between them: (1) $\mathbf{w}_i'\mathbf{P}_A\mathbf{1} = 0$ and (2) $\mathbf{w}_i'\mathbf{P}_A\mathbf{w}_j = 0$ for $i \neq j$. In this case, the partial regression coefficients are simplified to

$$\beta_i = \frac{\mathbf{G}_A'\mathbf{P}_A\mathbf{w}_i}{\mathbf{w}_i'\mathbf{P}_A\mathbf{w}_i}, \tag{8}$$

and there is no correlation between β_1 and β_2.

The genetic effects in vector $\boldsymbol{\beta}$ can be either *functional* (genotype-based) or *statistical* (gene-based) [11, 12]. Álvarez-Castro and Carlborg [11] gave the general constructs to design a matrix for the functional genetic effects ($\mathbf{W}_{F\cdot A}$)

$$\mathbf{W}_{F\cdot A} = \begin{bmatrix} 1 & 2-2p_A & -P_a^A \\ 1 & 1-2p_A & P_A^A + P_a^a \\ 1 & 0-2p_A & -P_a^A \end{bmatrix} \tag{9}$$

and for the statistical genetic effects ($\mathbf{W}_{S\cdot A}$)

$$\mathbf{W}_{S\cdot A} = \begin{bmatrix} 1 & 2-2p_A & -\dfrac{P_a^a P_a^A}{P_a^a p_A + P_A^A p_a} \\[3mm] 1 & 1-2p_A & \dfrac{2P_a^a P_a^A}{P_a^a p_A + P_A^A p_a} \\[3mm] 1 & 0-2p_A & -\dfrac{P_a^A P_a^A}{P_a^a p_A + P_A^A p_a} \end{bmatrix}, \tag{10}$$

where $2p_A$ is the expected content of allele A across all genotypes in the reference population. The second column (vector $\mathbf{w}_1$) in $\mathbf{W}_{F \cdot A}$ and $\mathbf{W}_{S \cdot A}$ are the deviations of the observed content of allele A in individual genotypes [AA, Aa, and aa] from the expected values. The third column (vector $\mathbf{w}_2$) in $\mathbf{W}_{S \cdot A}$ differs from its counterpart in $\mathbf{W}_{F \cdot A}$ because a further condition of orthogonality between $\mathbf{w}_1$ and $\mathbf{w}_2$ ($\mathbf{w}_1' \mathbf{P}_A \mathbf{w}_2 = 0$) is imposed. Once the genetic-effect design matrix is constructed, the genetic effects are calculated as $\boldsymbol{\beta}_{F \cdot A} = \mathbf{W}_{F \cdot A}^{-1} \mathbf{G}_A$ or $\boldsymbol{\beta}_{S \cdot A} = \mathbf{W}_{S \cdot A}^{-1} \mathbf{G}_A$. It should be noted that the regression coefficients in $\boldsymbol{\beta}_A = \begin{bmatrix} \beta_0 & \beta_1 & \beta_2 \end{bmatrix}'$ can be either functional or statistical genetic effects, but the usual convention [7, 11, 18] is that the functional genetic effects are labeled by Latin letters $\boldsymbol{\beta}_{F \cdot A} = \begin{bmatrix} m & a_A & d_d \end{bmatrix}'$, and the statistical genetic effects are labeled by Greek letters $\boldsymbol{\beta}_{S \cdot A} = \begin{bmatrix} \mu & \alpha_A & \delta_d \end{bmatrix}'$.

The general expression of $\mathbf{W}_{F \cdot A}$ in Eq. 9 reduces to the genetic-effect design matrices for some well-known reference populations [6] by setting gene and genotypic frequencies to specific values, e.g., the F_∞ population ($p_A = p_a = \frac{1}{2}$; $P_A^A = P_a^a = \frac{1}{2}$ and $P_a^A = 0$), the F_2 population ($p_A = p_a = \frac{1}{2}$; $P_A^A = P_a^a = \frac{1}{4}$ and $P_a^A = \frac{1}{2}$), or the unweighted (UW) hypothetical population ($p_A = p_a = \frac{1}{2}$; $P_A^A = P_a^A = P_a^a = \frac{1}{3}$):

$$\mathbf{W}_{F_\infty \cdot A} = \begin{bmatrix} 1 & 1 & 0 \\ 1 & 0 & 1 \\ 1 & -1 & 0 \end{bmatrix}, \quad \mathbf{W}_{F_2 \cdot A} = \begin{bmatrix} 1 & 1 & -\frac{1}{2} \\ 1 & 0 & \frac{1}{2} \\ 1 & -1 & -\frac{1}{2} \end{bmatrix}, \quad \text{and } \mathbf{W}_{UW \cdot A} = \begin{bmatrix} 1 & 1 & -\frac{1}{3} \\ 1 & 0 & \frac{2}{3} \\ 1 & -1 & -\frac{1}{3} \end{bmatrix}. \quad (11)$$

More generally, we can have any other population as a reference population. For example, the design matrices for a monomorphic (MONO) population of one homozygote, aa ($p_A = 0$; $p_a = 1$; $P_A^A = P_a^A = 0$ and $P_a^a = 1$), F_1 population ($p_A = p_a = \frac{1}{2}$; $P_A^A = P_a^a = 0$ and $P_a^A = 1$), and backcross (BC, $Aa \times aa$) population ($p_A = \frac{1}{4}$; $p_a = \frac{3}{4}$; $P_a^A = P_a^a = \frac{1}{2}$ and $P_A^A = 0$) are

$$\mathbf{W}_{MONO \cdot A} = \begin{bmatrix} 1 & 2 & 0 \\ 1 & 1 & 1 \\ 1 & 0 & 0 \end{bmatrix}, \quad \mathbf{W}_{F_1 \cdot A} = \begin{bmatrix} 1 & 1 & -1 \\ 1 & 0 & 0 \\ 1 & -1 & -1 \end{bmatrix}, \quad \text{and } \mathbf{W}_{BC \cdot A} = \begin{bmatrix} 1 & 1\frac{1}{2} & -\frac{1}{2} \\ 1 & \frac{1}{2} & \frac{1}{2} \\ 1 & -\frac{1}{2} & -\frac{1}{2} \end{bmatrix}. \quad (12)$$

This indirect approach has been extended to the case of multiple alleles at a locus [13, 14].

A linear model (6) for a single locus forms a basis for extension to multiple loci [for details, *see* [11]]. With L (≥ 2) independent loci in a general HWD population ($\mathbf{\Omega}_{AB\ldots L} = 0$ and $\mathbf{P}_{AB\ldots L} = \mathbf{\Psi}_{AB\ldots L}$), there are $(q+1)$ parameters in $\boldsymbol{\beta}$ with the mean of a reference population (β_0) plus $q = 3^L - 1$ genetic effects ($\beta_1, \beta_2, \ldots, \beta_q$). Thus,

$$\mathbf{G}_{AB\cdots L} = \mathbf{1}\beta_0 + \sum_{i=1}^{q} \mathbf{w}_i \beta_i \qquad (13)$$

where $\mathbf{G}_{AB\ldots L}$ is the $[(q+1) \times 1]$ vector of L-locus genotypic values that can be conveniently arranged as a series of direct products of L vectors of single-locus genotypic values, $G_{AB\ldots L} = \mathbf{G}_L \otimes \cdots \otimes \mathbf{G}_B \otimes \mathbf{G}_A = [G_{AB\ldots L}{}^{AB\cdots L} \ G_{aB\ldots L}{}^{AB\cdots L} \ \cdots \ G_{ab\ldots l}{}^{ab\cdots l}]'$ corresponding to the genotypes $[AABB\cdots LL \, AaBB\cdots LL \cdots aabb\cdots ll]$, $\beta_0 = \mathbf{G}'_{AB\ldots L} \mathbf{\Psi}_{AB\ldots L} \mathbf{1}$ is the mean of a given reference population, $\mathbf{w}_i$ is the $(i+1)$th column vector of the $(q+1) \times (q+1)$ design matrix ($\mathbf{W}_{AB\ldots L} = [\mathbf{1} \, \mathbf{w}_1 \cdots \mathbf{w}_q]$), and β_i is the coefficient of partial regression of $\mathbf{G}_{AB\ldots L}$ on $\mathbf{w}_i$,

$$\beta_i = \frac{\mathbf{G}'_{AB\cdots L} \mathbf{\Psi}_{AB\cdots L} \mathbf{w}_i - \left(\mathbf{G}'_{AB\cdots L} \mathbf{\Psi}_{AB\cdots L} \mathbf{1}\right)\left(\mathbf{w}'_i \mathbf{\Psi}_{AB\cdots L} \mathbf{1}\right)}{\mathbf{w}'_i \mathbf{\Psi}_{AB\cdots L} \mathbf{w}_i - \left(\mathbf{w}'_i \mathbf{\Psi}_{AB\cdots L} \mathbf{1}\right)^2}. \qquad (14)$$

if $\mathbf{w}_i$'s are not orthogonal contrasts; otherwise,

$$\beta_i = \frac{\mathbf{G}'_{AB\cdots L} \mathbf{\Psi}_{AB\cdots L} \mathbf{w}_i}{\mathbf{w}'_i \mathbf{\Psi}_{AB\cdots L} \mathbf{w}_i} \qquad (15)$$

if $\mathbf{w}_i$'s are orthogonal contrasts. The L-locus design matrix for the reference population ($\mathbf{W}_{AB\ldots L}$) can be obtained from a series of direct products of L single-locus design matrices, i.e., $\mathbf{W}_{AB\ldots L} = \mathbf{W}_L \otimes \cdots \otimes \mathbf{W}_B \otimes \mathbf{W}_A$. The L-locus genetic-effect vector $\boldsymbol{\beta}_{AB\ldots L}$ of order $(q+1) \times 1$ (β_0 to β_q) can be arranged according to the sequence of the terms obtained from a series of direct products of single-locus vectors:

$$\boldsymbol{\beta}_{AB\cdots L} = \boldsymbol{\beta}_L \otimes \cdots \otimes \boldsymbol{\beta}_B \otimes \boldsymbol{\beta}_A = \begin{bmatrix} 1 & \beta_A & \cdots & \beta_{2A}\beta_{2B}\cdots\beta_{2L} \end{bmatrix}'.$$

3.2 Direct Approach: Functional Genetic Effects

The key step of the indirect approach is to construct the genetic-effect design matrix (**W**-matrix). However, it appears that the genetic-effect design variables are defined on an ad hoc basis, and there is no general rule that governs the construction of these design variables. For example, the single-locus design matrices for the F_∞, F_2, and UW reference populations differ only in the definition of the dominance-effect design variable [cf. Eq. 11] but there is no clear indication of how this variable would vary from one reference population to another. This masks the simple, fundamental fact that the genetic effects defined from comparisons among the same genotypes in a given population should be the same regardless of which reference population is used. This point can be more explicitly illustrated by inspecting the inverses of the genetic-effect design matrices, for example those given in Eq. 11 for the F_∞, F_2, and UW models:

$$\mathbf{W}_{F_\infty \cdot A}^{-1} = \begin{bmatrix} \frac{1}{2} & 0 & \frac{1}{2} \\ \frac{1}{2} & 0 & -\frac{1}{2} \\ -\frac{1}{2} & 1 & -\frac{1}{2} \end{bmatrix}, \quad \mathbf{W}_{F_2 \cdot A}^{-1} = \begin{bmatrix} \frac{1}{4} & \frac{1}{2} & \frac{1}{4} \\ \frac{1}{2} & 0 & -\frac{1}{2} \\ -\frac{1}{2} & 1 & -\frac{1}{2} \end{bmatrix} \quad \text{and}$$

$$\mathbf{W}_{UW \cdot A}^{-1} = \begin{bmatrix} \frac{1}{3} & \frac{1}{3} & \frac{1}{3} \\ \frac{1}{2} & 0 & -\frac{1}{2} \\ -\frac{1}{2} & 1 & -\frac{1}{2} \end{bmatrix} \tag{16}$$

We immediately recognize that the gene action models defined under the F_∞, F_2, and UW models have the same definitions for genetic effects (the second and third rows of the $\mathbf{W}^{-1}$-matrices) but differ only in the means of different reference populations (the first row of the $\mathbf{W}^{-1}$-matrices). Thus, a generalization of these specific models is given by taking the inverse of the design matrix (9) as shown in Álvarez-Castro and Carlborg [11],

$$\mathbf{W}_{F \cdot A}^{-1} = \begin{bmatrix} P_A^A & P_a^A & P_a^a \\ \frac{1}{2} & 0 & -\frac{1}{2} \\ -\frac{1}{2} & 1 & -\frac{1}{2} \end{bmatrix}, \tag{17}$$

so that the functional genetic effect at locus A is $\boldsymbol{\beta}_A = \mathbf{W}_{F \cdot A}^{-1} \mathbf{G}_A$:

$$\begin{bmatrix} m_A \\ a_A \\ d_A \end{bmatrix} = \begin{bmatrix} P_A^A & P_a^A & P_a^a \\ \frac{1}{2} & 0 & -\frac{1}{2} \\ -\frac{1}{2} & 1 & -\frac{1}{2} \end{bmatrix} \begin{bmatrix} G_A^A \\ G_a^A \\ G_a^a \end{bmatrix}. \tag{18}$$

The functional genetic effects in Eq. 18 can be obtained directly from the mean of the reference population and two simple comparisons among genotypic values. The mean of the reference is obviously dependent on the genotypic frequencies of that population:

$$m_A = \mathbf{G}_A^{'} \mathbf{P}_A \mathbf{1} = P_A^A G_A^A + P_a^A G_a^A + P_a^a G_a^a. \tag{19}$$

For example, the means of three well-known reference populations, F_∞, F_2, and UW are

$$m_A = \begin{cases} \frac{1}{2}\left(G_A^A + G_a^a\right), & F_\infty\left(P_A^A = P_a^a = \frac{1}{2}; P_a^A = 0\right) \\ \frac{1}{4}\left(G_A^A + 2G_a^A + G_a^a\right), & F_2\left(P_A^A = 2P_a^A = P_a^a = \frac{1}{4}\right) \\ \frac{1}{3}\left(G_A^A + G_a^A + G_a^a\right), & UW\left(P_A^A = P_a^A = P_a^a = \frac{1}{3}\right) \end{cases}$$

The two genotypic comparisons lead to the usual definitions of functional additive and dominance effects [5, 7]: (1) the functional additive effect is half the difference between the values of two homozygotes,

$$a_A = \left(G_A^A - G_a^a\right) / 2 \tag{20}$$

and (2) the functional dominance effect is the deviation of the heterozygote value to the average of the values of the two homozygotes,

$$d_A = G_a^A - \left(G_A^A + G_a^a\right) / 2. \tag{21}$$

Putting Eqs. 19–21 together in matrix form, Eq. 18 is reproduced.

Now consider that a trait is controlled by genes at a pair of independent loci. We can obtain the general form of the two-locus genetic effects for any reference population from the knowledge of the genetic-effect design matrices at the two individual loci,

$$\beta_{F \cdot AB} = \mathbf{W}_{F \cdot AB}^{-1} \mathbf{G}_{AB} = \left(\mathbf{W}_{F \cdot B}^{-1} \otimes \mathbf{W}_{F \cdot A}^{-1}\right) \mathbf{G}_{AB},$$

where $\mathbf{W}_{F \cdot A}^{-1}$(and thus $\mathbf{W}_{F \cdot B}^{-1}$) for locus A (and locus B) is given in Eq. 17 and

$$\mathbf{W}_{F.AB}^{-1} = \mathbf{W}_{F.B}^{-1} \otimes \mathbf{W}_{F.A}^{-1}$$

$$= \begin{bmatrix} P_{.B}^{.B} & P_{.b}^{.B} & P_{.b}^{.b} \\ \frac{1}{2} & 0 & -\frac{1}{2} \\ -\frac{1}{2} & 1 & -\frac{1}{2} \end{bmatrix} \otimes \begin{bmatrix} P_{A.}^{A.} & P_{a.}^{A.} & P_{a.}^{a.} \\ \frac{1}{2} & 0 & -\frac{1}{2} \\ -\frac{1}{2} & 1 & -\frac{1}{2} \end{bmatrix}$$

$$= \begin{bmatrix}
P_{A.}^{A.}P_{.B}^{.B} & P_{a.}^{A.}P_{.B}^{.B} & P_{a.}^{a.}P_{.B}^{.B} & P_{A.}^{A.}P_{.b}^{.B} & P_{a.}^{A.}P_{.b}^{.B} & P_{a.}^{a.}P_{.b}^{.B} & P_{A.}^{A.}P_{.b}^{.b} & P_{a.}^{A.}P_{.b}^{.b} & P_{a.}^{a.}P_{.b}^{.b} \\
\frac{1}{2}P_{.B}^{.B} & 0 & -\frac{1}{2}P_{.B}^{.B} & \frac{1}{2}P_{.b}^{.B} & 0 & -\frac{1}{2}P_{.b}^{.B} & \frac{1}{2}P_{.b}^{.b} & 0 & -\frac{1}{2}P_{.b}^{.b} \\
-\frac{1}{2}P_{.B}^{.B} & P_{.B}^{.B} & -\frac{1}{2}P_{.B}^{.B} & -\frac{1}{2}P_{.b}^{.B} & P_{.b}^{.B} & -\frac{1}{2}P_{.b}^{.B} & -\frac{1}{2}P_{.b}^{.b} & P_{.b}^{.b} & -\frac{1}{2}P_{.b}^{.b} \\
\frac{1}{2}P_{A.}^{A.} & \frac{1}{2}P_{a.}^{A.} & \frac{1}{2}P_{a.}^{a.} & 0 & 0 & 0 & -\frac{1}{2}P_{A.}^{A.} & -\frac{1}{2}P_{a.}^{A.} & -\frac{1}{2}P_{a.}^{a.} \\
\frac{1}{4} & 0 & -\frac{1}{4} & 0 & 0 & 0 & -\frac{1}{4} & 0 & \frac{1}{4} \\
-\frac{1}{4} & \frac{1}{2} & -\frac{1}{4} & 0 & 0 & 0 & \frac{1}{4} & -\frac{1}{2} & \frac{1}{4} \\
-\frac{1}{2}P_{A.}^{A.} & -\frac{1}{2}P_{a.}^{A.} & -\frac{1}{2}P_{a.}^{a.} & P_{A.}^{A.} & P_{a.}^{A.} & P_{a.}^{a.} & -\frac{1}{2}P_{A.}^{A.} & -\frac{1}{2}P_{a.}^{A.} & -\frac{1}{2}P_{a.}^{a.} \\
-\frac{1}{4} & 0 & \frac{1}{4} & \frac{1}{2} & 0 & -\frac{1}{2} & -\frac{1}{4} & 0 & \frac{1}{4} \\
\frac{1}{4} & -\frac{1}{2} & \frac{1}{4} & -\frac{1}{2} & 1 & -\frac{1}{2} & \frac{1}{4} & -\frac{1}{2} & \frac{1}{4}
\end{bmatrix} \quad (22)$$

Just as for one-locus models, the two-locus genetic effects given in Eq. 22 are obtained directly from comparisons among one-locus and two-locus genotypes. Thus, the additive and dominance effects at locus A (a_A and d_A) and locus B (a_B and d_B) are simply comparisons among the *marginal* means of the three genotypes at each of the two loci,

$$\begin{aligned}
a_A &= \tfrac{1}{2}\left(\bar{G}_{A.}^{A.} - \bar{G}_{a.}^{a.}\right) \\
d_A &= \bar{G}_{a.}^{A.} - \tfrac{1}{2}\left(\bar{G}_{A.}^{A.} + \bar{G}_{a.}^{a.}\right) \\
a_B &= \tfrac{1}{2}\left(\bar{G}_{.B}^{.B} - \bar{G}_{.b}^{.b}\right) \\
d_B &= \bar{G}_{.b}^{.B} - \tfrac{1}{2}\left(\bar{G}_{.B}^{.B} + \bar{G}_{.b}^{.b}\right)
\end{aligned} \quad (23)$$

where $\bar{G}_{v.}^{u.}$ and $\bar{G}_{.z}^{.y}$ are the marginal means of genotype uv ($uv = AA$, Aa or aa) at locus A and genotype yz ($yz = BB$, Bb or bb) at locus B, respectively

$$\begin{aligned}
\bar{G}_{v.}^{u.} &= P_{.B}^{.B}G_{vB}^{uB} + P_{.b}^{.B}G_{vb}^{uB} + P_{.b}^{.b}G_{vb}^{ub} \\
\bar{G}_{.z}^{.y} &= P_{A.}^{A.}G_{Az}^{Ay} + P_{a.}^{A.}G_{az}^{Ay} + P_{a.}^{a.}G_{az}^{ay}
\end{aligned} \quad (24)$$

Clearly, $\bar{G}_{v.}^{u.}$ and $\bar{G}_{.z}^{.y}$ are not the same across different reference populations because the genotypic frequencies at individual loci in different reference populations are different. For example, the marginal mean values of genotype uv ($uv = AA$, Aa, or aa) at locus A for the three reference populations (F_∞, F_2, UW) are

$$\bar{G}_{v\cdot}^{u\cdot} = \begin{cases} \frac{1}{2}\left(G_{vB}^{uB} + G_{vb}^{ub}\right), & F_{\infty} \\ \frac{1}{4}\left(G_{vB}^{uB} + 2G_{vb}^{uB} + G_{vb}^{ub}\right), & F_2 \\ \frac{1}{3}\left(G_{vB}^{uB} + G_{vb}^{uB} + G_{vb}^{ub}\right), & UW \end{cases} \tag{25}$$

Inserting the expressions $\bar{G}_{v\cdot}^{u\cdot}$ and $\bar{G}_{\cdot z}^{\cdot y}$ in Eq. 24 into Eq. 23 gives

$$a_A = \frac{1}{2}\left(P_{\cdot B}^{\cdot B}G_{AB}^{AB} + P_{\cdot b}^{\cdot B}G_{Ab}^{AB} + P_{\cdot b}^{\cdot b}G_{Ab}^{Ab} - P_{\cdot B}^{\cdot B}G_{aB}^{AB} - P_{\cdot b}^{\cdot B}G_{ab}^{aB} - P_{\cdot b}^{\cdot b}G_{ab}^{ab}\right)$$

$$d_A = P_{\cdot B}^{\cdot B}G_{aB}^{AB} + P_{\cdot b}^{\cdot B}G_{ab}^{AB} + P_{\cdot b}^{\cdot b}G_{ab}^{Ab} - \frac{1}{2}\begin{pmatrix} P_{\cdot B}^{\cdot B}G_{AB}^{AB} + P_{\cdot b}^{\cdot B}G_{Ab}^{AB} + P_{\cdot b}^{\cdot b}G_{Ab}^{Ab} + P_{\cdot B}^{\cdot B}G_{aB}^{aB} \\ + P_{\cdot b}^{\cdot B}G_{ab}^{aB} + P_{\cdot b}^{\cdot b}G_{ab}^{ab} \end{pmatrix}$$

$$a_B = \frac{1}{2}\left(P_{A\cdot}^{A\cdot}G_{AB}^{AB} + P_{a\cdot}^{A\cdot}G_{aB}^{AB} + P_{a\cdot}^{a\cdot}G_{aB}^{aB} - P_{A\cdot}^{A\cdot}G_{Ab}^{AB} - P_{a\cdot}^{A\cdot}G_{ab}^{Ab} - P_{a\cdot}^{a\cdot}G_{ab}^{ab}\right) \tag{26}$$

$$d_B = P_{A\cdot}^{A\cdot}G_{Ab}^{AB} + P_{a\cdot}^{A\cdot}G_{ab}^{AB} + P_{a\cdot}^{a\cdot}G_{ab}^{aB} - \frac{1}{2}\begin{pmatrix} P_{A\cdot}^{A\cdot}G_{AB}^{AB} + P_{a\cdot}^{A\cdot}G_{aB}^{AB} + P_{a\cdot}^{a\cdot}G_{aB}^{aB} + P_{A\cdot}^{A\cdot}G_{Ab}^{Ab} \\ + P_{a\cdot}^{A\cdot}G_{ab}^{Ab} + P_{a\cdot}^{a\cdot}G_{ab}^{ab} \end{pmatrix}$$

The four epistatic effects can be obtained simply by applying the following rules: (1) take the products of $a_A \times a_B$, $a_A \times d_B$, $d_A \times a_B$ and $d_A \times d_B$, where a_A, d_A, a_B, and d_B are given in Eq. 23; (2) expand these products into individual products of the marginal means, $\bar{G}_{v\cdot}^{u\cdot} \times \bar{G}_{\cdot z}^{\cdot y}$; and (3) replace the products of $\bar{G}_{v\cdot}^{u\cdot} \times \bar{G}_{\cdot z}^{\cdot y}$ by the corresponding two-locus genotypic values G_{vz}^{uy}. Following these rules, we obtain the four epistatic effects:

$$aa = \frac{1}{4}\left(G_{AB}^{AB} - G_{Ab}^{Ab} - G_{aB}^{aB} + G_{ab}^{ab}\right)$$

$$ad = \frac{1}{4}\left(-G_{AB}^{AB} + 2G_{Ab}^{AB} - G_{Ab}^{Ab} + G_{aB}^{aB} - 2G_{ab}^{aB} + G_{ab}^{ab}\right)$$

$$da = \frac{1}{4}\left(-G_{AB}^{AB} + 2G_{aB}^{AB} - G_{aB}^{aB} + G_{Ab}^{Ab} - 2G_{ab}^{Ab} + G_{ab}^{ab}\right) \tag{27}$$

$$dd = \frac{1}{4}\left(G_{AB}^{AB} - 2G_{Ab}^{AB} + G_{Ab}^{Ab} - 2G_{aB}^{AB} + 4G_{ab}^{AB} - 2G_{ab}^{Ab} + G_{aB}^{aB} - 2G_{ab}^{aB} + G_{ab}^{ab}\right)$$

It is evident from Eq. 27 that since the epistatic effects arise from direct comparisons among two-locus genotypic values and do not involve marginal genotypic means at individual loci, no application of the averaging rules is required and the epistatic effects stay the same across all reference populations.

Finally, the two-locus mean of a reference population is given by

$$m_{F\cdot AB} = \mathbf{G}_{AB}^{'}\Psi_{AB}\mathbf{1} = \sum P_{v\cdot}^{u\cdot} P_{\cdot z}^{\cdot y} G_{vz}^{uy} \tag{28}$$

which may take different values, depending on which reference population is used, similar to the mean of the one-locus models. For example, the means for the F_∞, F_2, and UW reference populations are

$$
m_{F \cdot AB} = \begin{cases}
\frac{1}{4}\left(G_{AB}^{AB} + G_{Ab}^{Ab} + G_{aB}^{aB} + G_{ab}^{ab}\right), & F = F_\infty \\
\frac{1}{16}\left(G_{AB}^{AB} + 2G_{Ab}^{AB} + G_{Ab}^{Ab} + 2G_{aB}^{AB} + 4G_{ab}^{AB} + 2G_{ab}^{Ab} + G_{aB}^{aB} + 2G_{ab}^{aB} + G_{ab}^{ab}\right), & F = F_2 \\
\frac{1}{9}\left(G_{AB}^{AB} + G_{Ab}^{AB} + G_{Ab}^{Ab} + G_{aB}^{AB} + G_{ab}^{AB} + G_{ab}^{Ab} + G_{aB}^{aB} + G_{ab}^{aB} + G_{ab}^{ab}\right), & F = UW
\end{cases}
\tag{29}
$$

To extend to cases of three or more uncorrelated loci is straightforward. For three independent loci (A, B, and C) with two alleles at each locus, a total of 27 parameters are needed for a complete description of $(3 \times 3 \times 3)$ genotypes for each gene action model. These 27 parameters consist of the mean of a reference population ($m_{R \cdot ABC}$), six one-locus effects including three additive (a_A, a_B, and a_C) and three dominance (d_A, d_B, and d_C) effects, 12 two-locus epistatic effects ($aa_{AB} = a_A \times a_B$, $ad_{AB} = a_A \times d_B$, $da_{AB} = d_A \times a_B$, $dd_{AB} = d_A \times d_B$, $aa_{AC} = a_A \times a_C$, $ad_{AC} = a_A \times d_C$, $da_{AC} = d_A \times a_C$, $dd_{AC} = d_A \times d_C$, $aa_{BC} = a_B \times a_C$, $ad_{BC} = a_B \times d_C$, $da_{BC} = d_B \times a_C$ and $dd_{BC} = d_B \times d_C$), and eight three-locus epistatic effects ($aaa = a_A \times a_B \times a_C$, $aad = a_A \times a_B \times d_C$, $ada = a_A \times d_B \times a_C$, $add = a_A \times d_B \times d_C$, $daa = d_A \times a_B \times a_C$, $dad = d_A \times a_B \times d_C$, $dda = d_A \times d_B \times a_C$, and $ddd = d_A \times d_B \times d_C$). The three-locus functional genetic effects can be obtained directly from the knowledge of single-locus design matrices:

$$
\boldsymbol{\beta}_{R \cdot ABC} = \mathbf{W}_{R \cdot ABC}^{-1} \mathbf{G}_{ABC} = \left(\mathbf{W}_{R \cdot C}^{-1} \otimes \mathbf{W}_{R \cdot B}^{-1} \otimes \mathbf{W}_{R \cdot A}^{-1}\right) \mathbf{G}_{ABC}
\tag{30}
$$

where $\boldsymbol{\beta}_{R \cdot ABC} = \boldsymbol{\beta}_{1 \cdot C} \otimes \boldsymbol{\beta}_{1 \cdot B} \otimes \boldsymbol{\beta}_{1 \cdot A}$ with the first element of the vector is replaced by the mean of the reference population ($m_{R \cdot ABC}$), the sequence of the vector elements explained earlier, and $\mathbf{G}_{ABC} = \mathbf{G}_C \otimes \mathbf{G}_B \otimes \mathbf{G}_A$.

3.3 Direct Approach: Statistical Genetic Effects

Like functional genetic effects given in Eq. 18, the statistical genetic effects can be written out directly as well. For example, at locus A, $\boldsymbol{\beta}_A = \mathbf{W}_{S \cdot A}^{-1} \mathbf{G}_A$ or

$$
\begin{bmatrix} \mu_A \\ \alpha_A \\ \delta_A \end{bmatrix} = \begin{bmatrix} P_A^A & P_a^A & P_a^a \\ \tau_A & (\tau_a - \tau_A) & -\tau_a \\ -\frac{1}{2} & 1 & -\frac{1}{2} \end{bmatrix} \begin{bmatrix} G_A^A \\ G_a^A \\ G_a^a \end{bmatrix},
\tag{31}
$$

where $\mathbf{W}_{S \cdot A}^{-1}$ is the inverse of the genetic-effect design matrix ($\mathbf{W}_{S \cdot A}$) that is given in Eq. 10 when the indirect two-step approach is used.

It should be noted that Eq. 31 is the same as Eq. 18 except that the statistical additive genetic effect is redefined. Unlike the functional additive effect, the statistical additive effect in Eq. 31 is a weighted average of the two differences at locus A, corresponding to the two comparisons among three genotypes, $G_A^A - G_a^A$ and $G_a^A - G_a^a$. Each of these differences measures the effect of substituting a by A, but each measure is made in the presence of a different allele. Thus, the statistical additive effect is given by

$$
\begin{aligned}
\alpha_A &= \tau_A \left(G_A^A - G_a^A \right) + \tau_a \left(G_a^A - G_a^a \right) \\
&= \tau_A G_{A\cdot}^{A\cdot} + \left(\tau_a - \tau_A \right) G_{a\cdot}^{A\cdot} - \tau_a G_{a\cdot}^{a\cdot}
\end{aligned}
\tag{32}
$$

where τ_A and τ_a are the weighting factors that need to be determined, unless the allele frequency is one half as with functional genetic effects.

Equation 32 is a general definition of the statistical additive effect and thus can take different forms, depending on specific measures of the statistical additive effects. Here we examine Fisher's [22, 23] definitions of the average excess and average effect of allele substitution and their relationships [24]. Such examination allows the weighting factors to be determined as well. Fisher [23] defined the average excess of an allele at a locus as the difference by which the average over genotypes carrying the allele exceeds the average over genotypes carrying the alternative allele. Thus, the average excess of allele A over allele a at locus A is given by

$$
\begin{aligned}
\alpha_{EX\cdot A} &= \frac{P_A^A G_A^A + \frac{1}{2} P_a^A G_a^A}{P_A^A + \frac{1}{2} P_a^A} - \frac{\frac{1}{2} P_a^A G_a^A + P_a^a G_a^a}{\frac{1}{2} P_a^A + P_a^a} \\
&= \frac{P_A^A}{p_A} G_A^A + \frac{1}{2} \left(\frac{P_a^A}{p_A} - \frac{P_a^A}{p_a} \right) G_a^A - \frac{P_a^a}{p_a} G_a^a \\
&= \left(p_A + f_A p_a \right) G_A^A + \left(p_a - p_A \right) \left(1 - f_A \right) G_a^A - \left(p_a + f_A p_A \right) G_a^a
\end{aligned}
\tag{33}
$$

where P_v^u and G_v^u are the frequency and value of genotype uv ($uv = AA$, Aa, or aa), respectively, and p_u ($= P_u^u + \frac{1}{2} P_v^u$) is the frequency of allele u ($u = A$ or a) with $p_A + p_a = 1$ and f_A as Wright's [25] fixation index at locus A with $-1 \leq f_A \leq +1$. The departure from Hardy-Weinberg equilibrium ($f_A = 0$) may arise from demographic and evolutionary forces including selection, random drift, and nonrandom mating. The last relation in Eq. 33 is obtained by expressing the genotypic frequencies in terms of allele frequencies and the fixation index:

$$
P_v^u = \begin{cases} p_u^2 + p_u \left(1 - p_u \right) f_A, & \text{If } u = v \\ 2 p_u \left(1 - p_u \right) \left(1 - f_A \right), & \text{If } u \neq v \end{cases}
\tag{34}
$$

It is evident from Eqs. 32 and 33 that $\tau_i = p_i + (1 - p_i)f_A$ for $i = A, a$. Fisher [23] also defined the average effect of allele A as the linear regression coefficient of genotypic values ($y = G_A^A, G_a^A$, or G_a^a) on the gene content, i.e., the number of A alleles in the genotype ($x = 2, 1$ or 0):

$$
\begin{aligned}
\alpha_{EF \cdot A} &= \frac{\mathrm{Cov}(y,x)}{\mathrm{Var}(x)} \\
&= \frac{2P_A^A G_A^A + P_a^A G_a^A - 2p_A\left(P_A^A G_A^A + P_a^A G_a^A + P_a^a G_a^a\right)}{4P_A^A + P_a^A - 4p_A^2} \\
&= \frac{\left(p_A + f_A p_a\right)G_A^A + \left(p_a - p_A\right)\left(1 - f_A\right)G_a^A - \left(p_a + f_A p_A\right)G_a^a}{1 + f_A}
\end{aligned}
\tag{35}
$$

where the identity of $\tau_i = [p_i + (1 - p_i)f_A]/(1 + f_A)$ for $i = A, a$ is readily identified.

A comparison of Eqs. 33 and 35 reveals a well-known relationship between the average excess and average effect [23, 26, 27]:

$$
\alpha_{EF \cdot A} = \alpha_{EX \cdot A} / \left(1 + f_A\right).
$$

In an HWE population ($f_A = 0$), the two measures of the statistical additive effect are identical:

$$
\alpha_{EF \cdot A} = \alpha_{EX \cdot A} = p_A G_A^A + \left(p_a - p_A\right)G_a^A - p_a G_a^a
\tag{36}
$$

which reduces to a more familiar form of $a + (p_a - p_A)d$ [7] if the functional genetic effects are in place of the actual genotypic values, $a_A = (G_A^A - G_a^a)/2$ and $d_A = G_a^A - (G_A^A + G_a^a)/2$.

Extensions to two or more independent loci are straightforward. For example, for two independent loci, A and B with arbitrary gene and genotypic frequencies ($p_A \neq p_a$ and $p_B \neq p_b$), nine parameters, eight statistical genetic effects, and the population mean (μ_{AB}), are required for the complete description of nine genotypes. The eight effects include the statistical additive and dominance effects at locus A (α_A and δ_A) and locus B (α_B and δ_B), and four types of nonallelic (epistatic) interactions between the two loci: additive × additive ($\alpha\alpha_{AB}$), additive × dominance ($\alpha\delta_{AB}$), dominance × additive ($\delta\alpha_{AB}$), and dominance × dominance ($\delta\delta_{AB}$). The single-locus results can be calculated in the same way as in Eq. 23 for the functional genetic effects, but they are based on the marginal means over all genotypic values at another locus. Thus, the average excess at locus A is

$$
\alpha_A = \tau_A G_A^{A \cdot} + \left(\tau_a - \tau_A\right)G_a^{A \cdot} - \tau_a G_a^{a \cdot}
\tag{37}
$$

where $\tau_i = p_i + (1 - p_i)f_A$ and the marginal mean for genotype uv ($G_v^{u \cdot}$) is over the genotypic values at locus B, $G_v^{u \cdot} = \Sigma P_{\cdot z}^{\cdot y} G_{vz}^{uy}$. Inserting these marginal means into Eq. 37, we have

$$
\alpha_A = \tau_A P_{\cdot B}^{\cdot B} G_{AB}^{AB} + \left(\tau_a - \tau_A\right) P_{\cdot B}^{\cdot B} G_{aB}^{AB} - \tau_a P_{\cdot B}^{\cdot B} G_{aB}^{AB} + \tau_A P_{\cdot B}^{\cdot B} G_{Ab}^{AB} + \left(\tau_a - \tau_A\right) P_{\cdot b}^{\cdot B} G_{ab}^{AB}
$$
$$
-\tau_a P_{\cdot B}^{\cdot B} G_{ab}^{aB} + \tau_A P_{\cdot b}^{\cdot b} G_{Ab}^{AB} + \left(\tau_a - \tau_A\right) P_{\cdot b}^{\cdot b} G_{ab}^{Ab} - \tau_a P_{\cdot b}^{\cdot b} G_{ab}^{ab}. \tag{38}
$$

The other effects can be similarly obtained.

We streamline such calculations as follows. Each of the nine effects is the sum of the products of the two-locus coefficients (Table 2) and the corresponding two-locus genotypic values. The nine sets of two-locus coefficients in Table 2 are simply the elements in appropriate rows of the direct product of $\mathbf{W}_A^{-1}$ and $\mathbf{W}_B^{-1}$, the inverted design matrices for loci A and B, i.e.,

$$
\mathbf{W}_{AB}^{-1} = \mathbf{W}_B^{-1} \otimes \mathbf{W}_A^{-1}
$$

$$
= \begin{bmatrix} P_B^B & P_b^B & P_b^b \\ \tau_B & \tau_b - \tau_B & -\tau_b \\ -\tfrac{1}{2} & 1 & -\tfrac{1}{2} \end{bmatrix} \otimes \begin{bmatrix} P_A^A & P_a^A & P_a^a \\ \tau_A & \tau_a - \tau_A & -\tau_a \\ -\tfrac{1}{2} & 1 & -\tfrac{1}{2} \end{bmatrix}
$$

$$
= \begin{bmatrix}
P_A^A P_B^B & P_a^A P_B^B & P_a^a P_B^B & P_A^A P_b^B & P_a^A P_b^B & P_a^a P_b^B & P_A^A P_b^b & P_a^A P_b^b & P_a^a P_b^b \\
\tau_A P_B^B & (\tau_a - \tau_A)P_B^B & \tau_a P_B^B & \tau_A P_b^B & (\tau_a - \tau_A)P_b^B & \tau_a P_b^B & \tau_A P_b^b & (\tau_a - \tau_A)P_b^b & \tau_a P_b^b \\
-\tfrac{1}{2}P_B^B & P_B^B & -\tfrac{1}{2}P_B^B & -\tfrac{1}{2}P_b^B & P_b^B & -\tfrac{1}{2}P_b^B & -\tfrac{1}{2}P_b^b & P_b^b & -\tfrac{1}{2}P_b^b \\
\tau_B P_A^A & \tau_B P_a^A & \tau_B P_a^a & (\tau_b - \tau_B)P_A^A & (\tau_b - \tau_B)P_a^A & (\tau_b - \tau_B)P_a^a & -\tau_b P_A^A & -\tau_b P_a^A & -\tau_b P_a^a \\
\tau_A \tau_B & (\tau_a - \tau_A)\tau_B & -\tau_a \tau_B & \tau_A(\tau_b - \tau_B) & (\tau_a - \tau_A)(\tau_b - \tau_B) & -\tau_a(\tau_b - \tau_B) & -\tau_A \tau_b & -(\tau_a - \tau_A)\tau_b & \tau_a \tau_b \\
-\tfrac{1}{2}\tau_B & \tau_B & -\tfrac{1}{2}\tau_B & -\tfrac{1}{2}(\tau_b - \tau_B) & \tau_b - \tau_B & -\tfrac{1}{2}(\tau_b - \tau_B) & \tfrac{1}{2}\tau_b & -\tau_b & \tfrac{1}{2}\tau_b \\
-\tfrac{1}{2}P_A^A & -\tfrac{1}{2}P_a^A & -\tfrac{1}{2}P_a^a & P_A^A & P_a^A & P_a^a & -\tfrac{1}{2}P_A^A & -\tfrac{1}{2}P_a^A & -\tfrac{1}{2}P_a^a \\
-\tfrac{1}{2}\tau_A & -\tfrac{1}{2}(\tau_a - \tau_A) & \tfrac{1}{2}\tau_a & \tau_A & \tau_a - \tau_A & -\tau_a & -\tfrac{1}{2}\tau_A & -\tfrac{1}{2}(\tau_a - \tau_A) & \tfrac{1}{2}\tau_a \\
\tfrac{1}{4} & -\tfrac{1}{2} & \tfrac{1}{4} & -\tfrac{1}{2} & 1 & -\tfrac{1}{2} & \tfrac{1}{4} & -\tfrac{1}{2} & \tfrac{1}{4}
\end{bmatrix} \tag{39}
$$

where $\tau_A = p_A + p_a f_A$, $\tau_a = p_a + p_A f_A$, $\tau_B = p_B + p_b f_B$, and $\tau_b = p_b + p_B f_B$. Let the nine two-locus statistical genetic effects be appropriately arranged in vector $\boldsymbol{\beta}_{AB}$, i.e., $\boldsymbol{\beta}_{AB} = [\mu_{AB}\ \alpha_A\ \delta_A\ \alpha_B\ \delta_B\ \alpha\alpha_{AB}\ \alpha\delta_{AB}\ \delta\alpha_{AB}\ \delta\delta_{AB}]'$. Then we obtain the general form of the two-locus statistical genetic effects in matrix form:

$$
\boldsymbol{\beta}_{AB} = \mathbf{W}_{AB}^{-1}\mathbf{G}_{AB} = \left(\mathbf{W}_B^{-1} \otimes \mathbf{W}_A^{-1}\right)\mathbf{G}_{AB}, \tag{40}
$$

where $\mathbf{G}_{AB} = \mathbf{G}_B \otimes \mathbf{G}_A = [G_{AB}^{AB}\ G_{aB}^{AB}\ G_{aB}^{aB}\ G_{Ab}^{AB}\ G_{ab}^{AB}\ G_{ab}^{aB}\ G_{Ab}^{Ab}\ G_{ab}^{Ab}G_{ab}^{ab}]'$ corresponding to genotypes [$AABB$ $AaBB$ $aaBB$ $AABb$ $AaBb$ $aaBb$ $Aabb$ $Aabb$ $aabb$]. The arrangements in vectors $\boldsymbol{\beta}_{AB}$ and $\mathbf{G}_{AB}$ follow the rules developed in Álvarez-Castro and Carlborg [11]. As a check, calculate the statistical additive effect at locus A as given in Eq. 38, the coefficients associated with individual genotypic values should be identical to those in the second row of matrix (39).

In general, with L independent loci, $3^L - 1$ genetic effects plus the mean of a reference population ($\mu_{AB...L}$) are needed for the complete description of 3^L possible genotypes. The key to directly calculating these multilocus effects is to find out the inverse of the genetic-effect design matrix, $\mathbf{W}_{AB...L}^{-1}$. Generalizing the two-locus case for the multi-locus case, $\mathbf{W}_{AB...L}^{-1}$ is obtained from the reverse-order

Table 2
Coefficients associated with two-locus genotypic values and their marginal means for calculating statistical genetic effects

Genotype	AA	Aa	aa	Total
Mean				
BB	$P_{A\cdot}^A P_{\cdot B}^{\cdot B}$	$P_{a\cdot}^A P_{\cdot B}^{\cdot B}$	$P_{a\cdot}^a P_{\cdot B}^{\cdot B}$	$P_{\cdot B}^{\cdot B}$
Bb	$P_{A\cdot}^A P_{\cdot b}^{\cdot B}$	$P_{a\cdot}^A P_{\cdot b}^{\cdot B}$	$P_{a\cdot}^a P_{\cdot b}^{\cdot B}$	$P_{\cdot b}^{\cdot B}$
bb	$P_{A\cdot}^A P_{\cdot b}^{\cdot b}$	$P_{a\cdot}^A P_{\cdot b}^{\cdot b}$	$P_{a\cdot}^a P_{\cdot b}^{\cdot b}$	$P_{\cdot b}^{\cdot b}$
Total	$P_{A\cdot}^A$	$P_{a\cdot}^A$	$P_{a\cdot}^a$	
Additive effect at locus A				
BB	$\tau_A P_{\cdot B}^{\cdot B}$	$(\tau_a - \tau_A) P_{\cdot B}^{\cdot B}$	$-\tau_a P_{\cdot B}^{\cdot B}$	$P_{\cdot B}^{\cdot B}$
Bb	$\tau_A P_{\cdot b}^{\cdot B}$	$(\tau_a - \tau_A) P_{\cdot b}^{\cdot B}$	$-\tau_a P_{\cdot b}^{\cdot B}$	$P_{\cdot b}^{\cdot B}$
bb	$\tau_A P_{\cdot b}^{\cdot b}$	$(\tau_a - \tau_A) P_{\cdot b}^{\cdot b}$	$-\tau_a P_{\cdot b}^{\cdot b}$	$P_{\cdot b}^{\cdot b}$
Total	τ_A	$\tau_a - \tau_A$	$-\tau_a$	
Dominance effect at locus A				
BB	$-\frac{1}{2} P_{\cdot B}^{\cdot B}$	$P_{\cdot B}^{\cdot B}$	$-\frac{1}{2} P_{\cdot B}^{\cdot B}$	$P_{\cdot B}^{\cdot B}$
Bb	$-\frac{1}{2} P_{\cdot b}^{\cdot B}$	$P_{\cdot b}^{\cdot B}$	$-\frac{1}{2} P_{\cdot b}^{\cdot B}$	$P_{\cdot b}^{\cdot B}$
bb	$-\frac{1}{2} P_{\cdot b}^{\cdot b}$	$P_{\cdot b}^{\cdot b}$	$-\frac{1}{2} P_{\cdot b}^{\cdot b}$	$P_{\cdot b}^{\cdot b}$
Total	$-\frac{1}{2}$	1	$-\frac{1}{2}$	
Additive effect at locus B				
BB	$P_{A\cdot}^A \tau_B$	$P_{a\cdot}^A \tau_B$	$P_{a\cdot}^a \tau_B$	τ_B
Bb	$P_{A\cdot}^A (\tau_b - \tau_B)$	$P_{a\cdot}^A (\tau_b - \tau_B)$	$P_{a\cdot}^a (\tau_b - \tau_B)$	$\tau_b - \tau_B$
bb	$-P_{A\cdot}^A \tau_b$	$-P_{a\cdot}^A \tau_b$	$-P_{a\cdot}^a \tau_b$	$-\tau_b$
Total	$P_{A\cdot}^A$	$P_{a\cdot}^A$	$P_{a\cdot}^a$	
Dominance effect at locus B				
BB	$-\frac{1}{2} P_{A\cdot}^A$	$-\frac{1}{2} P_{a\cdot}^A$	$-\frac{1}{2} P_{a\cdot}^a$	$-\frac{1}{2}$
Bb	$P_{A\cdot}^A$	$P_{a\cdot}^A$	$P_{a\cdot}^a$	1
bb	$-\frac{1}{2} P_{A\cdot}^A$	$-\frac{1}{2} P_{a\cdot}^A$	$-\frac{1}{2} P_{a\cdot}^a$	$-\frac{1}{2}$
Total	$P_{A\cdot}^A$	$P_{a\cdot}^A$	$P_{a\cdot}^a$	
Additive × additive				
BB	$\tau_A \tau_B$	$(\tau_a - \tau_A)\tau_B$	$-\tau_a \tau_B$	τ_B
Bb	$\tau_A(\tau_b - \tau_B)$	$(\tau_a - \tau_A)(\tau_b - \tau_B)$	$-\tau_a(\tau_b - \tau_B)$	$\tau_b - \tau_B$
bb	$-\tau_A \tau_b$	$-(\tau_a - \tau_A)\tau_b$	$\tau_a \tau_b$	$-\tau_b$
Total	τ_A	$\tau_a - \tau_A$	$-\tau_a$	

(continued)

Table 2
(continued)

Genotype	AA	Aa	aa	Total
Additive × dominance				
BB	$-\frac{1}{2}\tau_A$	$-\frac{1}{2}(\tau_a - \tau_A)$	$-\frac{1}{2}\tau_a$	$-\frac{1}{2}$
Bb	τ_A	$(\tau_a - \tau_A)$	$-\tau_a$	1
bb	$-\frac{1}{2}\tau_A$	$-\frac{1}{2}(\tau_a - \tau_A)$	$-\frac{1}{2}\tau_a$	$-\frac{1}{2}$
Total	τ_A	$\tau_a - \tau_A$	$-\tau_a$	
Dominance × additive				
BB	$-\frac{1}{2}\tau_B$	τ_B	$-\frac{1}{2}\tau_B$	τ_B
Bb	$-\frac{1}{2}(\tau_b - \tau_B)$	$(\tau_b - \tau_B)$	$-\frac{1}{2}(\tau_b - \tau_B)$	$\tau_b - \tau_B$
bb	$\frac{1}{2}\tau_b$	$-\tau_b$	$\frac{1}{2}\tau_b$	$-\tau_b$
Total	$-\frac{1}{2}$	1	$-\frac{1}{2}$	
Dominance × dominance				
BB	$\frac{1}{4}$	$-\frac{1}{2}$	$\frac{1}{4}$	$-\frac{1}{2}$
Bb	$-\frac{1}{2}$	1	$-\frac{1}{2}$	1
bb	$\frac{1}{4}$	$-\frac{1}{2}$	$\frac{1}{4}$	$-\frac{1}{2}$
Total	$-\frac{1}{2}$	1	$-\frac{1}{2}$	

[a] $\tau_A = p_A + p_a f_A$, $\tau_a = p_a + p_A f_A$, $\tau_B = p_B + p_b f_B$, and $\tau_b = p_b + p_B f_B$

direct products of L single-locus matrices, $\mathbf{W}_A^{-1}$ for locus A, and $\mathbf{W}_B^{-1}$ for locus $B,\ldots,$ and $\mathbf{W}_L^{-1}$ for locus L, i.e.,

$$\mathbf{W}_{AB\cdots L}^{-1} = \mathbf{W}_L^{-1} \otimes \cdots \otimes \mathbf{W}_B^{-1} \otimes \mathbf{W}_A^{-1}. \qquad (41)$$

where the single-locus inverted design matrices, e.g., $\mathbf{W}_A^{-1}$, are obtained directly without any statistical analysis (cf. Eq. 31).

3.4 Numerical Examples

To illustrate the various aspects of the models described above, we now analyze two numerical examples. The first example is taken from Table 2 of Cheverud and Routman [28]. Specifically, the data consists of adult body weight (grams) of 534 F_2 mice derived from an intercross of the LG/J and SM/J strains. A pair of microsatellite loci, *D10Mit10* (locus A) and *D11Mit64* (locus B), are significantly associated with adult body weight. There is strong suppressing epistasis between this pair of loci, because such epistasis leads to diminished additive genetic variance at intermediate allele frequencies. Cheverud and Routman [28] used the indirect approach to partitioning the genotypic values at these two loci into functional additive, dominance, and epistatic effects under the UW model. In Table 3, we show the use of our direct approach to such partitioning under the UW model, the F_2-metric model, and the F_∞-metric model. It should be noted that the genotypic frequencies at the individual loci are known for different reference populations under the UW, F_2, and F_∞ models (*see* Subheading 2). The numerical results in Table 3 confirm the following features as revealed in the above theoretical analysis. First, the genetic effects obtained under the UW model are identical to those from the use of the indirect approach by Cheverud and Routman [28]. Second, nonepistatic (additive and dominance) effects differ among the three models, as they are the functions of single-locus genotypic frequencies in the reference populations corresponding to the different models. Third, epistatic effects are independent of the genotypic frequencies in the reference populations and thus are the same across the different models.

The second example is taken from Table 8 of Doebley et al. [21] who identified two QTLs, *UMC107* (designated as locus *A*) and *BV302* (designated as locus *B*), control differences in plant and inflorescence architecture between cultivated maize (*Zea mays* ssp. *mays*) and teosinte (*Zea mays* ssp. *parviglumis*) in the BC$_3$F$_2$ population (Teosinte-M1L × Teosinte-M3L) derived from an original cross of Reventador maize × *parviglumis* teosinte. Of nine morphological and inflorescence traits measured, average length of vegetative internodes in the primary lateral branch (LBIL) has been repeatedly reanalyzed to demonstrate the magnitude of epistasis between loci *A* and *B* using the indirect approach [9, 17, 29], we show the use of our direct approach to estimate the functional genetic effects for three reference populations with known or assumed genotypic frequencies as well as the statistical genetic effects for three reference populations, where the genotypic frequencies are unknown but are estimable from the genotype count data. Of six models examined, three statistical genetic effects correspond to the reference populations under HWE and two HWD scenarios as given in Yang [9] and Álvarez-Castro and Carlborg [11], and three functional genetic effects correspond to the UW, F_2, and F_∞ models (Table 4). The direct approach is implemented using the algorithm as described in Table 2. For comparison, we also include the results from the use of the indirect approach for the six models. The results show that the direct and indirect approaches lead to identical estimates of genetic effects, when genotypic frequencies in the reference populations are known or the reference population is in HWE. In the presence of HWD, however, the estimates of statistical additive and additive-related epistatic genetic effects differ between the direct and indirect approaches. It is shown in Eqs. 33 and 35 that the direct approach gives the estimates of average excess where the indirect approach gives the estimates of average effects; the two effects are different by a factor of $(1 + f)$. Nonzero residuals (ε) arise from the presence of LD.

Table 3
Partitioning of genotypic values (GVs) into nonepistatic and epistatic components for adult body weight of mice at *D10Mit10* (locus *A*) and *D11Mit64* (locus *B*) under three gene action models (F$_\infty$, UW, and F$_2$)

Genotype	GV[a]	Nonepistatic values					Epistatic values			
		$m_{R\text{-}AB}$	a_A	d_A	a_B	d_B	aa_{AB}	ad_{AB}	da_{AB}	dd_{AB}
aabb	33.90	$P_{a\cdot}^{a\cdot}\cdot P_{\cdot b}^{\cdot b}$	$-\frac{1}{2}P_{a\cdot}^{a\cdot}$	$-\frac{1}{2}P_{a\cdot}^{a\cdot}$	$-\frac{1}{2}P_{\cdot b}^{\cdot b}$	$-\frac{1}{2}P_{\cdot b}^{\cdot b}$	$\frac{1}{4}$	$-\frac{1}{4}$	$-\frac{1}{4}$	$\frac{1}{4}$
Aabb	35.31	$P_{a\cdot}^{A\cdot}\cdot P_{\cdot b}^{\cdot b}$	$-\frac{1}{2}P_{a\cdot}^{A\cdot}$	$-\frac{1}{2}P_{a\cdot}^{A\cdot}$	0	$P_{\cdot b}^{\cdot b}$	0	$\frac{1}{2}$	0	$-\frac{1}{2}$
AAbb	35.84	$P_{A\cdot}^{A\cdot}\cdot P_{\cdot b}^{\cdot b}$	$-\frac{1}{2}P_{A\cdot}^{A\cdot}$	$-\frac{1}{2}P_{A\cdot}^{A\cdot}$	$\frac{1}{2}P_{\cdot b}^{\cdot b}$	$-\frac{1}{2}P_{\cdot b}^{\cdot b}$	$-\frac{1}{4}$	$-\frac{1}{4}$	$\frac{1}{4}$	$\frac{1}{4}$
aaBb	34.56	$P_{a\cdot}^{a\cdot}\cdot P_{\cdot b}^{\cdot B}$	0	$P_{a\cdot}^{a\cdot}$	$-\frac{1}{2}P_{\cdot b}^{\cdot B}$	$-\frac{1}{2}P_{\cdot b}^{\cdot B}$	0	0	$-\frac{1}{2}$	$-\frac{1}{2}$
AaBb	35.31	$P_{a\cdot}^{A\cdot}\cdot P_{\cdot b}^{\cdot B}$	0	$P_{a\cdot}^{A\cdot}$	0	$P_{\cdot b}^{\cdot B}$	0	0	0	1
AABb	35.35	$P_{A\cdot}^{A\cdot}\cdot P_{\cdot b}^{\cdot B}$	0	$P_{A\cdot}^{A\cdot}$	$\frac{1}{2}P_{\cdot b}^{\cdot B}$	$-\frac{1}{2}P_{\cdot b}^{\cdot B}$	0	0	$\frac{1}{2}$	$-\frac{1}{2}$
aaBB	35.27	$P_{a\cdot}^{a\cdot}\cdot P_{\cdot B}^{\cdot B}$	$\frac{1}{2}P_{a\cdot}^{a\cdot}$	$-\frac{1}{2}P_{a\cdot}^{a\cdot}$	$-\frac{1}{2}P_{\cdot B}^{\cdot B}$	$-\frac{1}{2}P_{\cdot B}^{\cdot B}$	$-\frac{1}{4}$	$-\frac{1}{4}$	$-\frac{1}{4}$	$\frac{1}{4}$
AaBB	35.91	$P_{a\cdot}^{A\cdot}\cdot P_{\cdot B}^{\cdot B}$	$\frac{1}{2}P_{a\cdot}^{A\cdot}$	$-\frac{1}{2}P_{a\cdot}^{A\cdot}$	0	$P_{\cdot B}^{\cdot B}$	0	$\frac{1}{2}$	0	$-\frac{1}{2}$
AABB	38.96	$P_{A\cdot}^{A\cdot}\cdot P_{\cdot B}^{\cdot B}$	$\frac{1}{2}P_{A\cdot}^{A\cdot}$	$-\frac{1}{2}P_{A\cdot}^{A\cdot}$	$\frac{1}{2}P_{\cdot B}^{\cdot B}$	$-\frac{1}{2}P_{\cdot B}^{\cdot B}$	$\frac{1}{4}$	$-\frac{1}{4}$	$\frac{1}{4}$	$\frac{1}{4}$
UW	Effect	35.60	0.85	-0.79	1.07	-0.14	0.44	-0.82	-1.01	0.74
F$_\infty$	Effect	35.99	1.12	-1.04	1.41	-0.38	0.44	-0.82	-1.01	0.74
F$_2$	Effect	35.47	0.71	-0.67	0.90	-0.01	0.44	-0.82	-1.01	0.74

[a] Genotypic values were calculated using 534 F$_2$ mice formed from an intercross of the LG/J (*AABB*) and SM/J (*aabb*) inbred strains [28]

Table 4
Direct versus indirect approaches to estimating genetic effects (β_0–β_8) under three statistical genetic effect models (HWD1, HWD2, and HWE) and three functional genetic effect models (F_2, F_∞, and UW) for two QTLs controlling average length of vegetative internodes in the primary lateral branch (LBIL)[a]

Genic effects	Direct approach						Indirect approach					
	HWD1	HWD2	HWE	F_2	F_∞	UW	HWD1	HWD2	HWE	F_2	F_∞	UW
β_0	56.005	56.005	56.290	56.888	57.150	57.978	56.005	56.005	56.290	56.888	57.150	57.978
β_1	13.311	13.311	14.918	15.100	24.250	18.150	14.339	14.339	14.918	15.100	24.250	18.150
β_2	-4.329	-4.383	-3.877	-3.925	5.100	-0.917	-4.329	-4.383	-3.877	-3.925	5.100	-0.917
β_3	18.430	18.430	19.375	19.475	17.600	18.850	19.525	19.525	19.375	19.475	17.600	18.850
β_4	-5.200	-5.219	-4.574	-5.625	3.400	-2.617	-5.200	-5.219	-4.574	-5.625	3.400	-2.617
β_5	2.815	2.815	3.149	2.700	2.700	2.700	3.213	3.213	3.149	2.700	2.700	2.700
β_6	-18.007	-18.070	-19.309	-18.300	-18.300	-18.300	-19.397	-19.465	-19.309	-18.300	-18.300	-18.300
β_7	3.730	3.777	3.974	3.750	3.750	3.750	3.951	4.001	3.974	3.750	3.750	3.750
β_8	-17.764	-18.050	-18.050	-18.050	-18.050	-18.050	-17.764	-18.050	-18.050	-18.050	-18.050	-18.050
ϵ	-0.321	-0.321	-0.606	-1.204	-1.466	-2.294	-0.321	-0.321	-0.606	-1.204	-1.466	-2.294

HWD1 and HWD2 are gene action models described in Yang [9] and Álvarez-Castro and Carlborg [11], respectively
[a]Data taken from Table 8 of Doebley et al. [21]

4   Notes

This chapter describes a direct approach to estimating functional and statistical genetic effects. This direct approach is possible because the genetic effects are either direct comparisons among genotypes or direct calculations of the difference between the weighted mean of genotypes carrying one allele and the weighted mean of genotypes carrying an alternate allele (known as Fisher's average excess). In other words, the direct approach estimates the genetic effects with no need to carry out statistical analyses. In contrast, the conventional indirect approach requires the construction of liner models and the use of regression analyses [6, 10, 12]. More importantly, the direct approach allows for a clearer and easier interpretation of the genetic effects being estimated.

The use of our direct approach reveals an important pattern with all gene action models, specifically that the genetic effects of highest order (highest order epistatic effects) are invariant with the different gene action models but all the lower-order genetic effects including the overall mean are not. Thus, when a single locus is considered, additive and dominance effects are the same across the different models but the mean is not. When two loci are considered jointly, the two-locus epistatic effects are the same across the models, but the mean, additive and dominance effects at each of the two loci are not. When three loci are considered jointly, the three-locus epistatic effects are the same across the tree models, but all lower order genetic effects including mean, additive, and dominance effects at individual loci and all pairwise epistatic effects are not, and so on. However, the commonly used two-step indirect approach for model comparisons (e.g., [6, 10, 12]) fails to show such clear patterns. In addition because our direct approach is based on only a few simple average rules, it eliminates some inadvertent errors of tracking down different coefficients for the genetic effects that cause some discrepancies between the same models.

Some authors (e.g., [6, 10, 12]) have emphasized the need for the invariant properties of gene action models (i.e., the genetic effects remain constant across different loci). However, such a quest may not be realistic. As shown in this chapter (Table 3) and elsewhere [6, 29], the definitions of genetic effects are inconsistent across different loci. The inconsistency is undesirable because not all QTLs are identifiable, and the estimated genetic effects of the identified QTLs are probably biased due to the "background" effects of unidentified QTLs, particularly when strong epistasis is suspected. It is true that the orthogonality property does ensure the consistency of QTL effects defined at individual QTLs and across multiple QTLs [6, 29], but such orthogonality may not be obtainable for many cases including the presence of multiple alleles per locus and the presence of LD.

The basic assumption for all gene action models when considering two or more loci jointly is that the loci are independent of each other (i.e., no LD). This assumption is required for a complete specification of all intra-locus and inter-locus genetic effects. For example, there are nine possible genotypes for the two-locus case but as shown in this chapter, a full model with the nine parameters is needed to completely describe additive and dominance effects at each locus, and the epistatic effects between the two loci. The same argument also applies to cases of three or more loci. Because of a unique relationship between the parameters and genotypic values in these full models, the genetic effects are the same regardless of whether or not the population is in gametic and zygotic LD [6, 29]. Thus, in this case, there is no direct way to assess the impacts of multilocus associations. However, following the approach of Yang [9], we are able to assess the effects of gametic and/or zygotic LD between different loci (the ε values in Table 4). In the presence of LD, this approach enables the partitioning of the genetic values into equilibrium and residual portions. The equilibrium portion encompasses intra- and inter-locus genetic effects defined under a particular model whereas the residual portion arises from the presence of LD. There is certainly a need for further investigation into these issues.

Acknowledgments

This research was supported in part by the Natural Sciences and Engineering Research Council of Canada grant OGP0183983.

References

1. Meuwissen THE, Hayes BJ, Goddard ME (2001) Prediction of total genetic value using genome-wide dense marker maps. Genetics 157(4):1819–1829

2. Goddard ME, Hayes BJ (2009) Mapping genes for complex traits in domestic animals and their use in breeding programmes. Nat Rev Genet 10(6):381–391. doi:10.1038/Nrg2575

3. Jannink JL, Lorenz AJ, Iwata H (2010) Genomic selection in plant breeding: from theory to practice. Brief Funct Genomics 9(2):166–177

4. Fisher RA (1918) The correlation between relatives on the supposition of Mendelian inheritance. Trans R Soc Edinb 52(2):399–433

5. Mather K, Jinks JL (1982) Biometrical genetics: the study of continuous variation. Chapman and Hall, London

6. Zeng Z-B, Wang T, Zou W (2005) Modeling quantitative trait loci and interpretation of models. Genetics 169(3):1711–1725

7. Falconer DS, Mackay TFC (1996) Introduction to quantitative genetics, 4th edn. Longman, New York

8. Van Der Veen J (1959) Tests of non-allelic interaction and linkage for quantitative characters in generations derived from two diploid pure lines. Genetica 30(1):201–232

9. Yang R-C (2004) Epistasis of quantitative trait loci under different gene action models. Genetics 167(3):1493–1505. doi:10.1534/genetics.103.020016

10. Cheverud J, Routman E (1995) Epistasis and its contribution to genetic variance components. Genetics 130:1455–1461

11. Álvarez-Castro JM, Carlborg Ö (2007) A unified model for functional and statistical epistasis and its application in quantitative trait loci analysis. Genetics 176(2):1151–1167

12. Cheverud JM (2000) Detecting epistasis among quantitative trait loci. In: Wolf JB, Brodie ED, Wade MJ (eds) Epistasis and the evolutionary process. Oxford University Press, New York, pp 58–81

13. Yang R-C, Alvarez-Castro JM (2008) Functional and statistical genetic effects with multiple alleles. Curr Top Genet 3:49–62

14. Álvarez-Castro JM, Yang R-C (2011) Multiallelic models of genetic effects and variance decomposition in non-equilibrium populations. Genetica 139:1119–1134

15. Cockerham CC (1954) An extension of the concept of partitioning hereditary variance for analysis of covariances among relatives when epistasis is present. Genetics 39(6):859–882

16. Kempthorne O (1969) An introduction to genetic statistics. Iowa State University Press, Ames, IA

17. Lynch M, Walsh B (1998) Genetics and analysis of quantitative traits. Sinauer Associates, Inc., Sunderland, MA

18. Wang T, Zeng Z-B (2006) Models and partition of variance for quantitative trait loci with epistasis and linkage disequilibrium. BMC Genet 7:9. doi:10.1186/1471-2156-7-9

19. Yang R-C (2000) Zygotic associations and multilocus statistics in a nonequilibrium diploid population. Genetics 155(3):1449–1458

20. SAS Institute (2011) SAS/IML 9.3 user's guide. SAS Institute, Inc., Cary, NC, USA

21. Doebley J, Stec A, Gustus C (1995) teosinte branched1 and the origin of maize: evidence for epistasis and the evolution of dominance. Genetics 141(1):333–346

22. Fisher R (1930) The genetical theory of natural selection. Oxford University Press, London, UK

23. Fisher RA (1941) Average excess and average effect of a gene substitution. Ann Eugenics 11:53–63

24. Álvarez-Castro JM, Yang R-C (2012) Clarifying the relationship between average excesses and average effects of allele substitutions. Front Genet 3:30. doi:10.3389/fgene.2012.00030

25. Wright S (1965) The interpretation of population structure by F-statistics with special regard to systems of mating. Evolution 19:395–420

26. Templeton AR (1987) The general relationship between average effect and average excess. Genet Res 49(1):69–70

27. Falconer D (1985) A note on Fisher's 'average effect' and 'average excess'. Genet Res 46:337–347

28. Cheverud JM, Routman EJ (1996) Epistasis as a source of increased additive genetic variance at population bottlenecks. Evolution 50(3):1042–1051. doi:10.2307/2410645

29. Kao CH, Zeng Z-B (2002) Modeling epistasis of quantitative trait loci using Cockerham's model. Genetics 160(3):1243–1261

Chapter 10

Capacitating Epistasis—Detection and Role in the Genetic Architecture of Complex Traits

Mats E. Pettersson and Örjan Carlborg

Abstract

Here, we discuss the potential role of capacitating epistasis in the genetic architecture of complex traits. Two alternative methods for identifying such gene-gene interactions in genetic association studies—mapping of variance controlling loci and the variance plane ratio (VPR) method—are introduced. An overview of the theoretical foundation of the methods is presented together with a discussion on their implementation and available software for performing these analyses. We conclude by highlighting a few examples of capacitating epistasis described in the literature and its potential impacts on the genetics of complex traits.

Key words Epistasis, Capacitation, Genetic analysis, Gene-gene interaction

1 Introduction

Definition

A common problem in the literature on epistasis is the lack of coherence in the nomenclature and the definitions for several terms. Currently, the meaning of many terms vary depending on the context in which they are used. To avoid such confusion in this chapter, we will start by providing a definition of the term "capacitating locus" as it is used in this chapter.

A capacitating locus is a locus that suppresses or enhances the effect of one or more other locus (loci) involved in the regulation of the studied phenotype. The effect(s) of capacitated locus (loci) on the phenotype thus depend on the genotype at the capacitating locus.

1.1 Capacitating Epistasis in the Literature

From the molecular point of view it is well known that biological systems are, by their very nature, dependent on interactions between molecules. There is simply no such thing as a protein acting in isolation—its effects are always dependent on context. In spite of this fact, genetic interactions have proven difficult to

Jason H. Moore and Scott M. Williams (eds.), *Epistasis: Methods and Protocols*, Methods in Molecular Biology, vol. 1253, DOI 10.1007/978-1-4939-2155-3_10, © Springer Science+Business Media New York 2015

detect in classic quantitative genetics studies. In recent years, however, the inclusion of molecular genetic data in quantitative genetic studies has prompted the development of new analytical approaches that have made detection of interactions possible, and it is now widely recognized that epistasis is an important part of the genetic architecture of complex traits [1, 2]. Although the reports on epistasis are increasing in the literature, the connection between inferred statistical epistasis in GWAS or linkage studies and molecular interactions in cells, organs, and organisms are still largely unknown.

In this chapter, we discuss a particular type of epistasis, capacitation, that we think is particularly promising for bridging the gap between statistical and molecular interactions. This view is based on the findings of capacitating gene interaction networks in numerous biological systems, on various levels of molecular detail. The evidence ranges from statistical epistasis in genetic linkage mapping populations [3] through expression profiling [4], protein-protein interactions [5], and gene knockouts [6] to detailed molecular studies of the contributions of individual genes to capacitation [7–9]. By approaching a particular type of epistasis from many different angles, there is greater potential to understand the common features in the molecular background of such effects. Here, we discuss the potential for detection of capacitating epistasis in population-based linkage or association studies and describe two methods, the vQTL/vGWAS and the Variance Plane Ratio (VPR) methods, to detect such interactions.

1.2 Statistical Inference of Capacitating Epistasis

Capacitating epistasis is difficult to infer statistically since, by definition, the loci involved only affect the trait in a subset of the population, which weakens the statistical signal. Thus, detection of capacitating epistasis is dependent on the availability of large datasets. Most QTL studies performed to date do, however, lack power to detect capacitating epistasis, and often epistasis in general, which could be one explanation why the phenomenon has hitherto not been found with the prevalence expected from molecular evidence. As the costs for genotyping decrease, we expect the sizes of available datasets will increase, especially in model organisms where the cost of generating large populations is lower. This will then increase the potential to explore capacitating epistasis in greater detail. Therefore, it is timely to present these methods for detecting capacitating epistasis in greater detail.

2 Methods—New Statistical Methods for Detecting Capacitating Epistasis

2.1 Genome-Scans for Detecting Variance Controlling Loci

Traditionally, quantitative genetics has focused on evaluating the contributions of differences between the phenotypic means of different genotypes within a locus to the phenotypic variance in studied populations. Considerably less attention has been paid to the genetic control of the environmental variance, i.e., differences in variance between genotypic classes at a locus [10]. Recently, the interest in exploring this type of genetic regulation has increased and new methods such as the variance-heterogeneity genome wide association (vGWAS) [11, 12] and variance-heterogeneity quantitative trait locus (vQTL) [13, 14] analyses have been developed for detecting variance controlling loci. These approaches have been used to analyze experimental data from a number of systems [15–17], and have clearly demonstrated the potential of these methods in complex trait genetic studies. The mechanisms underlying the identified variance controlling loci are still largely unknown, but epistasis is one of several likely contributors to these effects [11, 14, 15]. Interestingly, some recent work on the contribution of genetic variance control to gene expression in yeast has empirically shown that epistasis can, in fact, be a strong contributing mechanism to single locus variance heterogeneity [18]. Since a key property of capacitating epistasis is that one genotype at the capacitating locus suppresses the effect of other loci, the capacitating locus will always be expected to display a difference in variance between genotypes, thus making this approach highly suitable for performing genome-wide scans to detect such loci.

Estimation of the variance in a given sample is inherently less powerful than estimation of the mean in that sample, as the variance is a second moment measure and the mean is a first moment measure. However, in cases where epistasis is the underlying reason for an observed heterogeneity in the variance, unidimensional scans for differences in the variance of a locus are typically more powerful than multidimensional scans for epistatic effects on the mean due to the reduced need for multiple-testing correction. In addition, the computational efficiency of the analysis is also significantly improved. By using these methods to search for candidate epistatic loci, one could resort to one-dimensional genome-scans without needing to make assumptions about whether epistatic loci have marginal effects on the mean and explore potential epistatic effects in a much smaller subset of loci. When key capacitating loci have been detected in this way, the data can be stratified based on the genotypes at the master regulatory locus, and the existence of capacitated loci be inferred by comparisons of the corresponding mean-effect profiles in each subset.

Before proceeding to discuss methods for inferring capacitating epistasis, we will first present the underlying theory of the methods for detecting variance controlling loci in more detail, introduce two useful software tools for performing such analyses in experimental linkage or association-mapping datasets, and discuss some of the challenges to be addressed in the future.

2.1.1 Theory

Detection of variance heterogeneity requires a slight reformulation of the model used for genetic association. Traditionally, in a single-locus additive model, the phenotypic variance, V_P, is partitioned as

$$V_P = V_A + V_E$$

where V_A is the additive variance and V_E is the residual variance. This decomposition only accounts for the mean difference between genotypes. If, however, the phenotypic variance is instead decomposed into the variance due to the mean shift between genotypes, V_M, the variance due to the variance heterogeneity, V_V, and the remaining residual variance V_R, we obtain the following expression:

$$V_P = V_M + V_V + V_R$$

For inbred, or haploid, populations, $V_M = V_A$ and thus, $V_R \leq V_E$. Equality holds if and only if $V_V = 0$, which is to say V_V captures a part of V_E that is not stochastic noise, but actually a genetic contribution [15].

For identifying variance-controlling SNPs using vGWAS, the Brown-Forsythe test can be used. Brown-Forsythe tests for equality of group variances and is based on an ANOVA of the absolute deviation from the median. It is a robust test, even if the phenotypic distribution deviates from normality [11]. The test yields an F-statistic with $m-1$, $n-m$ degrees of freedom, where m is the number of genotypes and n is the number of individuals. For large n, the F-statistic can be approximated with a χ^2-statistic with $m-1$ degrees of freedom, allowing a straightforward calculation of nominal p-values. Unless stratification is too strong, genomic control can be used to correct for inflation of p-values. Users need to be aware that test statistics are biased when there are few observations (<100) in the genotype-class associated with the high-variance estimate [19]. Results obtained under such situations should be interpreted with caution (**Note 1–2**).

2.1.2 Methods and Implementation

There is software available for detecting variance-controlling genes in a GWAS setting, predominantly within the R statistical framework [20]. These include the vGWAS package [21] and the VariABEL extension to the widely used GenABEL package [22]. There are ongoing efforts to produce software for detecting vQTL (Lars Rönnegård, *personal communication*) to be implemented in the MapFastR R-package [23].

2.1.3 Future Challenges

Current methods are limited to estimating either the mean or the variance within a single analysis. This can lead to complications, as there is typically some correlation between the two, in particular when the range of mean differences is large enough to span an order of magnitude or more. Current best practice is to perform both types of scans and compare the outcomes, but simultaneous estimation would be preferable (**Note 3**). Another problem with current methodology is the inability to correct for more severe stratification; the genomic control approach is only appropriate when the entire genome follows the same pattern of stratification and when inflation is weak (**Note 2**).

2.2 Inferring Genetic Capacitation Using the VPR Approach

In order to gain biologically meaningful insights into the genetic architecture of an epistatic network, it is not sufficient to detect the interacting genes; one also need to describe and visualize the interactions within the network. This can be achieved using multidimensional genotype-phenotype maps (GP-maps), which allow for interaction patterns to be discerned. Drawing two-dimensional GP-maps is a common way to illustrate epistasis, and classic types of epistasis are usually illustrated with this approach [24]. Some attempts to use graphic representations of epistasis to understand underlying mechanisms have been made [25–27] as well as listing the complete enumerations of all possible types of two-dimensional maps [28]. Although no generally applicable methods for inferring the type of underlying mechanisms have been proposed, the usefulness of the graphical approach to functionally dissect epistasis is illustrated by its role in the identification of a genetic network involving capacitation in chicken [3].

To further explore the usefulness of methods based on studies of high-dimensional GP-maps for detection of epistasis, we have recently proposed the VPR method for inferring capacitating epistasis involving three loci [29]. In this section, we describe this method in more detail and illustrate examples of results that can be obtained when analyzing experimental data.

2.2.1 Theory

To illustrate the concept of a basic three-locus VPR analysis, we use an example of a triplet of biallelic loci. The first step in the analysis is to construct the three-locus GP map. This map can be graphically represented as a cube with 27 cells, where each cell corresponds to a unique three-locus genotype with an associated phenotype. The second step is then to decompose this cube into slices or planes, one for each genotype at the three loci. This yields three separate, conditional GP-maps for each locus—each with nine cells—that can be compared and analyzed for epistatic patterns. In short, the VPR method relies on detection of different shapes among these planes and capacitating loci are inferred as those leading to a large difference among the three conditional planes.

Let us now consider a simple theoretical example in more detail. Assume that a phenotype is governed by three loci—$Q1$, $Q2$, and $Q3$—where all loci are bi-allelic with alleles A/a, B/b, and C/c, respectively. Let $Q3$ be the conditioning locus. If neither locus has any effect, the planes corresponding to the "CC" and "cc" genotypes for $Q3$ will be flat across the genotypes of $Q1$ and $Q2$ and be at the same level (Fig. 1a). If $Q3$ and only $Q3$ has a marginal

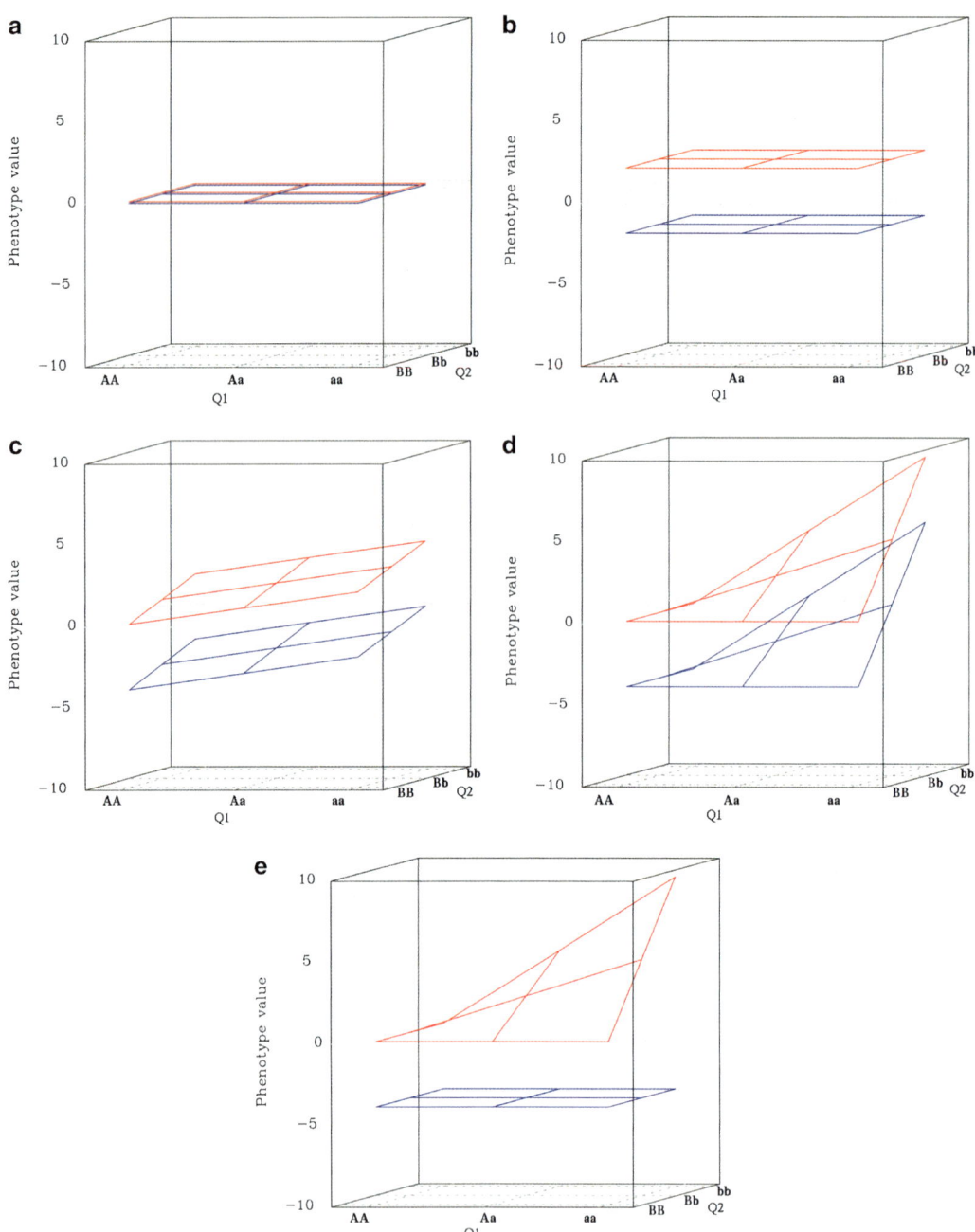

Fig. 1 Schematic description of genotype-phenotype planes across a three-locus GP map under various modes of gene-action. (**a**) Neither locus has any effect. (**b**) The conditioning locus has a main effect but does not interact with the other two loci. (**c**) All three loci have main effects but do not interact. (**d**) The conditioning locus has no effect, and the other two loci display synergistic epistasis. (**e**) The conditioning locus has a capacitating effect, and the two other loci display synergistic epistasis. This synergistic epistasis is, however, only visible in the CC background due to capacitation. The *red* and *blue* planes represent the CC and cc genotype classes at the conditioning locus, respectively. This figure is adapted from ref. 29 with permission

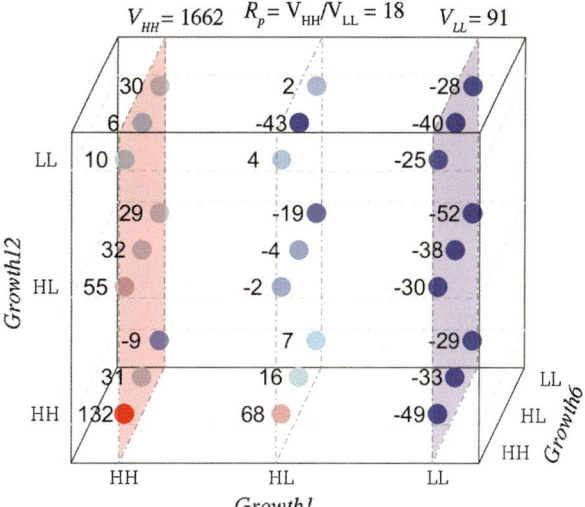

Fig. 2 Three locus GP-map for epistatic loci involved in the regulation of body-weight in an advanced intercross line of chicken. The figure shows the sex-corrected mean body-weight at 8 weeks of age for each of 27 possible three-locus genotypes, an empirical example of strong capacitating epistasis. The R_p-value (18) for *Growth1* is calculated by dividing the variance of the HH-plane (1,662) with that of the LL-plane (91). This figure is adapted from ref. 29 with permission, where a description of the underlying data and loci can be found

effect, the planes will still be flat and horizontal but have spatial separation due to the marginal effect (Fig. 1b). If all three loci have marginal effects and no interactions (Fig. 1c), or if *Q1* and *Q2* interact with each other but not with *Q3* (Fig. 1d), the planes will still have the same shape, but no longer be horizontal. If, however, *Q3* interacts with the other two loci, the shape of the planes will differ across the genotypes for *Q3*, and as a result of this they will have different within-plane variances (Fig. 1e). The difference in these within-plane variances is correlated with the strength of the interaction. In particular, if the locus displays a full capacitating effect, the difference in variance between the planes should be substantial, due to the plane(s) being flat, or close to flat, in the non-capacitated genotype(s). The ratio between the planes with the highest and lowest variance (R_p, in accordance with [29]) provides a quantitative measure of the capacitating epistatic effect of each locus in a given triplet.

An example of a result obtained by applying the VPR-method to experimental data is given in Fig. 2. Here, 8-week body-weight data from an advanced intercross line between two chicken lines with common ancestry that have been subject to more than 50 years of divergent selection [30] were studied. Three loci were selected for further study by previously identified effects on body-weight in an F_2 population between the divergently selected lines.

In this triplet, the R_p-value is 18 for *Growth1*, 4.3 for *Growth6*, and 5.0 for *Growth12*. The conclusion is that, in this triplet, *Growth1* is the main capacitating locus. Thus, the GP-map can be used to explore the interplay between interacting loci, something that is not easily attained with other methods.

2.2.2 Implementation

Calculation of R_p-values for a three-locus VPR analysis is straightforward: (1) use each locus as the conditioning locus in turn and (2), for each included locus, calculate the variances of the planes within each genotype at the conditioning locus and divide the variances for the planes with the maximum and the minimum variances. The method can thus, in principle, also be used for genome wide scanning as well. Even though it is an exhaustive three-dimensional scan, each calculation is very quick and scans including thousands of markers are feasible using a regular desktop computer. However, due to instability when all variances are small, the R_p-value should be combined with an absolute measure of variance in a scanning context.

2.2.3 Future Challenges

The VPR method can easily be extended for analyses in higher dimensions, including a fourth locus. With four loci, the planes become cubes extracted from a hypercube, but the conditioning on one locus in turn and comparisons of maximum and minimum plane (or here cube) variances remains the same. The major challenge with the approach in practice is to obtain a sufficient number of individuals in each *n*-locus genotype class to accurately estimate the phenotypic mean of all genotypes (**Note 4**). Large samples are needed to ensure that the least common genotypes, i.e., the multi-locus homozygotes, become increasingly rare with each locus added. Assuming even allele frequencies, Hardy-Weinberg equilibrium, and independence between the loci, four-locus homozygotes have a frequency of $0.25^4 \approx 0.004$. To obtain a reasonable estimate of a genotype value, approximately 20 individuals are needed, meaning that even for such optimal experimental designs pedigrees of about 5,000 individuals are needed. If any of the assumptions do not hold, the homozygote genotypes are even more rare and thus the method will likely not be of greater practical use for analysis of rare or linked variants.

2.3 Implications of Capacitating Epistasis for Complex Trait Genetics

2.3.1 Capacitation and Response to Selection

In the setting of a selection experiment or selection acting on a natural population, capacitating epistasis releases selectable genetic variation over time, making it possible to obtain a response to selection far outside the range of phenotypes in the original population. This change in the total selectable additive genetic variance results from a release of cryptic (or standing) genetic variation, which has been implicated as a source of selection induced genetic variation [31]; something that was experimentally validated initially in the Virginia chicken lines [3]. Thus, estimating the level of capacitating epistasis is useful to predict a population's potential for genetic change, either in selection programs or in adaptation to new environments [32].

The cryptic genetic variation suppressed by this type of epistasis can be substantial in a population and also has implications for both intermediate- and long-term selection responses in that population, because the marginal genetic (additive) effects of individual loci in a network involving capacitating epistasis will be dependent on the allele-frequencies at other interacting loci. In general, the genetic potential—i.e., the allele-frequency-independent ability of existing alleles in a population to change the phenotype—of the population is higher than the estimate of the heritability in a population where capacitation exists [32]. This as the heritability is dependent on the current allele-frequencies in the studied population, whereas the genetic potential is not.

2.3.2 Detection and Replication of Epistatic Networks

Replication and cloning of loci affecting complex trait loci becomes more challenging if they are part of a network involving capacitation. As the effects of the loci are context dependent, it is not possible to use traditional approaches to identify single, main effect loci and then replicate and clone them independently. Instead, it becomes necessary to first identify the interacting loci, which in most cases is difficult due to the costs of generating sufficiently large datasets, and then the dependencies between the interacting loci must be disentangled. Only when that is done is it possible to develop an appropriate strategy for simultaneous replication and cloning of the key loci. Due to the complexities and the cost involved, we can expect that the ability to identify and clone causative mutations using top-down approaches will impede progress in identifying new loci for multifactorial disorders where capacitation is important. Nevertheless, such endeavors are still necessary to first explore and later confirm the contributions of capacitation to complex traits, as the results so far indicate an important and general role for this mechanism in trait expression.

2.3.3 Evolutionary Consequences of Capacitation

The potential evolutionary impact of capacitating genetic interactions is considerable. Theoretical and simulation based work [10, 33, 34] have clearly shown that directional selection acts strongly on capacitating loci. In fact, in an epistatic network, the loci with the strongest capacitating effects will experience the strongest selection due to the prevalence of the variance inducing allele in extreme individuals (Fig. 3). Fixation of the capacitor releases cryptic variation, which allows a rapid phenotypic change in response to selection imposed by changing environments, for example. On the other hand, balancing selection, perhaps on a well-adapted population in a stable environment, will instead favor the suppressing allele, as that leads to reduced phenotypic variance, and thus less deviation from the optimum [34].

In light of this, it seems likely that genetic capacitation is an important evolutionary resource, especially on shorter time scales

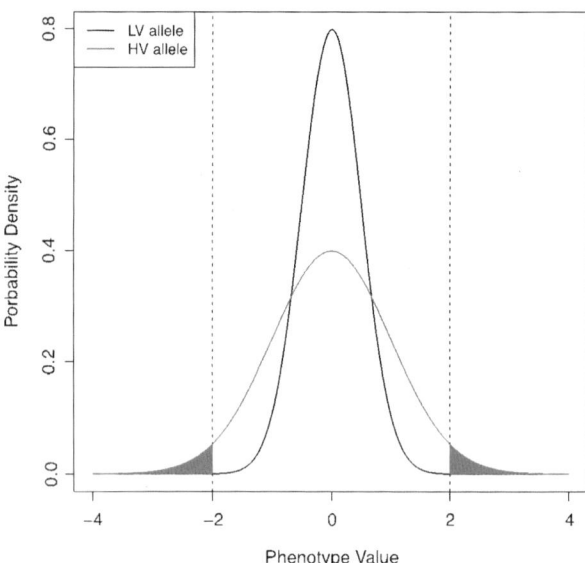

Fig. 3 Distributions illustrating the expected differential phenotypic variances across the genotypes at a variance-controlling capacitor locus. The distributions illustrate the effects of a locus where the capacitating high-variance allele (HV) releases variance of other capacitated loci. The *vertical lines* indicate selection thresholds that, if applied, would cause rapid fixation of the HV allele. This figure is adapted from ref. 34 with permission from John Wiley and Sons

in smaller populations. In addition to this, as illustrated by the Virginia line chicken selection experiment, it is also likely to be an important contributor in artificial selection programs in agricultural plants and animals. Interestingly, recent genomic studies have shown that most of the selection in the Virginia lines appears to have acted on variation present at the onset of the experiment, with few novel mutations identified [3, 35]. This implies that the release of cryptic variation was a major factor in allowing the selected populations to move far beyond the initial phenotypic range of the founder population.

3   Conclusions

In summary, capacitating epistasis is an important phenomenon in many biological systems, and needs to be taken into account when deciphering the genetic architecture of complex traits. Due to its inherent statistical properties, it is difficult to detect using standard analysis methods, and care needs to be taken at all steps of a study from the experiment design stage to the final analysis.

4   Notes

The following are some important tips for studying capacitating epistasis:

1. When interpreting the results from a vGWAS analysis, it is important to be aware of the risk of an increased false-positive rate if there are <100 individuals in the high-variance genotype class [19].

2. If working with highly stratified populations, there is a risk of p-value inflation that cannot be corrected for using genomic control. At present, there are no general vGWAS methods available that can deal with all such scenarios, although some recently proposed methods are able to accommodate this in specific scenarios [36].

3. At present, there are no methods available that simultaneously model the genetic effects on both the mean and the variance-heterogeneity. The best current practice is therefore to run separate analyses for mean- and variance-heterogeneity effects and combine the results afterwards to obtain a complete picture. However, when combined models are available, they will most likely be the method of choice.

4. The VPR method has the upside of being model-free, and assuming a particular epistatic pattern. However, the trade-off is that it can only be applied if all genotype class means can be accurately estimated. It is not advisable to use it if there are less than five individuals in any particular genotype class.

References

1. Andersson L, Georges M (2004) Domestic-animal genomics: deciphering the genetics of complex traits. Nat Rev Genet 5:202–212

2. Mackay TF, Lyman RF (2005) Drosophila bristles and the nature of quantitative genetic variation. Philos Trans R Soc B 360: 1513–1527

3. Carlborg Ö, Jacobsson L, Ahgren P et al (2006) Epistasis and the release of genetic variation during long-term selection. Nat Genet 38:418–420

4. Martin S, Söllner C, Charoensawan V et al (2010) Construction of a large extracellular protein interaction network and its resolution by spatiotemporal expression profiling. Mol Cell Proteomics 9:2654–2665

5. Yachie N, Saito R, Sugiyama N et al (2011) Integrative features of the yeast phosphoproteome and protein–protein interaction map. PLoS Comput Biol 7:e1001064

6. Gitter A, Siegfried Z, Klutstein M et al (2009) Backup in gene regulatory networks explains differences between binding and knockout results. Mol Syst Biol 5:276

7. Rutherford SL, Lindquist S (1998) Hsp90 as a capacitor for morphological evolution. Nature 396:336–342

8. Dworkin I, Palsson A, Birdsall K et al (2003) Evidence that Egfr contributes to cryptic genetic variation for photoreceptor determination in natural populations of *Drosophila melanogaster*. Curr Biol 13:1888–1893

9. Sangster T, Salathia N, Lee HN et al (2008) HSP90-buffered genetic variation is common in *Arabidopsis thaliana*. Proc Natl Acad Sci U S A 105:2969–2974

10. Hill WG, Zhang X-S (2004) Effects on phenotypic variability of directional selection arising through genetic differences in residual variability. Genet Res 83:121–132

11. Struchalin MV, Dehghan A, Witteman JC et al (2010) Variance heterogeneity analysis for detection of potentially interacting genetic loci: method and its limitations. BMC Genet 11:92

12. Paré G, Cook NR, Ridker PM et al (2010) On the use of variance per genotype as a tool to identify quantitative trait interaction effects: a report from the Women's Genome Health Study. PLoS Genet 6:e1000981

13. Jimenez-Gomez JM, Corwin JA, Joseph B et al (2011) Genomic analysis of QTLs and genes altering natural variation in stochastic noise. PLoS Genet 7:e1002295

14. Rönnegård L, Valdar W (2011) Detecting major genetic loci controlling phenotypic variability in experimental crosses. Genetics 188:435–447

15. Shen X, Pettersson M, Rönnegård L et al (2012) Inheritance beyond plain heritability: variance-controlling genes in *Arabidopsis thaliana*. PLoS Genet 8:e1002839

16. Hulse AM, Cai JJ (2012) Genetic variants contribute to gene expression variability in humans. Genetics 193:95–108

17. Yang J, Loos RJF, Powell JE et al (2012) FTO genotype is associated with phenotypic variability of body mass index. Nature 490:267–272

18. Nelson RM, Pettersson ME, Li X et al (2013) Variance heterogeneity in *Saccharomyces cerevisiae* expression data: trans-regulation and epistasis. PLoS One 8:e79507

19. Shen X, Carlborg Ö (2013) Beware of risk for increased false positive rates in genome-wide association studies for phenotypic variability. Front Genet 4:93

20. R Core Team (2012) R: a language and environment for statistical computing. R Foundation for Statistical Computing, Vienna, Austria. ISBN 3-900051-07-0

21. Shen X (2011) vGWAS: Variance Genome-wide Association. R package version 2011.11.01/r7. http://R-Forge.R-project.org/projects/vgwas/

22. Struchalin M (2011) VariABEL: testing of genotypic variance heterogeneity to detect potentially interacting SNP. R package version 0.9-0. http://CRAN.R-project.org/package=VariABEL

23. Nelson RM, Nettelblad C, Pettersson ME et al (2013) MAPfastR: quantitative trait loci mapping in outbred line crosses. G3 3:2147–2149

24. Carlborg Ö, Haley CS (2004) Epistasis: too often neglected in complex trait studies? Nat Rev Genet 5:618–625

25. Carlborg Ö, Kerje S, Schütz K et al (2003) A global search reveals epistatic interaction between QTL for early growth in the chicken. Genome Res 13:413–421

26. Carlborg Ö, Hocking PM, Burt DW et al (2004) Simultaneous mapping of epistatic QTL in chickens reveals clusters of QTL pairs with similar genetic effects on growth. Genet Res 83:197–209

27. Gjuvsland AB, Hayes BJ, Omholt SW et al (2007) Statistical epistasis is a generic feature of gene regulatory networks. Genetics 175:411–420

28. Hallgrímsdóttir IB, Yuster DS (2008) A complete classification of epistatic two-locus models. BMC Genet 9:17

29. Pettersson M, Besnier F, Siegel PB et al (2011) Replication and explorations of high-order epistasis using a large advanced intercross line pedigree. PLoS Genet 7:e1002180

30. Dunnington E, Siegel PB (1996) Long-term divergent selection for eight-week body weight in white Plymouth rock chicken. Poult Sci 75:1168–1179

31. Eitan Y, Soller M (2004) In: Wasser S (ed) Evolutionary theory and processes: modern horizons. Papers in honour of Eviatar Nevo. Kluwer Academic, Dordrecht, pp 153–176

32. LeRouzic A, Carlborg O (2008) Evolutionary potential of hidden genetic variation. Trends Ecol Evol 23:33–37 (Personal edition)

33. LeRouzic A, Siegel PB, Carlborg O (2007) Phenotypic evolution from genetic polymorphisms in a radial network architecture. BMC Biol 5:50

34. Pettersson M, Nelson R, Carlborg Ö (2012) Selection on variance-controlling genes: adaptability or stability. Evolution 66:3945–3949

35. Johansson AM, Pettersson ME, Siegel PB et al (2010) Genome-wide effects of long-term divergent selection. PLoS Genet 6:e1001188

36. Shen X, Forsberg S, Pettersson M, Sheng Z, Carlborg Ö (2013) Mutant epigenetic machinery mediates climate adaptation in *Arabidopsis thaliana*. arXiv:1310.4522 [q-bio.PE]

Chapter 11

Compositional Epistasis: An Epidemiologic Perspective

Etsuji Suzuki and Tyler J. VanderWeele

Abstract

Under Bateson's original conception, the term "epistasis" is used to describe the situation in which the effect of a genetic factor at one locus is masked by a variant at another locus. Epistasis in the sense of masking has been termed "compositional epistasis." In general, statistical tests for interaction are of limited use in detecting compositional epistasis. Using recently developed epidemiological methods, however, it has been shown that there are relations between empirical data patterns and compositional epistasis. These relations can sometimes be exploited to empirically test for certain forms of compositional epistasis, by using alternative nonstandard tests for interaction.

Using the counterfactual framework, we show conditions that can be empirically tested to determine whether there are individuals whose phenotype response patterns manifest epistasis in the sense of masking. Only under some very strong assumptions would tests for standard statistical interactions correspond to compositional epistasis. Even without such strong assumptions, however, one can still test whether there are individuals of phenotype response type representing compositional epistasis. The empirical conditions are quite strong, but the conclusions which tests of these conditions allow may be of interest in a wide range of studies. This chapter highlights that epidemiologic perspectives can be used to shed light on underlying mechanisms at the genetic, molecular, and cellular levels.

Key words Causality, Compositional epistasis, Counterfactual, Epidemiologic methods, Mechanistic interaction, Monotonicity assumptions, Potential outcomes, Statistical epistasis, Statistical models, Sufficient-cause framework

1 Introduction

In 1909, Bateson [1] used the term "epistasis" to describe the situation in which the effect of a genetic factor at one locus is masked by a variant at another locus. Bateson was specifically referring to the observation that, in some dihybrid crosses, not all possible phenotypic classes were observed and that some gene combinations resulted in novel phenotypes [2]. Since then, the use of the term epistasis has expanded to describe nearly any type of complex interaction among genetic loci. More recently, epistasis in the sense of masking has been termed "compositional epistasis" [2], and the terminology has been adopted in the literature [3–5] (*see* **Note 1**).

Jason H. Moore and Scott M. Williams (eds.), *Epistasis: Methods and Protocols*, Methods in Molecular Biology, vol. 1253, DOI 10.1007/978-1-4939-2155-3_11, © Springer Science+Business Media New York 2015

Meanwhile, several researchers have distinguished "statistical epistasis" from more biological forms of epistasis in the sense of masking or in the physical interaction of proteins [2–7]. Statistical epistasis is generally conceived of as a departure from additivity between the effects of two genetic factors in a statistical model under some link function so that correct specification of the model requires gene–gene interaction terms in the model [2]. It has then been argued that standard tests for statistical interaction or statistical epistasis are of limited use for elucidating epistasis in a more biological sense of the term [4, 6, 8].

In this chapter we focus on the concept of epistasis in the sense of masking and introduce recently developed epidemiologic methods to empirically test for certain forms of compositional epistasis using alternative nonstandard tests for interactions [9–12]. In the epidemiologic literature, formal thinking about causality has become mainstream, and the methodological advances discussed in this chapter were made possible by applying notions of causality in genetic association studies. To explain the underlying concept, we give a counterfactual exposition of compositional epistasis. We first consider a setting in which genotypes at both loci can be effectively considered binary. Then, we consider more general settings in which the genetic factors are considered to have three relevant levels.

2   Counterfactual Conception of Compositional Epistasis

As remarked above, under Bateson's original conception [1], epistasis is said to be present if variation of phenotype with genotype at one locus was only apparent among those with certain genotypes at the second locus. Those with other genotypes at the second locus would show no effect at the first, and the one being blocked is called hypostatic. Consider first a setting in which genotypes at both loci can effectively be considered binary.

Table 1 describes a potential phenotype pattern for a particular individual such that the effect of genotype at locus A is only present for the *B/B* variant; if the genotype at locus B is not *B/B* then the effect of the variation at locus A is not apparent. In other words, the effect of a genetic factor at locus A is masked when locus B is of the *b/b* or *b/B* genotype, and we say that locus B is epistatic to locus A. By symmetry, it is also the case that the effect of a genotype at locus B is only present for the *A/A* variant so the effect of the genetic factor at locus B is masked when locus A is of the *a/a* or *a/A* genotype, and we say that locus A is epistatic to locus B. Thus, the phenotype pattern in Table 1 is epistatic to both loci A and B.

In this simple setting, one way to conceive of epistasis is whether there are any individuals for whom the phenotype response pattern follows that in Table 1. In populations with heterogeneity

Table 1
An example of potential phenotypes for a particular individual that might result from different genotypes at two loci that exhibit compositional epistasis in the sense of making

Genotype at locus A	Genotype at locus B	
	b/b or *b/B*	*B/B*
a/a or *a/A*	0	0
A/A	0	1

Both loci A and B are epistatic in this phenotype pattern. Note that this phenotype pattern corresponds to response type 8 in Table 2

and for complex traits with non-Mendelian inheritance, the phenotype response patterns may vary between individuals, but we may be interested whether there are any individuals whose phenotype response patterns manifest such epistasis. Let X_1 be a binary indicator for a genotype at locus A (in the example, $X_1 = 0$ for genotype *a/a* or *a/A* and $X_1 = 1$ for genotype *A/A*); let X_2 be a binary indicator for a genotype at locus B (in the example, $X_2 = 0$ for genotype *b/b* or *b/B* and $X_2 = 1$ for genotype *B/B*). Let D be a binary indicator of phenotype, indicating the presence of some dichotomous trait. For each individual ω in the population let $D_{ij}(\omega)$ denote what the trait would have been if X_1 were i and if X_2 were j. In other words, for each individual ω, we could conceive of what might have happened to that individual had the genotype at each locus been something other than what it was. In the counterfactual framework [13], the random variable $D_{ij}(\omega)$ is referred to as potential outcomes for individual ω. For an individual ω who actually had $X_1 = i$ and $X_2 = j$, then the observed outcome D is that of the potential outcome $D_{ij}(\omega)$. For each individual ω, there would thus be four possible potential outcomes: $D_{11}(\omega)$, $D_{01}(\omega)$, $D_{10}(\omega)$, and $D_{00}(\omega)$. This form of thinking is justified if the outcome under the counterfactual settings of the factors are well defined, but this becomes subtle for genetic effects [14].

An epistatic interaction, in Bateson's original sense of masking, would be present if there were an individual for whom Table 1 describes the phenotype response pattern for that individual. In others words, by using the potential outcomes, an epistatic interaction shown in Table 1 would be present if there were an individual ω for whom

$$D_{11}(\omega) = 1, D_{01}(\omega) = 0, D_{10}(\omega) = 0, D_{00}(\omega) = 0. \qquad (1)$$

Note that, because there are four possible potential outcomes, individuals can be classified into 16 (i.e., 2^4) different phenotype response types [15], as enumerated in Table 2. The epistatic pattern

Table 2
Enumeration of 16 response types for phenotype *D* and corresponding potential outcomes

Type	$D_{11}(\omega)$	$D_{01}(\omega)$	$D_{10}(\omega)$	$D_{00}(\omega)$	$D_{11}(\omega) - D_{01}(\omega) - D_{10}(\omega) - D_{00}(\omega)$	$D_{11}(\omega) - D_{01}(\omega) - D_{10}(\omega)$	$D_{11}(\omega) - D_{01}(\omega) - D_{10}(\omega) + D_{00}(\omega)$
1	1	1	1	1	−2	−1	0
2[c,d]	1	1	1	0	−1	−1	−1
3[a,d]	1	1	0	1	−1	0	1
4[d]	1	1	0	0	0	0	0
5[b,c]	1	0	1	1	−1	0	1
6[d]	1	0	1	0	0	0	0
7[a,b,c,d]	1	0	0	1	0	1	2
8[c,d]	1	0	0	0	1	1	1
9[a,b]	0	1	1	1	−3	−2	−1
10[a,b,c,d]	0	1	1	0	−2	−2	−2
11[a]	0	1	0	1	−2	−1	0
12[a,d]	0	1	0	0	−1	−1	−1
13[b]	0	0	1	1	−2	−1	0
14[b,c]	0	0	1	0	−1	−1	−1
15[a,b]	0	0	0	1	−1	0	1
16	0	0	0	0	0	0	0

We let X_1 and X_2 be binary indicators for genotypes at loci A and B, respectively. We also let D be a binary indicator of phenotype, indicating the presence of some dichotomous trait. For each individual ω, we let $D_{ij}(\omega)$ denote what the trait would have been if X_1 were i and if X_2 were j

[a]Under the assumption of positive monotonicity of X_1 on D, these response types are excluded
[b]Under the assumption of positive monotonicity of X_2 on D, these response types are excluded
[c]Under the assumption of negative monotonicity of X_1 on D, these response types are excluded
[d]Under the assumption of negative monotonicity of X_2 on D, these response types are excluded

in Table 1 corresponds to response type 8 in Table 2. It is notable that, as shown in Table 2, the quantity $D_{11}(\omega) - D_{01}(\omega) - D_{10}(\omega) - D_{00}(\omega)$ can take a positive value only in the individual of response type 8. Thus, another way of asking whether there are individuals for whom relation (1) is satisfied is to ask whether there are individuals for whom

$$D_{11}(\omega) - D_{01}(\omega) - D_{10}(\omega) - D_{00}(\omega) > 0. \qquad (2)$$

In general, however, we are not able to infer what all of $D_{11}(\omega)$, $D_{01}(\omega)$, $D_{10}(\omega)$, and $D_{00}(\omega)$ would be for a particular individual ω; we can only observe either one of the four potential outcomes as an actual outcome and the other three potential outcomes are, by definition, not observable since they are counterfactual. This is indeed a "fundamental problem of causal inference" [16].

3 Empirical Tests for Compositional Epistasis

Although we cannot infer all the potential outcomes for a particular individual, in a genetic association study we might hope to be able to estimate the values of $E[D_{11}]$, $E[D_{01}]$, $E[D_{10}]$, and $E[D_{00}]$ on average for a population by carefully controlling for confounding by stratification and admixture. Since D is a dichotomous outcome, the mean $E[D_{ij}]$ is the risk $P[D_{ij}=1]$ of developing the outcome. If we let C denote a genetic marker for population substructure based on many loci [17–21], then if control for the marker suffices to control for confounding, we can estimate the average likelihood of the outcome D when $X_1 = i$ and $X_2 = j$ for those with genetic marker $C = c$ by

$$P\left[D_{ij} = 1 \mid C = c\right] \approx P\left[D = 1 \mid X_1 = i, X_2 = j, C = c\right]. \quad (3)$$

If marker C suffices to control for confounding, the average effects of genetic factors X_1 and X_2 on D can be estimated with data and relation (3) will hold [22]. We could thus test whether there are any individuals with $C = c$ for whom the response pattern is identical to Table 1 by testing

$$p_{11c} - p_{01c} - p_{10c} - p_{00c} > 0, \quad (4)$$

where $p_{ijc} = P[D = 1 \mid X_1 = i, X_2 = j, C = c]$. Provided the genetic marker C suffices to control for confounding by stratification and admixture so that relation (3) holds, then if for some value of the genetic marker we find that $p_{11c} - p_{01c} - p_{10c} - p_{00c} > 0$, there must be some individuals with $C = c$ for whom the response pattern is given by Table 1. Condition (4) is not the usual statistical test for interaction, but it can be tested empirically from the data to draw conclusions about whether there are at least some individuals with an epistatic response pattern. We discuss below the relation between condition (4) and the usual statistical tests for interactions. Note that the implication described above is one-way. If condition (4) is satisfied then there are individuals for whom $D_{11}(\omega) = 1$ but $D_{01}(\omega) = D_{10}(\omega) = D_{00}(\omega) = 0$; however, if condition (4) is not satisfied we cannot necessarily conclude that there are no individuals for whom $D_{11}(\omega) = 1$ but $D_{01}(\omega) = D_{10}(\omega) = D_{00}(\omega) = 0$. Condition (4) is a sufficient but not a necessary condition for the compositional epistasis given in Table 1.

Table 3
Other examples of compositional epistasis

	Genotype at locus B	
Genotype at locus A	b/b or b/B	B/B
A. (i.e., type 12)[a]		
a/a or a/A	0	1
A/A	0	0
B. (i.e., type 14)[a]		
a/a or a/A	0	0
A/A	1	0
C. (i.e., type 15)[a]		
a/a or a/A	1	0
A/A	0	0
D. (i.e., type 2)[a]		
a/a or a/A	0	1
A/A	1	1
E. (i.e., type 3)[a]		
a/a or a/A	1	1
A/A	0	1
F. (i.e., type 5)[a]		
a/a or a/A	1	0
A/A	1	1
G. (i.e., type 9)[a]		
a/a or a/A	1	1
A/A	1	0

Both loci A and B are epistatic in the seven phenotype patterns
[a]*See* Table 2

Using the same logic as that given above, we could empirically test for the other epistatic response patterns shown in Table 3. For example, we could test whether there are individuals for whom $D_{01}(\omega) = 1$ but $D_{11}(\omega) = D_{10}(\omega) = D_{00}(\omega) = 0$ (*see* Table 3A) by testing $p_{01c} - p_{11c} - p_{00c} - p_{10c} > 0$. We could test whether there are individuals for whom $D_{10}(\omega) = 1$ but $D_{11}(\omega) = D_{01}(\omega) = D_{00}(\omega) = 0$ (*see* Table 3B) by testing $p_{10c} - p_{11c} - p_{00c} - p_{01c} > 0$. Finally, we could test whether there are individuals for whom $D_{00}(\omega) = 1$ but $D_{11}(\omega) = D_{01}(\omega) = D_{10}(\omega) = 0$ (*see* Table 3C) by testing $p_{00c} - p_{01c} - p_{10c} - p_{11c} > 0$. Furthermore, by appropriately recoding the exposure of interest, we can also empirically test the other epistatic patterns shown in Tables 3D–G. Table 4 summarizes sufficient conditions for the eight epistatic patterns shown in Tables 1 and 3.

Table 4
Empirical tests for epistasis, with two relevant genetic variants at each locus

Form of epistasis	Assumption[a]	Sufficient condition[b]	Statistical testing[c]
Table 1 (i.e., type 8)[d]	No monotonicity assumption	$p_{11c} - p_{01c} - p_{10c} - p_{00c} > 0$	$\alpha_3 > 2\alpha_0$
	Either X_1 or X_2 with positive monotonicity	$p_{11c} - p_{01c} - p_{10c} > 0$	$\alpha_3 > \alpha_0$
	Both X_1 and X_2 with positive monotonicity	$p_{11c} - p_{01c} - p_{10c} + p_{00c} > 0$	$\alpha_3 > 0$
Table 3A (i.e., type 12)[d]	No monotonicity assumption	$p_{01c} - p_{11c} - p_{00c} - p_{10c} > 0$	$\alpha_3 < -2(\alpha_0 + \alpha_1)$
	Either $\bar{X}_1$ or X_2 with positive monotonicity	$p_{01c} - p_{11c} - p_{00c} > 0$	$\alpha_3 < -(\alpha_0 + \alpha_1)$
	Both $\bar{X}_1$ and X_2 with positive monotonicity	$p_{01c} - p_{11c} - p_{00c} + p_{10c} > 0$	$\alpha_3 < 0$
Table 3B (i.e., type 14)[d]	No monotonicity assumption	$p_{10c} - p_{11c} - p_{00c} - p_{01c} > 0$	$\alpha_3 < -2(\alpha_0 + \alpha_2)$
	Either X_1 or $\bar{X}_2$ with positive monotonicity	$p_{10c} - p_{11c} - p_{00c} > 0$	$\alpha_3 < -(\alpha_0 + \alpha_2)$
	Both X_1 and $\bar{X}_2$ with positive monotonicity	$p_{10c} - p_{11c} - p_{00c} + p_{01c} > 0$	$\alpha_3 < 0$
Table 3C (i.e., type 15)[d]	No monotonicity assumption	$p_{00c} - p_{01c} - p_{10c} - p_{11c} > 0$	$\alpha_3 < -2(\alpha_0 + \alpha_1 + \alpha_2)$
	Either $\bar{X}_1$ or $\bar{X}_2$ with positive monotonicity	$p_{00c} - p_{01c} - p_{10c} > 0$	$\alpha_0 + \alpha_1 + \alpha_2 < 0$
	Both $\bar{X}_1$ and $\bar{X}_2$ with positive monotonicity	$p_{00c} - p_{01c} - p_{10c} + p_{11c} > 0$	$\alpha_3 > 0$
Table 3D (i.e., type 2)[d]	No monotonicity assumption	$(1 - p_{00c}) - (1 - p_{01c}) - (1 - p_{10c}) - (1 - p_{11c}) > 0$	$\alpha_3 > 2(1 - \alpha_0 - \alpha_1 - \alpha_2)$
	Either X_1 or X_2 with positive monotonicity	$(1 - p_{00c}) - (1 - p_{01c}) - (1 - p_{10c}) > 0$	$1 - \alpha_0 - \alpha_1 - \alpha_2 < 0$
	Both X_1 and X_2 with positive monotonicity	$(1 - p_{00c}) - (1 - p_{01c}) - (1 - p_{10c}) + (1 - p_{11c}) > 0$	$\alpha_3 < 0$
Table 3E (i.e., type 3)[d]	No monotonicity assumption	$(1 - p_{10c}) - (1 - p_{11c}) - (1 - p_{00c}) - (1 - p_{01c}) > 0$	$\alpha_3 > 2(1 - \alpha_0 - \alpha_2)$
	Either $\bar{X}_1$ or X_2 with positive monotonicity	$(1 - p_{10c}) - (1 - p_{11c}) - (1 - p_{00c}) > 0$	$\alpha_3 > 1 - \alpha_0 - \alpha_2$
	Both $\bar{X}_1$ and X_2 with positive monotonicity	$(1 - p_{10c}) - (1 - p_{11c}) - (1 - p_{00c}) + (1 - p_{01c}) > 0$	$\alpha_3 > 0$
Table 3F (i.e., type 5)[d]	No monotonicity assumption	$(1 - p_{01c}) - (1 - p_{11c}) - (1 - p_{00c}) - (1 - p_{10c}) > 0$	$\alpha_3 > 2(1 - \alpha_0 - \alpha_1)$
	Either X_1 or $\bar{X}_2$ with positive monotonicity	$(1 - p_{01c}) - (1 - p_{11c}) - (1 - p_{00c}) > 0$	$\alpha_3 > 1 - \alpha_0 - \alpha_1$
	Both X_1 and $\bar{X}_2$ with positive monotonicity	$(1 - p_{01c}) - (1 - p_{11c}) - (1 - p_{00c}) + (1 - p_{10c}) > 0$	$\alpha_3 > 0$
Table 3G (i.e., type 9)[d]	No monotonicity assumption	$(1 - p_{11c}) - (1 - p_{01c}) - (1 - p_{10c}) - (1 - p_{00c}) > 0$	$\alpha_3 < 2(\alpha_0 - 1)$
	Either $\bar{X}_1$ or $\bar{X}_2$ with positive monotonicity	$(1 - p_{11c}) - (1 - p_{01c}) - (1 - p_{10c}) > 0$	$\alpha_3 < \alpha_0 - 1$
	Both $\bar{X}_1$ and $\bar{X}_2$ with positive monotonicity	$(1 - p_{11c}) - (1 - p_{01c}) - (1 - p_{10c}) + (1 - p_{00c}) > 0$	$\alpha_3 < 0$

[a]We let X_1 and X_2 be binary indicators for genotypes at loci A and B, respectively. We also let $\bar{X}_1$ and $\bar{X}_2$ denote the complements of X_1 and X_2, respectively, in the terminology of events

[b]We let p_{ijc} denote $P[D=1 | X_1 = i, X_2 = j, C = c]$

[c]Within each stratum $C = c$, we consider a statistical model of the following form: $P[D=1 | X_1 = x_1, X_2 = x_2] = \alpha_0 + \alpha_1 x_1 + \alpha_2 x_2 + \alpha_3 x_1 x_2$

[d]See Table 2

In some cases, we might be willing to assume that genotype $X_1 = 1$ (as compared with $X_1 = 0$) never prevents the outcome. Therefore, if $D_{00}(\omega) = 1$ then it is also the case that $D_{10}(\omega) = 1$, and if $D_{10}(\omega) = 0$ then it must also be the case that $D_{00}(\omega) = 0$. Similarly, if $D_{01}(\omega) = 1$ then it is also the case that $D_{11}(\omega) = 1$, and if $D_{11}(\omega) = 0$ then it must also be the case that $D_{01}(\omega) = 0$. In such cases, $X_1 = 1$ (as compared with $X_1 = 0$) is never preventive in that it has either a neutral or causative effect on all individuals. This assumption has been referred to as positive monotonicity [23, 24]. Stated more succinctly, we may say that X_1 has a positive monotonic effect on D if for all individuals in the population, $D_{1j}(\omega) \geq D_{0j}(\omega)$ for $j = 0, 1$. Similarly, we say that X_2 has a positive monotonic effect on D if for all individuals in the population, $D_{i1}(\omega) \geq D_{i0}(\omega)$ for $i = 0, 1$.

Monotonicity assumptions are strong assumptions insofar as they make reference to all individuals in the population. Empirical data can sometimes be used to invalidate such monotonicity assumptions; whenever a particular genetic variant is such that it makes the outcome more likely in some populations but less likely in others, its monotonicity assumption does not hold. Unfortunately, however, monotonicity assumptions can never be empirically verified with data, since they make reference to all of the potential outcomes for each individual in a population under each possible combination of the factors and we only observe the outcome D under one particular setting. One would thus generally have to rely on knowledge of the biology itself to reasonably make these monotonicity assumptions. In some cases, monotonicity of X_1 or X_2 might hold within certain strata of a genetic marker C but not in others.

When positive monotonicity assumptions do hold, we can test for the compositional epistasis shown in Table 1 by testing a condition weaker than condition (4). Suppose, for example, that X_1 had a positive monotonic effect on D then if it were the case that $D_{10}(\omega)$ were 0 then it must also be the case that $D_{00}(\omega)$ is 0. Thus if X_1 had a positive monotonic effect on D and there were individuals for whom $D_{11}(\omega) = 1$ and $D_{01}(\omega) = D_{10}(\omega) = 0$, then we could also conclude for such individuals that $D_{00}(\omega) = 0$ by positive monotonicity and thus that relation (1) held for such individuals (i.e., that an epistatic interaction as given in Table 1 was present). Indeed, as shown in Table 2, under the assumption of positive monotonicity of X_1 on D, the individuals of response types 3, 7, 9, 10, 11, 12, and 15 are excluded, and individuals of response types 1, 2, 4, 5, 6, 8, 13, 14, and 16 may remain. It is notable that, among these nine response types, the quantity $D_{11}(\omega) - D_{01}(\omega) - D_{10}(\omega)$ can take a positive value if and only if individual ω is of response type 8 (*see* Table 2). Thus, if some genetic marker C suffices to control for confounding by stratification and admixture, we could test whether there were individuals for whom the response pattern is given by Table 1 within stratum $C = c$ by testing.

$$p_{11c} - p_{01c} - p_{10c} > 0. \tag{5}$$

Note that condition (5) is a weaker condition than condition (4); condition (5) does not require subtracting p_{00c}. When we can assume that the effect of X_1 on D is positive monotonic, we can test this weaker condition instead. By symmetry, if X_2, rather than X_1, has a positive monotonic effect on D, then condition (5) can also be used to test whether there are individuals for whom the response pattern is given by Table 1. We have seen then that if either X_1 or X_2 has a positive monotonic effect on D then we can use the weaker condition (5), rather than condition (4), to test for epistasis shown in Table 1. Once again, however, condition (5) does not correspond to a standard statistical test for interaction.

Finally suppose that both X_1 and X_2 have positive monotonic effects on D. Suppose $X_1 = 1$ (as compared with $X_1 = 0$) never prevented the outcome for any individual and $X_2 = 1$ (as compared with $X_2 = 0$) never prevented the outcome for any individual. Stated another way, we are supposing that $D_{ij}(\omega)$ is non-decreasing in i and j, i.e., $D_{ij}(\omega) \geq D_{i'j'}(\omega)$ for $\forall \omega$ whenever $i \geq i'$ and $j \geq j'$. Under this assumption, the individuals of response types 3, 5, 7, and 9 through 15 are excluded, and individuals of response types 1, 2, 4, 6, 8, and 16 may remain [24]. Among these six response types, the quantity $D_{11}(\omega) - D_{01}(\omega) - D_{10}(\omega) + D_{00}(\omega)$ takes a positive value if and only if individual ω is of response type 8 (*see* Table 2). Then, if some genetic marker C suffices to control for confounding by stratification and admixture, we could test whether there were individuals for whom the response pattern is given by Table 1 within stratum $C = c$ by testing

$$p_{11c} - p_{01c} - p_{10c} + p_{00c} > 0. \tag{6}$$

Note that condition (6) is weaker than conditions (4) or (5), because we are adding back the term p_{00c}. Condition (6) is how additive interaction is often ordinarily assessed in statistical models; condition (6) essentially examines whether the effects of X_1 and X_2 combined are greater than the sum of the effects of X_1 and X_2 considered separately. However, condition (6) will only imply the presence of individuals with the epistatic response pattern in Table 1 if one can assume that both X_1 and X_2 have positive monotonic effects on D. In other words, if there are any individuals with $C = c$ for whom the outcome would be present if $X_1 = 0$ but for whom it would not be present if $X_1 = 1$ (or similarly for whom the outcome would be present if $X_2 = 0$ but for whom it would not be present if $X_2 = 1$) then the positive monotonicity assumptions would be violated and one could not use condition (6) to test for epistatic response pattern shown in Table 1. As remarked above, even if these monotonicity assumptions are violated, condition (4) or (5) could be used to test for epistasis in Table 1; however, condition (6), which is the usual test for interaction, only gives a test for epistasis under strong positive monotonicity assumptions for both genetic factors. With regard to the seven epistatic patterns in Table 3, we can analogously use weaker conditions under the assumptions of some monotonic effects. These weaker conditions are summarized in Table 4 (*see* **Notes 2** and **3**).

3.2 Relation to Statistical Models

In this section we briefly relate the empirical tests for compositional epistasis to standard tests for interactions in statistical models. When two binary genetic variants are considered, a statistical model of the following form is sometimes used to test for a statistical interaction:

$$P[D = 1 \mid X_1 = x_1, X_2 = x_2] = \alpha_0 + \alpha_1 x_1 + \alpha_2 x_2 + \alpha_3 x_1 x_2 \quad (7)$$

To control for confounding by stratification and admixture, one can fit a separate model like (7) within each stratum $C = c$ of some genetic marker. Statistical interaction is then often assessed by testing whether $\alpha_3 > 0$.

Note that we can express conditions (4–6) in terms of the coefficients of the statistical model (7). For example, when neither X_1 nor X_2 had positive monotonic effects (i.e., we make no assumptions about monotonicity), we can test for the presence of the epistatic pattern in Table 1 by testing condition (4). Within each stratum $C = c$ of some genetic marker, condition (4) can be expressed in terms of the coefficients of statistical model (7) as

$$\begin{aligned} p_{11c} - p_{10c} - p_{01c} - p_{00c} &= (\alpha_0 + \alpha_1 + \alpha_2 + \alpha_3) - (\alpha_0 + \alpha_1) - (\alpha_0 + \alpha_2) - \alpha_0 \\ &= \alpha_3 - 2\alpha_0 > 0. \end{aligned} \quad (8)$$

Thus, even without making any assumptions about monotonicity we can test for such epistasis by testing whether $\alpha_3 > 2\alpha_0$. In a similar manner, condition (5) can be expressed in terms of the coefficients of statistical model (7) as $\alpha_3 > \alpha_0$. Thus, if at least one of X_1 and X_2 had positive monotonic effects on the outcome, we can test for such epistasis by testing whether $\alpha_3 > \alpha_0$. These are nonstandard tests for interaction, but, when satisfied, allow for conclusions to be drawn not just about statistical interactions but about epistatic response patterns. Finally, we saw above that condition (6) can be used to test for the epistatic pattern in Table 1 only under the strong assumption that both X_1 and X_2 have positive monotonic effects on the outcome. Trivially, testing whether $\alpha_3 > 0$ corresponds to a test of condition (6).

As noted above, these tests are sufficient conditions for compositional epistasis, but not necessary. If the conditions are satisfied, there are at least some individuals with response patterns manifesting epistasis in Table 1. If the conditions are not satisfied, there may or may not be individuals with response patterns exhibiting epistasis in Table 1; we cannot tell from the data. Table 4 summarizes statistical testing for the eight epistatic patterns in terms of the coefficients of statistical model (7) (*see* **Note 4**).

3.3 Testing for Compositional Epistasis in Case-Control Studies

Many analyses of interaction use data from a case-control study. In such case-control studies, risks like p_{11c}, p_{01c}, p_{10c}, and p_{00c} cannot, in general, be estimated, but odds ratios (ORs) for the effects of genetic factors can be estimated. Thus in such studies, logistic regression is often used, which for interaction analyses may take the form of

$$\text{logit}\left(P\left[D = 1 \mid X_1 = x_1, X_2 = x_2\right]\right) = \beta_0 + \beta_1 x_1 + \beta_2 x_2 + \beta_3 x_1 x_2 \quad (9)$$

Model (9) can be used to calculate ORs comparing the odds of the outcome when both X_1 and X_2 are present to when both are absent (denoted by $\text{OR}_{11} = \exp(\beta_1 + \beta_2 + \beta_3)$), the odds when $X_1 = 1$ and $X_2 = 0$ to when both are absent (denoted by $\text{OR}_{10} = \exp(\beta_1)$) and the odds when $X_1 = 0$ and $X_2 = 1$ to when both are absent (denoted by $\text{OR}_{01} = \exp(\beta_2)$). When the outcome is rare these ORs approximate the corresponding relative risks (RRs), denoted by RR_{11}, RR_{10}, and RR_{01}, respectively. Although we cannot test conditions (4, 5, or 6) directly using risks, we could divide these conditions by p_{00c}. Then, condition (4) becomes

$$\text{RR}_{11c} - \text{RR}_{01c} - \text{RR}_{10c} - 1 > 0. \quad (10)$$

Condition (5) becomes

$$\text{RR}_{11c} - \text{RR}_{01c} - \text{RR}_{10c} > 0. \quad (11)$$

Condition (6) becomes

$$\text{RR}_{11c} - \text{RR}_{01c} - \text{RR}_{10c} + 1 > 0. \quad (12)$$

Under the assumption that the outcome is rare, these conditions could be tested using the estimated ORs instead. Thus even in a case-control study one can potentially test for epistasis in settings in which the outcome is rare. The quantity $\text{RR}_{11c} - \text{RR}_{01c} - \text{RR}_{10c} + 1$ is sometimes described as the "relative excess risk due to interaction" or RERI [25]. The three conditions given above could thus be written, respectively, as RERI > 2, RERI > 1, and RERI > 0. Statistical tests and confidence intervals for RERI are given elsewhere [26]. Thus, in a case-control study with a rare outcome, to test whether there were individuals for whom $D_{11}(\omega) = 1$ and $D_{01}(\omega) = D_{10}(\omega) = D_{00}(\omega) = 0$ within stratum $C = c$, RERI > 2 suffices if no monotonicity assumptions are made; RERI > 1 suffices if one of X_1 or X_2 is assumed to have a positive monotonic effect; and RERI > 0 suffices if both X_1 and X_2 can be assumed to have positive monotonic effects.

Alternatively, it can also be shown that when the outcome is rare, conditions (4, 5, or 6) will be satisfied if we have $\beta_3 > \log(3)$, $\beta_3 > \log(2)$, or $\beta_3 > 0$, respectively, provided that the main effects β_1 and β_2 are nonnegative. See VanderWeele [9] for a proof.

However, these are stronger conditions than $(10\text{--}12)$ above, and thus conditions $(10\text{--}12)$ should be used instead whenever possible. In case-only studies only β_3 is estimable and so the conditions $\beta_3 > \log(3)$, $\beta_3 > \log(2)$, or $\beta_3 > 0$ may be of use.

3.4 More General Settings

The remarks above also apply to settings in which the genetic factors are considered to have three relevant levels. Consider now a setting in which at loci A and B there are three distinct relevant genotypes: a/a, a/A, and A/A at locus A and b/b, b/B, and B/B at locus B. We let V_1 and V_2 be variables with three levels indicating the genotype at loci A and B, respectively (e.g., $V_1 = 0$ for a/a, $V_1 = 1$ for a/A, $V_1 = 2$ for A/A and $V_2 = 0$ for b/b, $V_2 = 1$ for b/B, $V_2 = 2$ for B/B). Once again, let D be a binary indicator of phenotype, indicating the presence of some dichotomous trait. For individual ω, we let $D_{ij}(\omega)$ denote what the trait would be if V_1 were i and if V_2 were j. Again let C denote a genetic marker for population substructure and suppose that the marker suffices to control for confounding by stratification and admixture so that $P[D_{ij} = 1 | C = c] \approx P[D = 1 | V_1 = i, V_2 = j, C = c]$. As before we let $p_{ijc} = P[D = 1 | V_1 = i, V_2 = j, C = c]$. We can then consider a variety of response patterns that would constitute instances of epistasis. Table 5 shows four examples among them.

Table 5
Examples of potential phenotypes for the effects of genotypes at two loci exhibiting epistasis, with three relevant genetic variants at each locus

Genotype at locus A	Genotype at locus B		
	b/b	*b/B*	*B/B*
A			
a/a	0	0	0
a/A	0	0	0
A/A	0	0	1
B			
a/a	0	0	0
a/A	0	0	1
A/A	0	0	1
C			
a/a	0	0	0
a/A	0	0	0
A/A	0	1	1
D			
a/a	0	0	0
a/A	0	1	1
A/A	0	1	1

Both loci A and B are epistatic in the four phenotype patterns

First, we consider the response pattern in Table 5A. Note that the effect of genetic variation at locus A is only apparent when the genotype is B/B at locus B, and the effect of the genetic factor at locus A is masked when locus B is of the b/b or b/B genotype. In other words, it does not matter what value locus A takes when locus B is of the b/b or b/B genotype, and we say that locus B is epistatic to locus A. Similarly, the effect of genetic variation at locus B is only apparent when the genotype is A/A at locus A, and the effect of the genetic factor at locus B is masked when locus A is of the a/a or a/A genotype. In other words, it does not matter what value locus B takes when locus A is of the a/a or a/A genotype, and we say that locus A is epistatic to locus B. Thus, the phenotype pattern in Table 5A is epistatic to both loci A and B.

For simplicity now assume that the effects of both V_1 and V_2 on D are positively monotonic so that whenever $i \geq i'$ and $j \geq j'$, we have $D_{ij}(\omega) \geq D_{i'j'}(\omega)$ for $\forall \omega$. Then, by arguments similar to those given above, there must be individuals with genetic marker $C = c$ who have the response patterns given by Table 5A if it is the case that

$$p_{22c} - p_{21c} - p_{12c} + p_{11c} > 0. \qquad (13)$$

In other words, if control for genetic marker C suffices to control for confounding by stratification and admixture, and if condition (13) holds, then there must be some individuals with genetic marker $C = c$ who have a phenotype pattern in Table 5A. Thus, even when genotypes are considered to have three possible relevant variants rather than two, we can once again test for compositional epistasis empirically. An analogous discussion applies to three other examples of compositional epistasis in Tables 5B–D, and their sufficient conditions are shown in Table 6.

The test represented by condition (13) presupposed that the effects of V_1 and V_2 on D were positively monotonic. Table 6 also shows sufficient conditions for the epistatic interaction in Table 5 without monotonicity assumptions or when only one of V_1 or V_2 has a monotonic effect on D. In the Appendix Table, we show a total of 18 possible phenotype patterns for the effects of genotypes under the assumption of positive monotonicity of both V_1 and V_2 on D. We do not show two non-epistatic patterns with all phenotype responses equal to 0 or 1.

4 Relation to Sufficient-Cause Framework

The empirical tests for compositional epistasis are related to but stronger than the notion of "synergism" in Rothman's sufficient-cause framework [27]. Rothman conceptualized causation as a collection of different causal mechanisms, each sufficient to bring about the outcome. These causal mechanisms are called

Table 6
Empirical tests for epistasis, with three relevant genetic variants at each locus

Form of epistasis	Assumption[a]	Sufficient condition[b]
Table 5A	No monotonicity assumption	$p_{22c} - p_{21c} - p_{20c} - p_{12c} - p_{11c} - p_{10c} - p_{02c} - p_{01c} - p_{00c} > 0$
	V_1 with positive monotonicity	$p_{22c} - p_{21c} - p_{20c} - p_{12c} > 0$
	V_2 with positive monotonicity	$p_{22c} - p_{12c} - p_{02c} - p_{21c} > 0$
	Both V_1 and V_2 with positive monotonicity	$p_{22c} - p_{21c} - p_{12c} + p_{11c} > 0$
Table 5B	No monotonicity assumption	Not available
	V_1 with positive monotonicity	$p_{12c} - p_{21c} - p_{20c} - p_{02c} > 0$
	V_2 with positive monotonicity	Not available
	Both V_1 and V_2 with positive monotonicity	$p_{12c} - p_{21c} - p_{02c} + p_{01c} > 0$
Table 5C	No monotonicity assumption	Not available
	V_1 with positive monotonicity	Not available
	V_2 with positive monotonicity	$p_{21c} - p_{12c} - p_{02c} - p_{20c} > 0$
	Both V_1 and V_2 with positive monotonicity	$p_{21c} - p_{12c} - p_{20c} + p_{10c} > 0$
Table 5D	No monotonicity assumption	Not available
	V_1 with positive monotonicity	Not available
	V_2 with positive monotonicity	Not available
	Both V_1 and V_2 with positive monotonicity	$p_{11c} - p_{20c} - p_{02c} + p_{00c} > 0$

[a]We let V_1 and V_2 be variables with three levels indicating the genotype at loci A and B, respectively
[b]We let p_{ijc} denote $P[D = 1 | V_1 = i, V_2 = j, C = c]$

"sufficient causes," and each sufficient cause would consist of a minimal set of conditions or "component causes" such that, whenever all the component causes for a particular causal mechanism were present, the mechanism would operate and the outcome would inevitably occur. Such mechanisms or sufficient causes might require the absence or presence of two or more particular factors of interest, X_1 and X_2, along with other possibly unknown factors. Synergism, or causal co-action, would be said to be present between X_1 and X_2 if there were a sufficient cause that required both X_1 and X_2 to operate.

VanderWeele and Robins [23, 28] formalized Rothman's sufficient cause framework and introduced the notion of a sufficient cause interaction to describe a form of mechanistic interaction [29]. A sufficient cause interaction is present when for some individuals, the outcome occurs if both of two exposures are present but not if only one or the other is present. For example, a sufficient cause interaction is present if there are individuals for whom $D_{11}(\omega) = 1$ and $D_{01}(\omega) = D_{10}(\omega) = 0$ in the population; $D_{00}(\omega)$ can be either 1 or 0. Response patterns of this type imply synergism in

Rothman's sufficient-cause framework [23, 28], which can be readily understood by explicating the link between the sufficient-cause model and the counterfactual model [24, 30, 31]. A complete enumeration of 512 risk-status types and 16 response types under two binary causes and a binary outcome is shown by Suzuki et al. [24].

The notion of an epistatic response pattern such as that given in Table 1 is stronger than that of a sufficient cause interaction, because the response pattern in Table 1 requires that $D_{00}(\omega) = 0$ [32]. Indeed, for binary exposures an epistatic interaction in the sense of masking is an even stronger notion of mechanistic interaction than a sufficient cause interaction [29]. If at least one of the two genetic factors has a monotonic effect on the outcome D, then the concepts of an epistatic interaction and a sufficient cause interaction between two factors coincide. If neither of the two factors has a monotonic effect on the outcome, then an epistatic interaction is a stronger condition than a sufficient cause interaction. Statistical tests for sufficient cause interactions have been described elsewhere [23, 28, 33–35], and these statistical tests could also be used for epistatic interactions if at least one of the two factors has a monotonic effect on the outcome.

5 Conclusions and Discussion

By using the counterfactual framework, we have derived conditions that can be tested empirically for detecting whether there are individuals whose phenotype response patterns manifest epistasis in the sense of masking. Only under some very strong assumptions would tests for standard statistical interactions correspond to compositional epistasis. Even without such strong assumptions, however, one can still test whether there are individuals of phenotype response type representing compositional epistasis. The empirical conditions are quite strong, but the conclusions which tests of these conditions allow may be of interest in a wide range of studies. For example, the empirical tests were used to detect possible gene-gene interaction between *HLA-DRB1* and R620W *PTPN22* alleles on anti-CCP-positive rheumatoid arthritis [11]. Although the results were essentially illustrative, there was evidence for a certain form of compositional epistasis, even without any assumption on the monotonicity of the two genetic factors.

The empirical tests require control for a genetic marker, denoted by C, to eliminate confounding by stratification and admixture so that the associations observed between the genes of interest and the outcome at least approximately correspond to the true causal effects of these genes. The tests will only be valid to the extent that this approximation holds. Other genetic or environmental factors could be included in C to attempt to better control for confounding. When C contains multiple factors, more sophisticated statistical techniques may be desirable to allow for multivariate control [35–38].

The goal of most biomedical studies for complex diseases in genetic associations, gene–gene interactions, and gene–environment interactions is to uncover causal relationships [39]. In this regard, epidemiologic methods have evolved as a strategy to uncover the structure of the data and eliminate all non-causative explanations for an observed association. The recent advances in causal inference have provided deeper insight into the nature of etiology [40], and the concepts of mediation and mechanism have been clarified in the sufficient-cause framework [41, 42]. It has also been shown that one can infer operating mediation and mechanism from empirical data [43]. Epidemiologic perspectives have been used to shed light on the underlying mechanisms at the genetic, molecular, and cellular levels [44] (*see* **Note 5**).

Finally, it should be noted that, although compositional epistasis is arguably a more biological notion of epistasis than is statistical epistasis, even compositional epistasis does not necessarily imply "functional epistasis"—molecular interactions of one protein with another [2]. The recent developments of epidemiologic methods have demonstrated that much insight can be gathered from simplified representations of the complex biological reality. However, it is also important to acknowledge the limitations of what can be learned from these methods. It is hoped that the empirical tests for compositional epistasis will be employed in future analyses of genetic data to allow somewhat deeper insight into the presence and nature of gene–gene interactions.

6 Notes

The following are some salient points in this chapter:

1. Under Bateson's original conception, the term "epistasis" is used to describe the situation in which the effect of a genetic factor at one locus is masked by a variant at another locus. Epistasis in the sense of masking has been termed "compositional epistasis."

2. In general, statistical tests for interaction are of limited use in detecting compositional epistasis. However, there are relations between empirical data patterns and compositional epistasis, and these relations can sometimes be exploited to empirically test for certain forms of compositional epistasis, by using alternative non-standard tests for interaction.

3. By using the counterfactual framework, we have derived conditions that can be tested empirically for detecting whether there are individuals whose phenotype response patterns manifest compositional epistasis.

4. Only under some very strong assumptions would tests for standard statistical interactions correspond to compositional

epistasis. Even without such strong assumptions, however, one can still test whether there are individuals of phenotype response type representing compositional epistasis.

5. Epidemiologic perspectives can be used to shed light on underlying mechanisms at the genetic, molecular, and cellular levels.

Appendix Table Potential Phenotype Patterns Under the Assumption of Positive Monotonicity of both V_1 and V_2 on D

Genotype at locus A	Genotype at locus B		
	b/b	b/B	B/B
A			
a/a	0	0	0
a/A	0	0	0
A/A	0	0	1
B			
a/a	0	0	0
a/A	0	0	1
A/A	0	0	1
C			
a/a	0	0	0
a/A	0	0	0
A/A	0	1	1
D			
a/a	0	0	0
a/A	0	1	1
A/A	0	1	1
E			
a/a	0	0	1
a/A	0	0	1
A/A	0	0	1
F			
a/a	0	0	0
a/A	0	0	0
A/A	1	1	1
G			
a/a	0	0	1
a/A	0	1	1
A/A	0	1	1
H			
a/a	0	1	1
a/A	0	1	1
A/A	0	1	1

(continued)

(continued)

Genotype at locus A	Genotype at locus B		
	b/b	*b/B*	*B/B*
I			
a/a	0	0	0
a/A	0	1	1
A/A	1	1	1
J			
a/a	0	0	1
a/A	0	1	1
A/A	1	1	1
K			
a/a	0	1	1
a/A	0	1	1
A/A	1	1	1
L			
a/a	0	0	0
a/A	1	1	1
A/A	1	1	1
M			
a/a	0	0	1
a/A	1	1	1
A/A	1	1	1
N			
a/a	0	1	1
a/A	1	1	1
A/A	1	1	1
O			
a/a	0	0	0
a/A	0	0	1
A/A	0	1	1
P			
a/a	0	0	1
a/A	0	0	1
A/A	0	1	1
Q			
a/a	0	0	0
a/A	0	0	1
A/A	1	1	1
R			
a/a	0	0	1
a/A	0	0	1
A/A	1	1	1

Both loci A and B are epistatic in the phenotype patterns A, B, C, D, J, K, M, N, O, and R
Locus A is epistatic to locus B in the phenotype patterns I and Q
Locus B is epistatic to locus A in the phenotype patterns G and P
Neither locus A nor locus B is epistatic in the phenotype patterns E, F, H, and L

References

1. Bateson W (1909) Mendel's principles of heredity. Cambridge University Press, Cambridge

2. Phillips PC (2008) Epistasis: the essential role of gene interactions in the structure and evolution of genetic systems. Nat Rev Genet 9: 855–867

3. Moore JH, Williams SM (2009) Epistasis and its implications for personal genetics. Am J Hum Genet 85:309–320

4. Cordell HJ (2009) Detecting gene-gene interactions that underlie human diseases. Nat Rev Genet 10:392–404

5. Steen KV (2012) Travelling the world of gene-gene interactions. Brief Bioinform 13:1–19

6. Cordell HJ (2002) Epistasis: what it means, what it doesn't mean, and statistical methods to detect it in humans. Hum Mol Genet 11:2463–2468

7. Moore JH, Williams SM (2005) Traversing the conceptual divide between biological and statistical epistasis: systems biology and a more modern synthesis. BioEssays 27:637–646

8. Cordell HJ, Clayton DG (2005) Genetic association studies. Lancet 366:1121–1131

9. VanderWeele TJ (2010) Epistatic interactions. Stat Appl Genet Mol Biol 9:1. doi:10.2202/1544-6115.1517

10. VanderWeele TJ (2010) Empirical tests for compositional epistasis. Nat Rev Genet 11:166

11. VanderWeele TJ, Laird NM (2011) Tests for compositional epistasis under single interaction-parameter models. Ann Hum Genet 75:146–156

12. VanderWeele TJ, Knol MJ (2011) Remarks on antagonism. Am J Epidemiol 173:1140–1147

13. Little RJ, Rubin DB (2000) Causal effects in clinical and epidemiological studies via potential outcomes: concepts and analytical approaches. Annu Rev Public Health 21:121–145

14. VanderWeele TJ, Hernán MA (2012) Causal effects and natural laws: towards a conceptualization of causal counterfactuals for nonmanipulable exposures, with application to the effects of race and sex. In: Berzuini C, Dawid P, Bernardinelli L (eds) Causality: statistical perspectives and applications. Wiley, Hoboken, NJ, pp 101–113

15. Greenland S, Poole C (1988) Invariants and noninvariants in the concept of interdependent effects. Scand J Work Environ Health 14: 125–129

16. Holland PW (1986) Statistics and causal inference. J Am Stat Assoc 81:945–960

17. Pritchard JK, Rosenberg NA (1999) Use of unlinked genetic markers to detect population stratification in association studies. Am J Hum Genet 65:220–228

18. Pritchard JK, Stephens M, Rosenberg NA, Donnelly P (2000) Association mapping in structured populations. Am J Hum Genet 67: 170–181

19. Price AL, Patterson NJ, Plenge RM, Weinblatt ME, Shadick NA, Reich D (2006) Principal components analysis corrects for stratification in genome-wide association studies. Nat Genet 38:904–909

20. Hoggart CJ, Parra EJ, Shriver MD, Bonilla C, Kittles RA, Clayton DG et al (2003) Control of confounding of genetic associations in stratified populations. Am J Hum Genet 72: 1492–1504

21. Satten GA, Flanders WD, Yang Q (2001) Accounting for unmeasured population substructure in case-control studies of genetic association using a novel latent-class model. Am J Hum Genet 68:466–477

22. Greenland S, Morgenstern H (2001) Confounding in health research. Annu Rev Public Health 22:189–212

23. VanderWeele TJ, Robins JM (2007) The identification of synergism in the sufficient-component-cause framework. Epidemiology 18:329–339

24. Suzuki E, Yamamoto E, Tsuda T (2011) On the link between sufficient-cause model and potential-outcome model. Epidemiology 22: 131–132

25. Rothman KJ (1986) Modern epidemiology. Little Brown and Company, Boston, MA

26. Richardson DB, Kaufman JS (2009) Estimation of the relative excess risk due to interaction and associated confidence bounds. Am J Epidemiol 169:756–760

27. Rothman KJ (1976) Causes. Am J Epidemiol 104:587–592

28. VanderWeele TJ, Robins JM (2008) Empirical and counterfactual conditions for sufficient cause interactions. Biometrika 95:49–61

29. VanderWeele TJ (2012) Invited commentary: assessing mechanistic interaction between coinfecting pathogens for diarrheal disease. Am J Epidemiol 176:396–399

30. Flanders WD (2006) On the relationship of sufficient component cause models with potential outcome (counterfactual) models. Eur J Epidemiol 21:847–853

31. VanderWeele TJ, Hernán MA (2006) From counterfactuals to sufficient component causes and vice versa. Eur J Epidemiol 21:855–858

32. VanderWeele TJ (2011) A word and that to which it once referred: assessing "biologic" interaction. Epidemiology 22:612–613

33. VanderWeele TJ (2010) Sufficient cause interactions for categorical and ordinal exposures with three levels. Biometrika 97:647–659

34. VanderWeele TJ (2009) Sufficient cause interactions and statistical interactions. Epidemiology 20:6–13

35. Vansteelandt S, VanderWeele TJ, Robins JM (2008) Multiply robust inference for statistical interactions. J Am Stat Assoc 103:1693–1704

36. Vansteelandt S, VanderWeele TJ, Robins JM (2012) Semiparametric tests for sufficient cause interaction. J Roy Stat Soc B 74:223–244

37. VanderWeele TJ, Vansteelandt S (2011) A weighting approach to causal effects and additive interaction in case-control studies: marginal structural linear odds models. Am J Epidemiol 174:1197–1203

38. VanderWeele TJ, Vansteelandt S, Robins JM (2010) Marginal structural models for sufficient cause interactions. Am J Epidemiol 171:506–514

39. Vansteelandt S, Lange C (2012) Causation and causal inference for genetic effects. Hum Genet 131:1665–1676

40. Suzuki E, Yamamoto E, Tsuda T (2012) On the relations between excess fraction, attributable fraction, and etiologic fraction. Am J Epidemiol 175:567–575

41. Hafeman DM (2008) A sufficient cause based approach to the assessment of mediation. Eur J Epidemiol 23:711–721

42. VanderWeele TJ (2009) Mediation and mechanism. Eur J Epidemiol 24:217–224

43. Suzuki E, Yamamoto E, Tsuda T (2011) Identification of operating mediation and mechanism in the sufficient-component cause framework. Eur J Epidemiol 26: 347–357

44. VanderWeele TJ, Asomaning K, Tchetgen Tchetgen EJ, Han Y, Spitz MR, Shete S et al (2012) Genetic variants on 15q25.1, smoking, and lung cancer: an assessment of mediation and interaction. Am J Epidemiol 175: 1013–1020

Chapter 12

Identification of Genome-Wide SNP–SNP and SNP–Clinical Boolean Interactions in Age-Related Macular Degeneration

Carlos Riveros, Renato Vimieiro, Elizabeth G. Holliday,
Christopher Oldmeadow, Jie Jin Wang, Paul Mitchell,
John Attia, Rodney J. Scott, and Pablo A. Moscato

Abstract

We propose here a methodology to uncover modularities in the network of SNP–SNP interactions most associated with disease. We start by computing all possible Boolean binary SNP interactions across the whole genome. By constructing a weighted graph of the most relevant interactions and via a combinatorial optimization approach, we find the most highly interconnected SNPs. We show that the method can be easily extended to find SNP/environment interactions. Using a modestly sized GWAS dataset of age-related macular degeneration (AMD), we identify a group of only 19 SNPs, which include those in previously reported regions associated to AMD. We also uncover a larger set of loci pointing to a matrix of key processes and functions that are affected. The proposed integrative methodology extends and overlaps traditional statistical analysis in a natural way. Combinatorial optimization techniques allow us to find the kernel of the most central interactions, complementing current methods of GWAS analysis and also enhancing the search for gene–environment interaction.

Key words Epistasis, Machine learning, Association studies, Combinatorial optimization, GPU-based methods, Gene–environment

---

1 Introduction

It is now recognized that genome-wide association studies (GWAS) have fallen short in finding the inherited components of complex diseases [1]. For many highly heritable traits the associated loci found through GWAS account for a relatively small and sometimes minute fraction of the assumed heritable content. The missing heritability is conjectured to be hidden in (a) rarer variants, difficult to spot with current technologies; (b) a much larger number of variants with smaller effects; (c) interaction between variants; and (d) interaction between variants and environment. Although the adoption of full genome sequencing technologies is assumed to overcome the relative scarcity of proxy markers available today with

Jason H. Moore and Scott M. Williams (eds.), *Epistasis: Methods and Protocols*, Methods in Molecular Biology, vol. 1253, DOI 10.1007/978-1-4939-2155-3_12, © Springer Science+Business Media New York 2015

commercial "whole-genome" scanning platforms (and thus the need for imputation of non-measured SNPs), it will be accompanied by a flood of data with a whole new set of methodological problems requiring new analytical approaches.

Traditional GWAS statistical methods analyze large categorical datasets by testing individual significance of each variable with abstraction of the rest. That is, each SNP is tested for significance independently of the other SNPs being measured, typically through *logistic regression* models [2] or *logic regression* models [3]. In the traditional framework, pairwise and higher order interactions are searched for using multivariate versions of the logistic or logic regression models, for examples *see* [2, 4, 5], or *log-linear* models, for example *see* BOOST [6]. Alternative nonparametric methods to analyze SNP interactions have been proposed, in particular *multifactor dimensionality reduction* (MDR) [7, 8] has been successfully employed in a number of cases.

It must be noted that a complete search for interactions in typical GWAS with any of the mentioned methodologies is a computationally demanding task. Recent analyses are starting to make use of modern computing architectures such as the general purpose *Graphics Processor Units* (GPUs) to obtain remarkable speedups through massive parallel computation. Some of the methods above (logistic and logic regression, MDR) [9, 10] have been translated, and others (log-linear models) [11] adapted to take advantage of these novel platforms with impressive results.

Age-related macular degeneration (AMD) is exemplary of a complex disease that provided excellent results from early GWAS; direct-to-consumer genetic testing companies, *deCODE* and *23andMe*, now regularly report on biomarkers for AMD. The two most conspicuous genes are CFH (1q32, Complement Factor H; Aliases: *AHUS1, ARMD4, ARMS1, CFHL3, FH, FHL1, HF, HF1, HF2, H factor 1, HUS, MGC88246*) [12–14] and ARMS2 (age-related maculopathy susceptibility 2; aliases: *Age-related maculopathy susceptibility protein 2, LOC387715*). The identification of CFH stemmed from results of a genome-wide screen study of 116,204 single-nucleotide polymorphisms (SNPs) in a fairly small study (96 cases and 50 controls) [13]. It is perhaps interesting to remark that a study of this size would likely not be funded today if we consider a priori its power. A larger follow-up study confirmed CFH as a susceptibility locus [12]. Meanwhile, Rivera et al. [15] also identified ARMS2/LOC387715 as the second major susceptibility locus. A number of studies around the world subsequently replicated CFH and ARMS2 as susceptibility loci [16–21], and, as sample sizes grew, additional loci were found. A recent meta-analysis, through imputation of six million SNPs in 2,594 advanced AMD cases and 4,134 controls of European ancestry, has now confirmed 14 genetic loci associated with AMD [21].

However, relying on ever-increasing sample sizes and meta-analyses can be an expensive and blunt exercise to identify loci, and it does not address the possibility that epistatic interactions may be missed. Investigators require new methods to identify loci of interest, and it has been suggested that screening for SNP–SNP interactions may be a way of identifying individual SNPs that are good candidates as well as detecting SNPs that may be missed when focusing on genetic main effects only [22]. It is therefore plausible that checking for SNP/environment interactions, not only SNP–SNP interactions, may also identify potential candidate loci.

AMD is the leading cause of irreversible blindness in the elderly (in developed nations) and the third major cause of blindness in all ages around the world [23]. Epidemiology studies have revealed that environmental factors have definite roles in AMD [24–26]. Cigarette smoking has been shown to be a major contributing factor [27, 28], and many other environmental exposures [29, 30] and clinical conditions were identified as possibly related to AMD [31, 32]. Little work has been done in investigating the role of gene-environment interactions for this disease, and certainly not on a genome-wide scale.

We previously presented a computational approach to analyze GWAS [33] by combining evidence from different methods of GWAS analysis to reach consensus. We also introduced a novel method to explicitly compute all possible SNP×SNP interactions in a reasonable time-frame. Here we describe the interaction scan method in more detail, and, for the first time, present a topological graph analysis of the most significant interactions, complementary to our previous analysis. We expand the analysis of AMD considering SNP×SNP and SNP×clinical/environment interactions. In [33] we focused on regions based on aggregation of a huge number of interactions to produce a rank, a situation most akin to traditional individual SNP analysis; whereas here the analysis focuses on topmost interactions by identification of key graph structures and strongest graph edges.

Our methodology can be briefly summarized as follows: each genotyped SNP is considered as a set of three binary variables, one for each possible state or number of copies of the minor allele (0,1,2). Pairwise interaction between any two binary variables is expressed by their combinations through all possible Boolean functions. Association of the interaction results with case/control status is tested by means of the exact statistic. As we exhaust all possible forms of combining binary interactions, the search for "interesting" interactions effectively becomes model-agnostic.

The process can be regarded as the computation of the complete binary interaction graph between SNPs. Managing and/or storing such graph is unfeasible; we extract from it two (related) types of information: summary SNP information measures (degree or weighted degree centrality) and summary graph information (we retain only a very large number of the most significant interactions—still a minute fraction of the total number of interactions). We use this information to assign individual SNP significance and to search for interaction structure among all SNPs by combinatorial optimization graph analysis tools and ranking and selection of interactions. Centrality results show good agreement, in general, with individual SNP significance assigned by standard statistical methods (e.g., logistic regression), and we refer the reader to our previous paper for analysis of results in the context of AMD.

We extend here our analysis of AMD with the same dataset to uncover the SNPs participating in the set of most significant interactions. Specifically, we look for SNPs belonging to the largest fully connected sets of vertices among the strongest interactions (or *clique*), which is a combinatorial optimization problem. What makes the SNPs in *cliques* interesting is both a high interaction significance and the fact that each of those SNPs is significantly interacting with *all* other SNPs in the structure, *regardless of individual SNP association with disease*. The analysis is also extended to the interactions between SNPs and clinical/environmental variables. Variables are discretized and association is tested as for SNP–SNP interactions.

Full details of the binary Boolean interaction scan and graph analysis methodology, and results obtained from its application to AMD are presented and discussed throughout Subheading 3. Our implementation of the method has been made available at https://github.com/cibm/gwa-clique-central. Additional expanded results and supporting material files are available in the CIBM wiki at https://cibm.newcastle.edu.au/doku.php?id=public:amdchapter:start, hereafter referred to as the *supporting site*. Subheading 2 briefly describe the main characteristics of the AMD dataset used, its demographics and quality control procedures used.

2   Materials

We have applied the binary Boolean interaction scanning technique presented here to a modestly sized GWAS of 315 individuals diagnosed with AMD and 1,014 paired healthy controls from the Blue Mountains Eye Study (BMES) cohort. A total of 541,094 genotyped SNPs remained after quality control and removal of chromosome 23 SNPs to create a sex-agnostic dataset. Details of demographics, genotyping, and quality control are given in our previous publication [33].

2.1   Data Collection: Ethics

The BMES study was conducted according to the recommendations of the Declaration of Helsinki. Ethical approval obtained from the Western Sydney Area Human Ethics Committee and written, informed consent was obtained from all participants.

2.2 Subjects and Genotyping	The Blue Mountains Eye Study (BMES) is a population-based survey of vision and common eye diseases in an urban population-based cohort, initially 49 years or older resident in the Blue Mountains region, west of Sydney, Australia. The sample is of predominantly European Caucasian ancestry; detailed recruitment methods have been extensively described elsewhere [34–36]. We refer the reader to these references and [33] for details on genotyping and quality control. SNPs with call rates <0.95, failing Hardy-Weinberg equilibrium (HWE) at P-value $<10^{-5}$, or with observed minor allele frequency MAF <0.01 were removed. Samples with call rate <0.95 and outliers in principal component analysis were also removed. The final dataset comprised 1,329 samples (315 affected, 1,014 unaffected) with observed genotypes for 554,484 SNPs.
2.3 Clinical Variables	Several environmental and clinical variables were measured for this cohort and screened against the genotyped SNPs for association with AMD, including smoker status, fibrinogen, white blood cell count (WBC), omega-3 intake, total omega-3 consumption, and the combination of fibrinogen and WBC (F*WBC). We refer to [30, 37, 38] for methodology of detection of those variables. Non-binary variables failed to pass the entropy filtering and minimum description length criterion for association with AMD [39], indicating weak or nonexistent direct association with case/control labels. With the exception of smoker status (already in binary form), variables were then discretized by mean, by median, by first and second tertile. Fibrinogen plasma level was also discretized with a clinically supplied threshold value (395 µg/ml), which is somewhat lower than the high tertile value (430). Discretized variables were then tested for binary interaction with SNPs. From the results, the best interaction P-value for the different discretization thresholds was used. The possible concurrent interaction between fibrinogen and white cell count was also tested by generating a pseudo-variable product of the former two, and discretizing as explained.

3 Methods

3.1 SNP–SNP Interaction as a Binary Pattern	Most genotyping platforms measure SNPs as *biallelic* variations. We considered each SNP variable as three different independent binary variables, denoted A_0, A_1, and A_2 for SNP A, where A_i indicates the SNP A being in state i. For a given individual (a *sample*), and ignoring copy number variation issues, only one of these binary variables can be *true*, corresponding to 2, 1, or 0 copies of one of the alleles (usually the "minor" variant) of SNP A. In a dataset, each binary variable is an array of zeros and ones (ones being "true" values and zeros being "false"), one per sample. The binary SNP variables satisfy the obvious condition:

$$A_0 \vee A_1 \vee A_2 = \text{true}$$

for every sample and every SNP A successfully genotyped, where $\vee$ is the *logical or binary OR* operation. This condition renders one of the binary SNP variables dependent on the other two, but we consider each binary variable as independent in the methodology presented here.

Any interaction between two SNPs A and B can be expressed as a binary function operator relating the result value to the two interacting operands. For two operands, there are only 16 different ways to combine binary values to obtain another binary value; we list those cases in Table 1. Two results are *trivial functions*:

Table 1
The 16 Boolean functions of two variables

Inputs					Boolean	Equiv.	Mnemonic
A	0	0	1	1			
B	0	1	0	1			
O_1	0	0	0	0	0	T	
O_2	0	0	0	1	$A \wedge B$	F_1	AND
O_3	0	0	1	0	$A \wedge \neg B$	F_2	ANDN
O_4	0	0	1	1	A	U	
O_5	0	1	0	0	$\neg A \wedge B$	F_3	NAND
O_6	0	1	0	1	B	U	
O_7	0	1	1	0	$A \otimes B$	F_5	XOR
O_8	0	1	1	1	$A \vee B$	$\neg F_4$	
O_9	1	0	0	0	$\neg (A \vee B)$	$\neg A \wedge \neg B$	F_4	NANDN
O_{10}	1	0	0	1	$\neg (A \otimes B)$	$\neg F_5$	
O_{11}	1	0	1	0	$\neg B$	U	
O_{12}	1	0	1	1	$\neg (\neg A \wedge B)$	$A \vee \neg B$	$\neg F_3$	
O_{13}	1	1	0	0	$\neg A$	U	
O_{14}	1	1	0	1	$\neg (A \wedge \neg B)$	$\neg A \vee B$	$\neg F_2$	
O_{15}	1	1	1	0	$\neg (A \wedge B)$	$\neg F_1$	
O_{16}	1	1	1	1	1	T	

On the left side of the table, the outcomes (denoted as O_1 to O_{16}) for the 16 possible functions between two binary inputs is shown. The column Boolean is a possible representation in terms of *logical negation* ($\neg$), AND ($\wedge$), OR ($\vee$) and *Exclusive OR* ($\otimes$). For some functions, another possible alternative representation is shown in the column Equiv.. The functions are either marked as *trivial* (T) (always true or false independent of the variables), *unary* (U) (depend only on one variable), or in terms of five unique binary functions F_1 to F_5

either always *true* (1) or always *false* (0), another four results are *unary functions*: always equal to the A input, or to the B input, or their negated forms. The remaining ten functions can be expressed by five unique Boolean functions, and their negated forms. The representation, in terms of logical disjunctions or conjunctions, is not unique. We have chosen to use the *logical AND* ($\wedge$) function with the different negated combinations of its arguments, plus the *Exclusive OR* ($\otimes$). The set of five Boolean functions exhaust all possible ways of combining two binary variables, modulo negation of the result.

3.2 Significance Measurement

In a case-control study, a variable is said to be "associated with the disease" when the difference of observed frequency of appearance of values in the two populations is above a certain user-defined confidence threshold based on the accumulated probability of observing such difference through chance alone. For binary variables, one very common confidence measure is the P-value, defined as obtaining a chance value, which is at least as extreme as the observed value. Any binary variable can be analyzed with this criterion, in particular, the result of applying a Boolean function to a pair of variables. A 2×2 contingency table can be constructed with the observed counts of samples per class with a value of 1:

	False (0)	True (1)	
Class 0	n_{00}	n_{01}	n_{0+}
Class 1	n_{10}	n_{11}	n_{1+}
	n_{+0}	n_{+1}	N

where n_{00} is sometimes named as the *number of true negatives*, n_{11} as the *number of true positives*, n_{01} as *number of false positives*, n_{10} as *number of false negatives*, n_{0+} and n_{1+} as *total number of samples in class 0 and 1*, respectively, n_{+0} and n_{+1} as *total number of negatives and positives*, respectively ("marginals"), and N as the *total number of samples*.

We chose to use the non-parametric Fisher exact test to evaluate the probability of such observed counts [40, 41]:

$$P\text{-value} = P\left(n_{11} \geq n_{11,\text{obs}} \mid \{n_{i+}\}, \{n_{+j}\}; \theta = 1\right)$$

where

$$P\left(n_{11} = k \mid \{n_{i+}\}, \{n_{+j}\}; \theta\right) = \frac{\binom{n_{1+}}{k}\binom{N - n_{1+}}{n_{+1} - k}\theta^{k}}{\sum_{u}\binom{n_{1+}}{u}\binom{N - n_{1+}}{n_{+1} - k}\theta^{u}}$$

is the conditional probability that $n_{11} = k$ for given marginals and odds ratio θ.

It must be emphasized that this particular measure of statistical significance can be replaced by others (for example, a χ^2 test), and this particular choice has the most impact on the computational cost, as it is evaluated for each combination of variables. For two classes with two outcomes, the Fisher exact test follows a binomial distribution; with a higher number of classes the multinomial distributions become more costly to evaluate. For contingency tables where some of the cells have very small numbers (as might be the case for most minor alleles of SNPs) Fisher exact test is more accurate than χ^2 [41].

The associated test for a 2×2 contingency table is invariant to class relabeling or to outcome relabeling; the latter meaning that the P-value is the same for a binary variable or its negated, which shows that only five Boolean functions are needed to evaluate all possible combinations of two binary variables. Interaction significance is determined by its resultant P-value.

Other useful measures are the *odds ratio* of each result, together with its standard deviation:

$$L = \log_{10}\left(\mathrm{OR}\right) = \log_{10}\left(\frac{n_{00} n_{11}}{n_{01} n_{10}}\right)$$

$$\mathrm{SD}(L) = \sqrt{\frac{1}{n_{00}} + \frac{1}{n_{01}} + \frac{1}{n_{11}} + \frac{1}{n_{10}}},$$

and Pearson's ϕ, or Matthews correlation coefficient M_{CC} [42, 43]:

$$M_{CC} = \frac{n_{00} n_{11} - n_{01} n_{10}}{\sqrt{n_{00} n_{11} n_{10} n_{01}}}$$

To compare the resultant P-value of a binary interaction (A,B) with the P-value of their components A, B, we resort to the ratio

$$P_{\mathrm{corr}} = \frac{P_{(A,B)}}{P_A P_B}$$

which is referred to as the *relative* or *corrected* P-value.

Values for the abovementioned measures are produced by our implementation in addition to the interaction P-value, and are provided in the detailed result listings available at the supporting site.

3.3 Interaction Graph

Each computed Boolean SNP pairwise interaction is an edge in the interaction graph in which each SNP variable is a vertex. Edges have associated weights, representing interaction "strength" or significance. We choose a monotonically increasing function of the (numerical) significance measure of the interaction between these two nodes, higher weights corresponding to more significant inter-actions. If a representation with implicit "geometrical" meaning is desired, the "interaction distance" between a pair of nodes is a monotonically *decreasing* function of the interaction significance. This definition leads to a family of interaction graphs depending on the particular choice of mapping function from significance to weight. As we are ultimately interested in each SNP as a unique entity, a further reduction is obtained assigning the three binary variables associated with each SNP to the same node. The induced graph represents the connectivity of the network of SNPs; if all interactions were equally significant, then the graph would be "flat" and all nodes would have, on average, the same importance.

Our working hypothesis is that the topology of the graph car-ries information. The use of centrality measures to characterize the graph nodes appears as a natural choice. If only interactions above a certain *threshold weight* are considered, the graph becomes more sparse as some edges are pruned. Eventually, for a high enough threshold, no interactions subsist and all nodes become discon-nected. This simple observation leads to the consideration of two immediate options for centrality as measure of node significance:

1. Degree centrality, where the centrality of a node is proportional to the number of interactions with weight greater than the selected threshold.

2. Weighted centrality, where the centrality of a node is propor-tional to the sum of weights of its adjacent edges.

It is evident that the restriction to a monotonous mapping between interaction significance and edge weight makes the first case invariant to the particular mapping function used, as any threshold can also be established as a threshold on significance, and it would not alter the count of interactions greater than the thresh-old. We have restricted to the simple case of proportionality between weights and $-\log_{10}(P)$ to retain ease of interpretation of edge weights and reduce numerical rounding error.

Noteworthy, degree centrality makes use of the aggregated effect of all interactions to infer individual SNP relevance, in con-trast to a statistical analysis in which the relevance is estimated independently of any other variable. If there are confounding or hidden factors their effects will influence all SNPs by the same amount on average, providing background noise against which true associations are searched. In the non-weighted centrality case, this suggests a threshold value such that all SNPs have at least some interactions, or the number of SNPs without interactions is rela-tively very small.

In addition to global measures as the degree centrality, other types of topological graph structures are of interest. Finding the largest subset of nodes such that each node is connected to all other nodes in the subset (a structure called a *clique*) is of particular interest. The search for cliques on an undirected connected graph is a difficult computational problem (more precisely, it belongs to the class of so called NP-hard combinatorial optimization problems [44]). In the interaction graph, cliques constitute the kernel, or the topologically more interconnected part of the network of binary interactions (*see* **Note 1**). The information conveyed by cliques is not necessarily the same as that provided by the node centrality, as some nodes of high centrality may gather small amounts of "evidence" through many interactions of relatively low significance, while looking for cliques among the most significant interactions ensures that each interaction is highly significant.

3.4 Method Implementation

For a GWAS, undertaking the exhaustive evaluation of all pairwise binary SNP Boolean interactions is a daunting task. A typical GWAS has 5×10^5 SNPs, measured for thousands of samples. An order of magnitude of the number of binary interactions evaluated is $nF \times \frac{1}{2} (nV \times nS)^2 = 5 \times (3 \times 5 \times 10^5)^2 \approx 10^{13}$, where nV is the number of variants per SNP, nF is the number of binary functions, and nS is the number of SNPs. Number of SNPs affects the computation time quadratically, while the number of samples affects it linearly. For the dataset analyzed here, a well tuned distributed implementation of the Boolean interaction scan procedure would take more than one month of processing time in a cluster of 128 cores.

We have implemented the procedure in CUDA [45] using General Purpose Graphical Processing Units (GPGPU); reducing computation of results for a GWAS dataset to 2–3 days in a single machine with four GPUs (specifically, a double Quad Core CPU machine with four nVidia C2050 GPUs—448 cores each; this architecture has also a much lower cost than a cluster of computers) (*see* **Note 2**). The most computationally expensive part is the computation of the associated *P*-value for each binary interaction, but for a given threshold it is possible to pre-evaluate if it has to be computed or not, significantly reducing the effective number of *P*-value evaluations (*see* **Note 3**).

Graph analysis, in particular clique determination (which is an NP-hard problem) and clique weight computation, was performed in R [46] using the `igraph` package [47] (*see* **Note 4**). Graphical production was done using a variety of packages: `venny` [48], `circos` [49], and R itself. We have made available a tool and scripts implementing the method at https://github.com/cibm/gwa-clique-central. The tool can handle up to four GPUs simultaneously, and is provided in source code form and as a `linux x64` binary. Sample input data, scripts, and a short description of use are provided as well.

3.5 Numerical Experiments

We have performed a series of numerical experiments to characterize the stability of the method results and the enhanced description of disease the results provide. A full discussion is out of the scope of

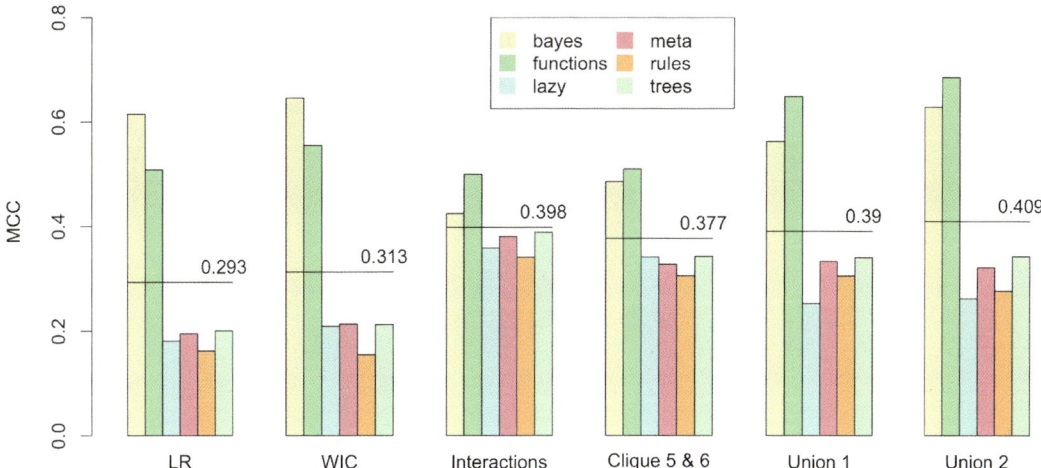

Fig. 1 Comparison of methodology performance. Average Matthews correlation coefficient (M_{cc}) values for each type of classifier on the datasets obtained with the different selection criteria (more details at the *supporting site*). *LR* stands for Logistic Regression, *WIC* for Weighted Interaction Count, *Union 1* is the union of the WIC, Interactions, and Clique 5 and 6 datasets, and *Union 2* is the union of the Logistic Regression, Weighted Interaction Count, and Interactions datasets. We observe the different average performance of classifiers according to type and selection criteria of SNPs and SNP–SNP interaction markers. This indicates that the use of a single classifier in GWAS may not be an optimal strategy and the approach of using ensemble classifier methods warrants further investigation

this work, complete details can be obtained at the supporting site. However, it is worth mentioning the performance of a wide range of classifiers on datasets consisting of the top results from logistic regression, weighted interaction count, cliques interactions, and unions of these sets, where we find that combining results from interactions and individual significance results improves classification accuracy; these results are summarized in Fig. 1.

3.5.1 Comparison with Boost

We have compared our results to the results produced by one of the standard epistasis analysis tools (GBOOST [11], a GPU-enabled implementation of Boost) and our algorithm. Although we observe a small amount (approximately 11 %) of interactions being shared with our relative *P*-value interaction weight results for the topmost 1,000 interactions, this is reduced to approximately 1 % when using the absolute *P*-value results (*see* Fig. 2). We emphasize that cliques are found on the interaction graph of absolute weights; there are no cliques of size >3 on the interaction graph constructed neither with the topmost 100, 200, or 300 K interactions with weights provided by Boost, nor with our relative *P*-value weights. This difference can be understood considering that the GBOOST model is designed to bring up pairs of SNPs for which the interaction term is more significant than the individual SNP association. It appears that using Boost for building a network,

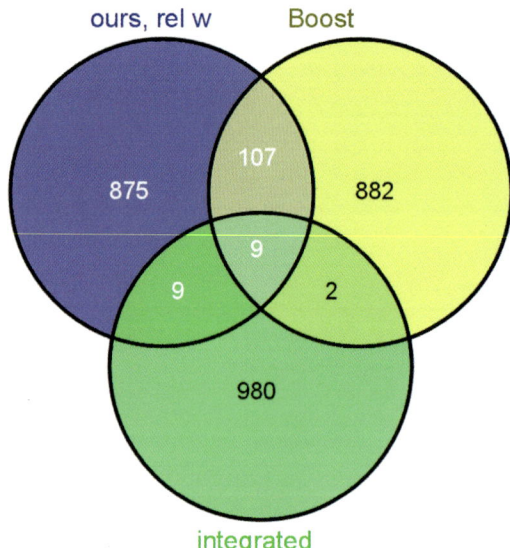

Fig. 2 Comparison of composition of results. Difference in composition of results for Boost, our integrated method, and the interaction network when relative weights are used, for the topmost 1,000 SNP–SNP interactions for each method. As explained in the text, structural information is only found in the integrated results

instead of our proposed method, leaves behind important densely connected groups of SNPs.

3.6 Results: Aggregate Information— Elucidating Individual SNPs

As previously discussed, one way to use the interaction data is to aggregate the information to assign a significance of association value to SNPs. This has been the subject of our previous work [33], where all interactions with P-value $<10^{-5}$ have been aggregated in an interaction graph, the weighted degree centrality for each SNP computed, and the SNPs ranked according to their centrality. This involved aggregating interaction P-values for 137,467,878 interactions.

3.7 Results: Ranking and Selection of Interactions— Elucidating Networks

The 100,000 most significant interactions range in P-value from 2.5×10^{-13} to 6.8×10^{-9}, and involve 54,497 different SNPs. The SNPs and neighboring genes for the 200 most significant interactions are listed in Table 2, with P-values ranging from 5×10^{-13} to 1.5×10^{-12}, while an expanded table with the most significant binary interactions up to a P-value of 10^{-9} is provided in the supporting site. Figure 3 shows the most significant interactions; the dominance of the CFH gene regions is evident; however, a small number of significant interactions between other regions are also noticeable, even among the topmost 1,000. It is encouraging to note that CFH and ARMS2/HTRA1 show up so strongly among the topmost interactions, despite the fact that they do not

Table 2
SNPs in the top 200 Boolean interactions

CHR	Position	SNP	P_{min}	Gene	Neighbor genes
chr1	18490269	rs11260939	11.021	IGSF21(+)	
	26275548	*rs2992068*	10.844		STMN1(−,42) MIR3917(−,43) PAFAH2(−,11)
	40439598	rs4313386	11.504		MFSD2A(+,4)
	68990480	rs12136612	10.936		DEPDC1(−,28)
	69008903	rs1924533	11.542		DEPDC1(−,46)
	108569904	rs12145813	10.876		
	156659768	rs7532302	10.887		ISG20L2(−,33) RRNAD1(+,38) MRPL24(−,47) BCAN(+,30) NES(−,13) CRABP2(−,10)
	168653857	rs7416358	11.147		DPT(−,11)
	168654873	rs10489358	11.191		DPT(−,10)
	168658880	rs1412335	11.503		DPT(−,6)
	196660261	rs10801555	10.859	**CFH(+)**	
	196674917	rs2019724	12.002	**CFH(+)**	
	196675571	rs6428357	11.717	**CFH(+)**	
	196693451	rs379489	11.989	**CFH(+)**	
	196702810	rs1329428	11.354	**CFH(+)**	**CFHR3(+,41)**
	196867233	*rs6685931*	12.321	**CFHR4(+)**	**CFHR2(+,46)**
	196937536	*rs6428379*	12.540		**CFHR4(+,50) CFHR2(+,9) CFHR5(+,9)**
	196941493	*rs6667243*	12.603		**CFHR2(+,13) CFHR5(+,5)**
	196984679	rs1332666	11.001		**CFHR5(+,6) F13B(−,24)**
	201241279	rs6687886	10.957		IGFN1(+,43) PKP1(+,11)
chr2	2021648	rs4853811	12.603	MYT1L(−)	
	2050818	rs4853952	12.508	MYT1L(−)	
	2132794	rs12614597	11.537	MYT1L(−)	
	2160605	rs3748985	11.036	MYT1L(−)	
	11162735	rs1349722	11.623		
	29533782	rs7599783	11.664	ALK(−)	
	38253013	*rs7595482*	12.423	FAM82A1(+)	CYP1B1(−,42)
	80112641	rs986211	11.022	CTNNA2(+)	

(continued)

Table 2
(continued)

CHR	Position	SNP	P_{min}	Gene	Neighbor genes
	107906065	rs758822	10.880		
	107931902	rs6755570	10.905		
	133526973	rs10165182	10.851	NCKAP5(−)	
	155188523	rs707025	11.161	GALNT13(+)	
	175904189	rs1153719	11.664		CHN1(−,34) ATF2(−,35)
chr3	73944436	rs6549566	10.878		
	105807750	rs767150	10.915		
	112771505	rs2399460	11.411		C3orf17(−,33)
	116579556	rs6438355	11.220		
	116579622	rs2373	11.301		
	132006118	rs1505829	10.903		ACPP(+,30)
	176623079	rs1387161	11.268		
	189343424	rs7610017	11.024		TP63(+,6)
chr4	54707431	rs1498833	11.598		
	54710952	rs10517313	10.889		
	74653151	rs16850039	11.123		IL8(+,44) CXCL6(+,49)
	166298451	rs1583645	11.256		SC4MOL(+,34) CPE(+,2) MIR578(+,9)
	174803471	rs10471202	10.969		
	178700699	rs3919788	10.868	LOC285501(+)	
	181440776	rs6843673	11.029		
chr5	6671959	rs39847	11.018		NSUN2(−,38) SRD5A1(+,2) PAPD7(+,43)
	7025264	rs11134199	12.002		
	7025546	rs4702421	11.989		
	7441829	rs7733194	11.163	ADCY2(+)	
	71137731	rs12659543	11.056		
	71180449	rs4704363	11.029		
	71667074	rs4507472	11.237		PTCD2(+,12)
	99944279	rs380360	11.563		FAM174A(+,22)
	99952114	rs492938	11.563		FAM174A(+,30)
	117492859	rs1382723	11.002		

(continued)

Table 2
(continued)

CHR	Position	SNP	P_{min}	Gene	Neighbor genes
	117533130	rs17354084	11.039		
	133434914	rs244660	10.989		TCF7(+,15)
	133455153	rs244693	12.154	TCF7(+)	SKP1(-,37)
	133514972	rs3776841	11.845		MIR3661(+,46) TCF7(+,31) SKP1(-,2) PPP2CA(-,17)
chr6	21037940	rs989969	10.886	CDKAL1(+)	
	23034890	rs17670031	11.656		
	30418662	rs4947284	10.852		HLA-E(+,39)
	75536940	rs17786760	10.929		
	75611839	rs12523819	11.080		
	165797945	rs11754977	11.296	PDE10A(-)	
	166468915	rs300614	11.761		
chr7	49739058	rs10263787	11.251		
	83300296	rs17517322	11.837		SEMA3E(-,22)
	127111144	rs1419409	10.955		
	127138903	rs954008	10.955		
	147784487	rs7809670	11.471	CNTNAP2(+)	
chr8	29697805	rs13270786	10.855		
	125044339	*rs4389918*	11.237	FER1L6(+)	
chr9	16918196	rs1547761	11.249		BNC2(-,47)
	19666510	rs7040903	11.471	SLC24A2(-)	
	79258969	rs4744800	11.241	PRUNE2(-)	
	79310431	rs505034	11.283	PRUNE2(-)	
	79315400	rs620321	11.779	PRUNE2(-)	
	79318677	rs561970	11.271	PRUNE2(-)	
	79318998	rs620552	11.251	PRUNE2(-)	
	79333056	rs890150	11.153	PRUNE2(-)	PCA3(+,46)
	81872880	rs10746589	11.354		
	115245875	rs7048129	11.046		HSDL2(+,11) KIAA1958(+,3)
	115371397	rs2676629	11.050	KIAA1958(+)	
	115375153	rs9886783	11.082	KIAA1958(+)	

(continued)

Table 2
(continued)

CHR	Position	SNP	P_{min}	Gene	Neighbor genes
	117062677	rs1626295	11.369	COL27A1(+)	AKNA(−,34) ORM1(+,23) ORM2(+,29)
	117679474	rs2418326	10.851	TNFSF8(−)	
	117697831	rs2181033	11.807		TNFSF8(−,5)
	119816239	rs10983479	11.829	ASTN2(−)	
	121325521	rs10818187	11.566		
	134503810	rs17148121	10.914	RAPGEF1(−)	
chr10	29542629	rs7909464	11.201		LYZL1(+,35)
	29991935	rs1752582	11.883	SVIL(-)	
	106349202	rs2183164	12.321		
	109395356	rs7914264	10.903		
	124215421	*rs3750847*	11.402	**ARMS2(+)**	**HTRA1(+,6)** PLEKHA1(+,24) MIR3941(+,39)
	124231464	*rs932275*	11.659	**HTRA1(+)**	PLEKHA1(+,40) **ARMS2(+,15)**
	133561098	rs9419649	11.022		FLJ46300(−,44)
chr11	6794304	rs555693	10.892		OR2AG2(−,4) OR2AG1(+,12) OR6A2(−,21)
	100510039	rs1393357	10.920		ARHGAP42(+,48)
chr12	43399128	rs2407183	10.887		
	53023812	rs686339	11.308		KRT1(−,45) KRT72(−,28) KRT73(−,11) KRT2(−,15)
	53027400	rs609307	11.308		KRT1(−,41) KRT72(−,32) KRT73(−,15) KRT2(−,11)
	115575730	*rs1828384*	11.161		
chr13	26035663	rs17082351	11.557	ATP8A2(+)	
	31993213	rs17076005	10.852		
	43270517	rs1169791	10.871		
	46516959	rs1080107	10.874		ZC3H1(3–9)
	46550443	rs4942460	11.131	ZC3H13(−)	
	62730174	rs2121505	11.321		
	74549101	rs4454843	11.593	KLF12(−)	
	109319559	rs17478892	11.040	MYO16(+)	
	112806993	rs9324263	11.019		

(continued)

Table 2
(continued)

CHR	Position	SNP	P_{min}	Gene	Neighbor genes
chr14	29992503	*rs1191555*	12.423		
	31059969	rs7154847	10.940	G2E3(+)	SCFD1(+,32)
	52326085	rs17124656	11.504		GNG2(+,1)
	94329276	rs10484040	11.764		C14orf86(−,42)
chr15	27597927	rs28671569	10.855	GABRG3(+)	
chr16	9708268	rs718396	11.088		
	9734782	rs1548853	10.940		
	13980348	rs8054265	12.111		ERCC4(+,34)
	22768821	rs1895494	11.782		
	58253460	rs2731783	11.253		CSNK2A2(−,22) CCDC113(+,30)
	58254730	rs2731780	11.533		CSNK2A2(−,23) CCDC113(+,29)
	58267472	rs2550333	10.829		CSNK2A2(−,36) CCDC113(+,16) PRSS54(−,46)
chr17	15300309	*rs2323661*	10.914		FAM18B2-CDRT4(−,39) CDRT4(−,39)
	55086808	rs4794679	11.809		COIL(−,48) SCPEP1(+,3) RNF126P1(+,36)
chr18	8899016	rs590770	10.824		
	39883347	rs717127	10.831		
	41980868	rs2925518	11.268		
	50192702	*rs9966348*	12.111	**DCC(+)**	
chr19	22639349	rs1230300	10.920		ZNF98(−,34)
chr20	19616837	rs4814877	10.952	SLC24A3(+)	
	41015030	*rs6030189*	11.254	PTPRT(−)	
	54938207	rs6024830	10.934	C20orf108(+)	CASS4(+,49) AURKA(−,6) CSTF1(+,29)
chr21	34532556	rs2834115	11.590		C21orf54(−,5)
	34800872	rs8130924	11.256	IFNGR2(+)	TMEM50B(−,4)
	34804966	rs1532	11.256	IFNGR2(+)	TMEM50B(−)
	34810007	rs8131980	10.883	TMEM50B(−)	IFNGR2(+,0)
	46023668	rs392686	10.989	TSPEAR(−)	KRTAP10-5(−,23) KRTAP10-6(−,11) KRTAP10-7(+,2) KRTAP10-8(+,8) KRTAP10-9(+,23) KRTAP10-10(+,34) KRTAP10-11(+,43) KRTAP10-3(-,45) KRTAP10-9(+,23) KRTAP10-4(+,29)

(continued)

Table 2
(continued)

CHR	Position	SNP	P_{min}	Gene	Neighbor genes
	46028252	rs8132500	11.081	TSPEAR(-)	KRTAP10-5(−,28) KRTAP10-6(−,16) KRTAP10-7(+,6) KRTAP10-8(+,4) KRTAP10-9(+,19) KRTAP10-10(+,29) KRTAP10-11(+,38) KRTAP12-4(−,46) KRTAP12-3(+,50) KRTAP10-3(−,50) KRTAP10-4(+,33)

For each SNP, the most significant SNP *P*-value is listed (chosen as the minimum *P*-value of all interactions involving that particular SNP). Refer to Fig. 3 for a pictorial view of interactions. SNPs in boldface appear in the union of the largest size cliques (Figure 4). Genes and neighbor genes up to a distance of 50 kbp of SNP are included: "+" or "−" indicates the strand, and the approximate distance of SNP to the 5′ or 3′ exon of gene is given in kbp. *P*-values are given as $\log_{10}(P)$

attain genome-wide significance in our dataset with traditional individual logistic regression analysis (LR *P*-value $= 4.75 \times 10^{-9}$) due to small sample size.

The interaction graph for the 54,497 SNPs highlighted above is sparse, with an average of 1.83 interactions per SNP. At the core is a much smaller subset of SNPs with four or more interactions per SNP. A search for cliques in the interaction graph of the 100,000 most significant interactions yields 21 cliques of size 6, 325 cliques of size 5, and 2,204 of size 4; there are no cliques of size larger than 6. Many of those cliques share some nodes; for instance, the 21 cliques of size 6 form a highly connected structure involving only 19 SNPs; they are schematically depicted in Fig. 4. Table 3 lists the SNPs, neighbor significant SNPs, and genes. It must be stressed that these 19 SNPs are highlighted due to the significance of their mutual pairwise interactions and constitute, in fact, a compact kernel of interaction—a structure sometimes called a *para-clique* (compare 19 vertices to 126, or 93 edges to 315 in the extreme case of 21 *disjoint* cliques of size 6). The complete list of cliques, detailing interactions and some clique characterizing values such as the clique sum of weights, minimum and maximum interaction weights, and clique composition can be obtained at the supporting site.

3.8 Results: Interaction Structure—Elucidating Epistatic and Gene-Environment Effects

A combined compact graphical representation of the interaction plus the individual significance results for the whole genome is presented in Fig. 5. The regions in agreement between the weighted interaction count and logistic regression methods have been marked as well, to better highlight the complementary nature of the information provided by the interactions. In fact, we show in Subheading 3.5 that the quantitative description of disease provided by individual significance is enhanced with the interaction

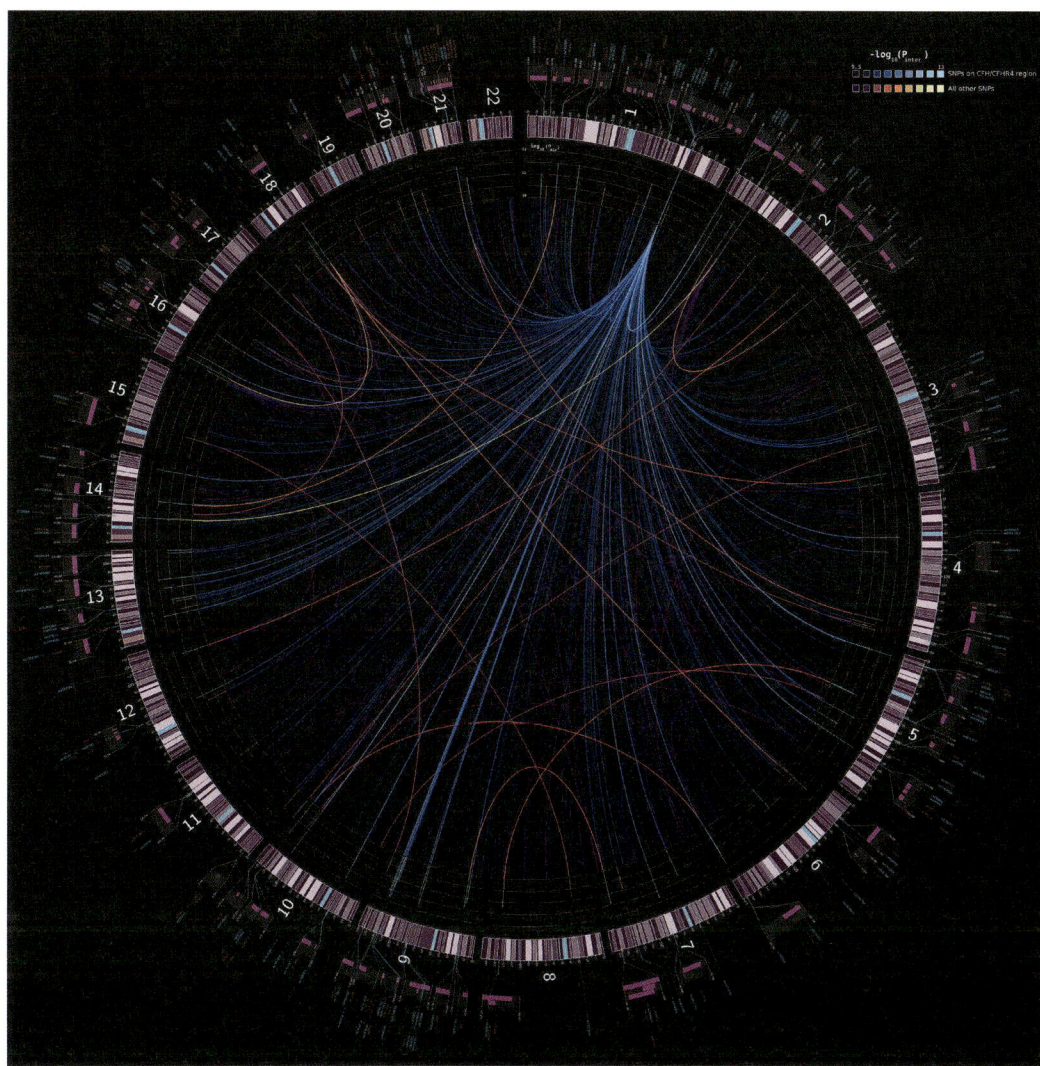

Fig. 3 Most significant Boolean interactions. The figure shows the genome-wide 1,000 most significant Boolean interactions. The genome is represented in chromosome—position order. All interactions involving at least one SNP in the CFH/CFHR4 region are shown in a *blue* scale, and all other interactions in a *red–yellow* scale. Color intensity is proportional to interaction weight, measured as $-\log_{10}(P)$ of the interaction (*see* Subheading 3 for calculation details). Genes and SNPs in Table 2 are shown in the zoomed sectors. The *inner bar plot* shows the value of the most significant interaction in which a particular SNP is involved (the values are also given in Table 2)

description. This is not a surprising result, as it is expected that some additional insight is contained in pairwise and higher order interactions.

Some of the SNP–SNP interactions noted above (Fig. 5) appear strong because of individual SNP main effects, e.g., CFH. However, others are showing combined effects that are orders of magnitude

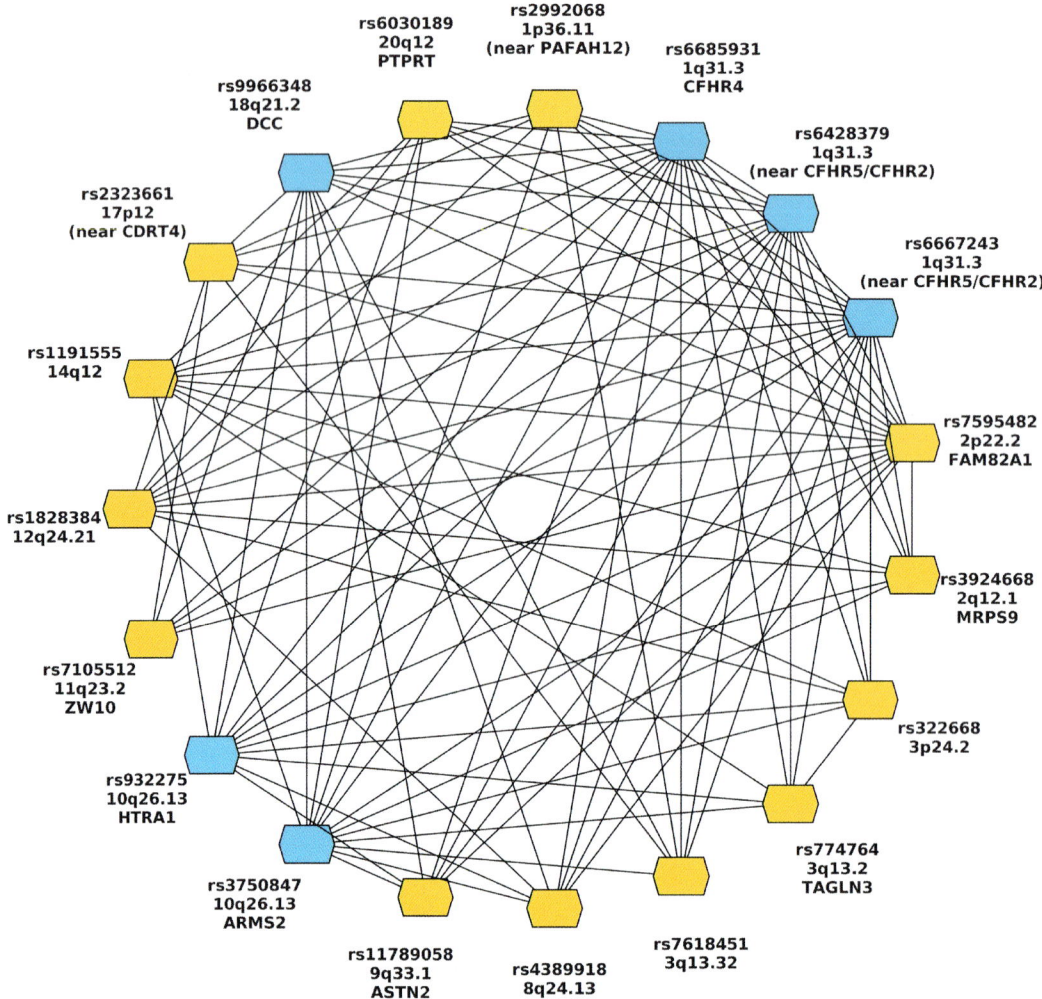

Fig. 4 The union of the largest size cliques from the 100,000 most significant interactions. It is interesting to remark that this subgraph contains the major regions previously associated with AMD (CFHR4/CFHR5, HTRA1/ARMS2) and at least one other gene associated in at least one study (Netrin-1 receptor DCC) [16]; these are shown in *blue*. There appears to be a correlation between the degree of the node and the previous validation of these regions in other studies; *rs6685931* (CFHR4) is connected to all other nodes (*rs6685931* is the SNP with the strongest individual association for this cohort [33]), *rs6428379* and *rs6667243* (near CFHR5/CFHR2) are connected to all but one, and the SNPs in ARMS2, HTRA1, and Netrin-1 receptor DCC have 13, 12, and 12 neighbors, respectively

greater than what could be expected on the basis of individual main effects, i.e., "purely epistatic" effects. The list of most significant interactions with P-values corrected by individual SNP significance (relative P-values), up to a P-value of 10^{-7}, is available at the supporting site.

To gain insight into the information being extracted by the interaction method, an enrichment analysis was performed with

Table 3
Significant regions from the union of all cliques of size 6

CHR	Position	SNP	P_6	P_{LR}	P_{WIC}	Gene	Neighbor genes
chr1	26275548	*rs2992068*	10.84	3.02	3.39		STMN1(–,42) MIR3917(–,43) PAFAH2(–,11)
	196619502	rs505102		5.64	4.31		KCNT2(–,42) **CFH(+,2)**
	196623337	rs7524776		3.25	2.47	CFH(+)	KCNT2(–,46)
	196642233	rs800292		5.86	4.35	CFH(+)	
	196660261	rs10801555		7.67	4.79	CFH(+)	
	196670839	rs10754199		7.50	4.59	CFH(+)	
	196674917	rs2019724		8.28	5.49	CFH(+)	
	196675571	rs6428357		8.13	4.98	CFH(+)	
	196693451	rs379489		8.32	6.03	CFH(+)	
	196702525	rs2284664		4.37	3.03	CFH(+)	**CFHR3(+,41)**
	196702810	rs1329428		7.99	5.24	CFH(+)	**CFHR3(+,41)**
	196823300	rs10922144		3.68	2.51		**CFHR1(+,22) CFHR4(+,34)**
	196838060	rs6657442		3.72	2.50		**CFHR1(+,37) CFHR4(+,19)**
	196867233	*rs6685931*	11.66	9.67	6.67	CFHR4(+)	**CFHR2(+,46)**
	196897088	rs4915318		4.40	3.07		**CFHR4(+,9) CFHR2(+,16) CFHR5(+,50)**
	196937536	*rs6428379*	10.21	8.30	5.02		**CFHR4(+,50) CFHR2(+,9) CFHR5(+,9)**
	196941493	*rs6667243*	10.84	7.72	4.89		**CFHR2(+,13) CFHR5(+,5)**
	196982154	rs7555070		3.51	3.18		**CFHR5(+,3)** F13B(–,26)
	196984679	rs1332666		6.71	4.55		**CFHR5(+,6)** F13B(–,24)
chr2	38245716	rs4670230		1.21	3.88	FAM82A1(+)	CYP1B1(–,49)
	38253013	*rs7595482*	12.42	1.46	4.83	FAM82A1(+)	CYP1B1(–,42)
	105648817	rs2376275		2.61	3.76		MRPS9(+,6)
	105656718	rs13419511		2.73	3.78	MRPS9(+)	
	105699038	*rs3924668*	10.61	2.69	3.71	MRPS9(+)	
chr3	25343615	*rs322668*	9.60	3.02	4.52		
	25349414	rs9809773		1.69	3.06		
	25349434	rs322694		2.34	3.42		
	111719493	*rs774764*	8.85	2.66	3.49	TAGLN3(+)	PHLDB2(+,24) ABHD10(+,7) TMPRSS7(+,39)

(continued)

**Table 3
(continued)**

CHR	Position	SNP	P_6	P_{LR}	P_{WIC}	Gene	Neighbor genes
	118198045	rs9849379		3.14	2.59		
	118202086	*rs7618451*	9.15	4.22	4.05		
chr8	125044339	*rs4389918*	9.39	3.00	4.09	FER1L6(+)	
chr9	119816239	rs10983479		2.51	3.58	ASTN2(−)	
	119826169	*rs11789058*	9.57	2.96	2.73	ASTN2(−)	
chr10	124215421	*rs3750847*	11.40	5.78	4.68	**ARMS2(+)**	PLEKHA1(+,24) **MIR3941(+,39) HTRA1(+,6)**
	124231464	*rs932275*	11.66	5.45	4.62	**HTRA1(+)**	PLEKHA1(+,40) **ARMS2(+,15)**
	124240785	rs2284666		2.53	3.07	**HTRA1(+)**	PLEKHA1(+,49) **ARMS2(+,24)**
chr11	113590575	rs7932415		3.85	4.01		TMPRSS5(−,14) ZW10(−,13)
	113619339	*rs7105512*	9.27	3.76	3.94	ZW10(−)	TMPRSS5(−,42) CLDN25(+,31) USP28(−,49)
	113620632	rs10891592		3.69	3.83	ZW10(−)	TMPRSS5(−,44) CLDN25(+,30) USP28(−,48)
	113679119	rs2465647		3.54	3.59	USP28(−)	ZW10(−,35) CLDN25(+,28)
	113696653	rs7111825		3.68	3.69	USP28(−)	CLDN25(+,45)
	113712526	rs7112957		3.39	3.47	USP28(−)	
	113753516	rs2097078		3.42	3.30		USP28(−,7) HTR3B(+,22)
chr12	115570046	rs35426		2.55	3.35		
	115575730	*rs1828384*	10.70	4.15	4.49		
chr14	29992503	*rs1191555*	12.42	4.25	4.30		
chr17	15300309	*rs2323661*	9.97	0.22	4.73		CDRT4(−,39) FAM18B2-CDRT4(−,39)
chr18	50192702	*rs9966348*	9.43	3.46	5.13	**DCC(+)**	
chr20	41015030	*rs6030189*	11.25	3.59	4.43	PTPRT(−)	
	41017932	rs2425483		2.56	3.27	PTPRT(−)	

The 19 SNPs found in the 21 cliques of size 6 among the best 100,000 binary interactions (SNP id shown in ***boldface italics***). A neighborhood of up to 100 kbp to each side of these SNPs was explored in the weighted interaction count and logistic regression results, for other SNPs with high significance. All SNPs with logistic or weighted interaction count P-value $<10^{-3}$ were selected (all SNPs but one are among the top 500 ranked in weighted interaction count). Genes at the SNP position or within 50 kbp are indicated: "+" or "−" indicates the strand, and the approximate distance of SNP to the 5′ or 3′ exon of gene is given in kbp. Genes in regions previously associated with AMD in other GWAS are indicated in *boldface*. P_6 denotes the best interaction P-value for the SNP in the cliques of size 6; P_{LR}, P_{WIC} are the logistic regression and weighted interaction count P-values, respectively. All P-values as $-\log_{10}(P)$

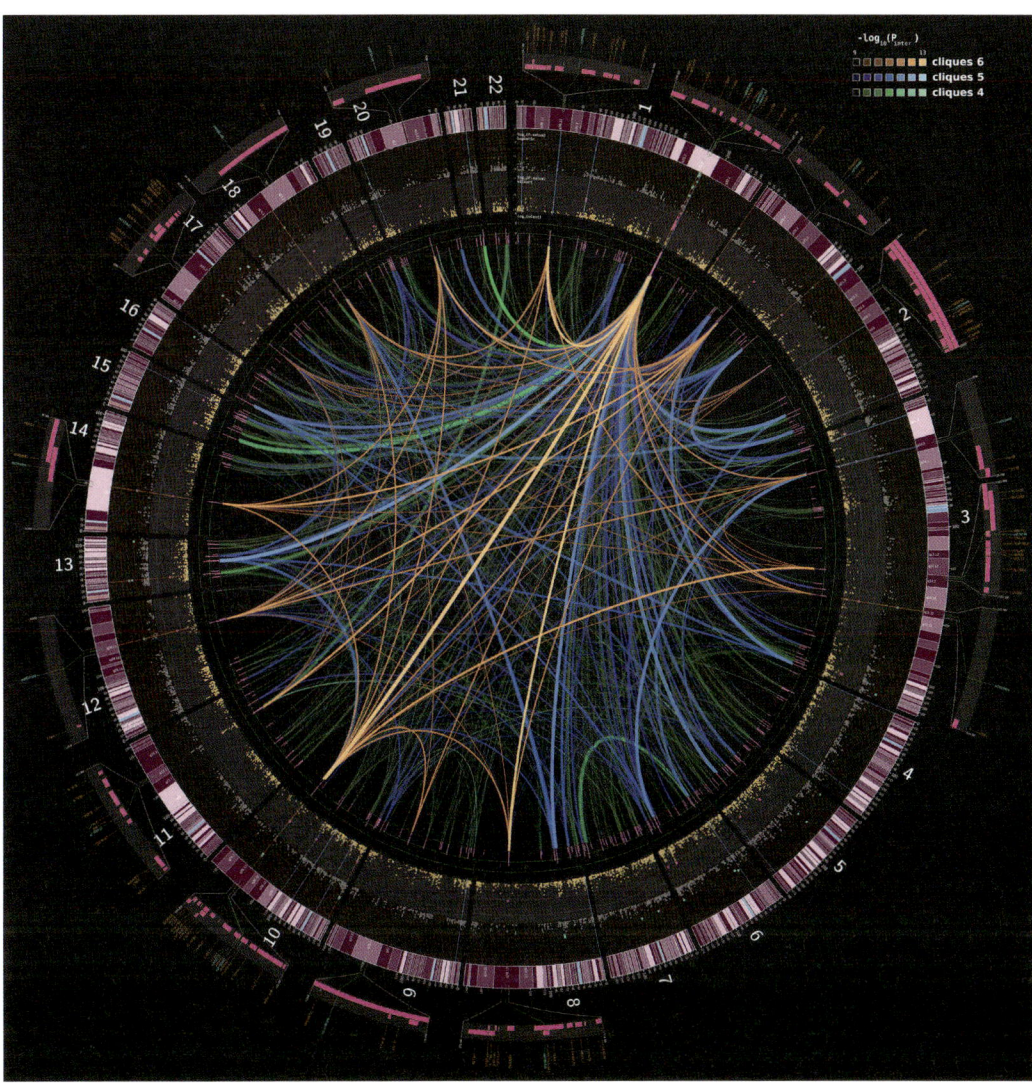

Fig. 5 Circular plot depicting main interactions. The plot shows on the sex-agnostic ideogram the interactions between the SNPs in the cliques of size 6 (*orange lines*), 5 (*blue lines*) and 4 (*green lines*); *see* text for details. The genome is represented in chromosome—position order. Line width and color intensity are proportional to interaction significance as measured by interaction *P*-value, from a *P*-value of 10^{-8} for the *thin lines* to 10^{-12} for the *thick* ones. The $-\log_{10}(P)$ of weighted degree centrality for each SNP in cliques of size 4, 5, and 6 is shown in the innermost circular band with *thin purple bars*, scale ranges from 10 to 10^4. In addition, the $-\log_{10}(P)$ of logistic regression analysis (*light orange dots*) and weighted interaction count analysis (*grey dots*) are shown; *P*-value scale ranges from 10^{-3} to 10^{-9} and values better than 10^{-5} have been enhanced as *colored square dots*. *Thin orange lines* crossing the Manhattan plots of logistic and weighted interaction count mark regions of at most 100 kbp in agreement between the above mentioned individual SNP significance methodologies. For each of the 16 regions comprising the 19 SNPs in the size 6 cliques (*see* Fig. 4), a *zoomed-in circular arc* shows the positions of SNPs and near genes

ToppGene [50] on the results from the topmost interactions and the cliques (enrichment analysis results available at the supporting site). While, as expected, results from the genes in the cliques of size 6 are clearly related to AMD, a different subjacent set of themes emerges from genes in Table 2 and becomes highly significant for the expanded lists of topmost interactions available at the supporting site; actin-binding proteins, genes related to cell adhesion, neurogenesis, axon guidance, Wnt, and cadherin signaling pathways appear strongly overrepresented.

In addition to SNP–SNP interactions, we also look at the pairwise gene–environment interactions. Figure 6 shows the main results in this respect, and the list of SNP×clinical/environment interactions can be obtained from the supporting site. We proceed to highlight the results from the regions in both Tables 2 and 3, discussing its possible relevance to AMD.

3.9 Discussion

We have described a method based on a genome-wide screen of SNP–SNP and SNP×environment Boolean function interactions and a combinatorial optimization search, tractable in a reasonable time. Our implementation in a multi-GPU processor machine takes 3–4 days to complete, compared to several months in a CPU-only cluster. In contrast to standard statistical approaches used to analyze GWAS (e.g., PLINK/BOOST) (1) our integrated method does not assume a particular model of interaction and explores them all and (2) the results are obtained from the totality of interactions through graph analysis and combinatorial optimization. This approach aims at capturing structure and patterns of co-occurrence across a larger number of explanatory variables. The data from our results can be used in three ways: (a) aggregated to obtain **individual** SNP significance; (b) ranking and selecting largest cliques among topmost interactions to highlight **networks** of SNPs frequently represented in disease; and (c) searching for interaction structure to reveal strongly **epistatic** or gene–environment effects not readily seen in individual SNP-based analyses. In addition, we have also shown how the method can incorporate environmental/clinical variables in the analysis (*see* **Note 5**).

3.9.1 CFH and HTRA1

All our interaction scan results, either as weighted interaction count, most significant interactions, or significant interactions in cliques, contain the already reported results from the CFH/CFHR4 (1q31.3, complement factor H/complement factor H-related 4) and ARMS2/HTRA1 (10q26.13, age-related maculopathy susceptibility 2/HtrA serine peptidase 1) regions. Despite the small size of our dataset and the fact that these loci do not reach genome-wide significance, our method consistently identifies these loci and confirms that looking for interactions may be a useful way of identifying true signals from noise in GWAS studies. Indeed, these methods may be used as alternatives to (or in conjunction with) "brute-force"

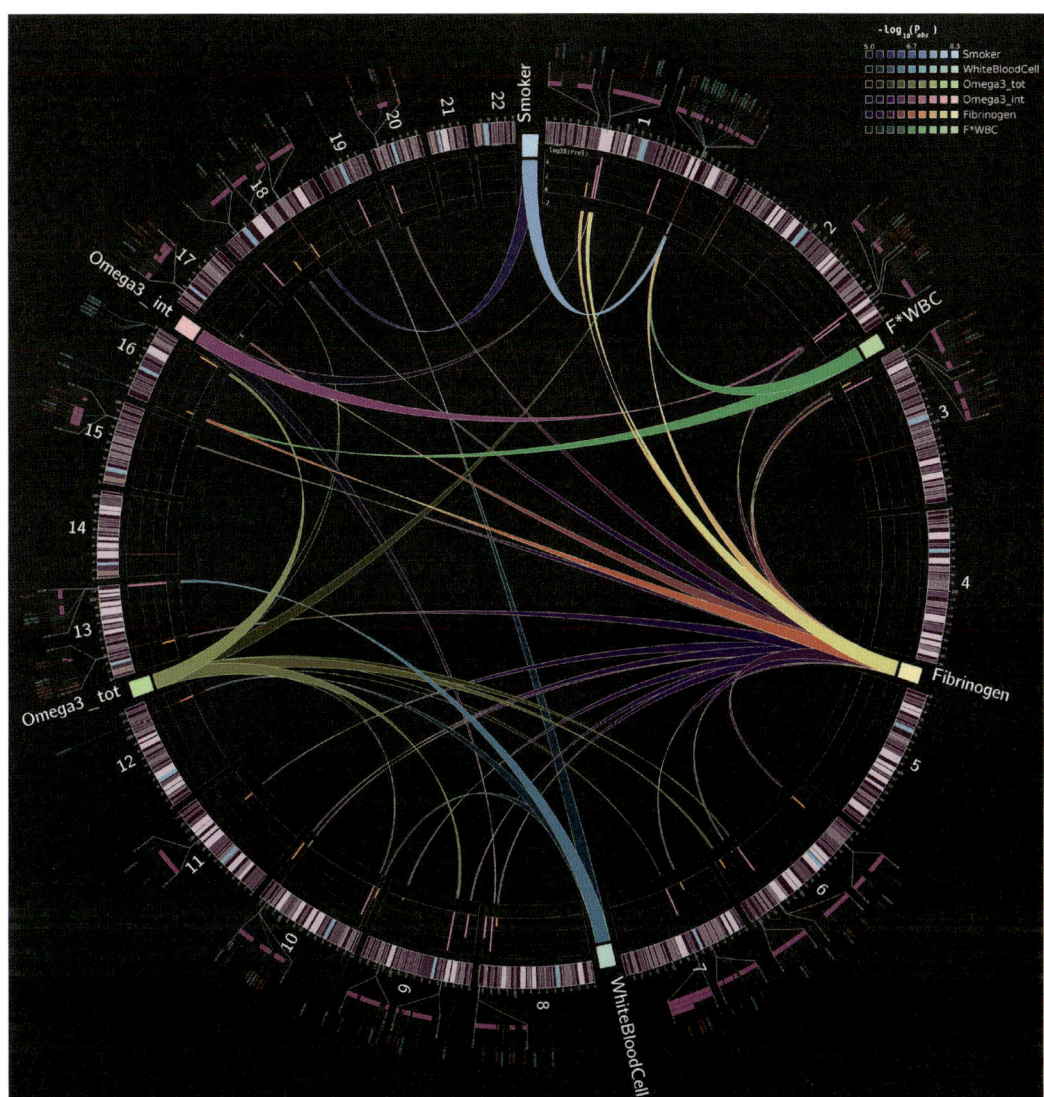

Fig. 6 Boolean Interactions between SNPs and clinical variables. The plot shows the significant interactions between six clinical variables and SNPs across the whole genome. Clinical variables tested are smoker status (*Smoker*), blood fibrinogen protein dosage (*Fibrinogen*), omega-3 intake (*Omega3_in*) and total omega-3 (*Omega3_tot*), white-cell blood count (*WBC*), and the product of WBC and Fibrinogen (*F*WBC*). The clinical variables have been discretized for the interaction analysis. Intensity of color and width of ribbons at the SNP end are proportional to $-\log_{10}(P_{abs})$. The *inner histogram* shows the relative significance once the SNP and

variable significance have been factored, in *purple* if $\log_{10}\left(\dfrac{P_{inter}}{P_{SNP}P_{var}}\right) > 3$ and in *light blue* if <0, *orange*

otherwise. *Superimposed thin orange lines* indicate regions of at most 100 kbp where logistic regression and weighted interaction count agree. SNPs and neighbor genes have been annotated on the zoomed regions. For each SNP—clinical variable pair, the best binary interaction is shown; *see* Subheadings 2.3 and 3 for details

pooling or very large sample sizes. As we mentioned in the introduction, there are several studies that linked CFH with AMD in a variety of different populations [15, 21, 51–78].

CFH is constitutively expressed by retinal pigment epithelial cells (RPE) [79] and regulated by the proinflammatory cytokines TNF-alpha and IL-6 [80].

3.9.2 New Loci

It is interesting to note that our method also reveals new loci that have not been detected even in the largest meta-analysis to date [81]. All these loci are plausible candidates based on their biology, and we list the most consistent hits and their known functions in Table 4. The results suggest that a small number of key molecular processes are being affected and acting as co-adjuvants to the observed main effects.

3.9.3 Netrin Receptors

As an example, we focus on one of our significant results. We find multiple significant interactions for four SNPs (*rs9966348, rs3862681, rs1143768, rs12953529*) in the Netrin-1 receptor gene DCC (18q21.2, deleted in colorectal carcinoma), and several other SNPs near the 5′ exon. The name "Netrin," which labels a family of proteins, has been derived from the Sanskrit "Netr," meaning "guide" [82]. Netrin-1 and its receptor DCC have roles in axonal guidance [83], neuronal migration and retinal development [84], steering of axons out of the retina [85], guidance of commissural neurons toward the floor plate [86], maintenance of axo-oligodendroglial paranodal junctions [87, 88], neurite growth, and retinal ganglion cell axon pathfinding in several species [89–91]. Together, Netrin-1 and its receptor seem to have a balancing role in apoptosis, as the Netrin-1 receptor DCC is blocked by the presence of Netrin-1. Recently, Shi et al. have shown that retinas in Netrin-1 receptor DCC (–/–) mice exhibit reduced ganglion cell layers [92], and Xu et al. and Tian et al. have shown its relation to retinal neovascularization [93, 94].

Indeed, we know of at least one previous study in which the Netrin-1 receptor DCC was related to AMD. Edwards et al. identified sixteen DCC polymorphisms (*see* Supplementary Table 3 of [16]), but these findings failed to replicate in a different cohort on six intronic Netrin-1 receptor DCC SNPs with allelic test P-values $<10^{-4}$. Interestingly, their replication of a coding C3 polymorphism (and concurrent replication of C3, CFH, and C2/BF in other studies) led them to share the view that the alternative complement pathway is implicated in AMD. In fact, the C-terminal domains of netrins and complement proteins C3, C4, and C5 are homologous with the N-terminal domains of tissue inhibitors of metalloproteases (TIMPs) [95]. Associations of polymorphisms for CFH, CFI, C2, C3, and TIMP3 have been reported in [51, 96], and TIMP3 is highly expressed in the human retinal pigment epithelium [97]. Together with other proteins that share this module, their common

Table 4
Significant loci identified by our method

SNPs	Cytoband	Gene	Function	Role
rs2992068	1p36.11	PAFAH2 (platelet-activating factor acetylhydrolase 2, 40 kDa)	This enzyme hydrolyzes oxidatively fragmented fatty acyl chains attached to phospholipids [105]	A similar enzyme, PLA2G7 (phospholipase A2, group VII; PAF-AH, Lp-PLA2) is implicated in dominant form of macular dystrophy (Benign Concentric Annular Macular Dystrophy [BCMAD]) [106, 105]
		STMN1 (stathmin 1)	Microtubule dynamics regulator	Neuroprotective [107, 108], Associated with retinal protection and regeneration [109–113]
rs7599783	2p23	ALK (anaplastic lymphoma receptor tyrosine kinase)	Receptor for neurotrophic growth factor pleiotrophin (PTN) [114, 115]	Angiogenesis in malignant cancers [116], in particular associated to VEGF, (vascular endothelial growth factor) and brain injury [117]
rs4670230 rs7595482	2p22.2	FAM82A1 (family w/ sequence similarity 82, member A1)	Regulator of microtubule. Antisense of CYP1B1	Glaucoma [118, 119], optic neuropathies of mitochondrial origin [120], pro-angiogenic in retinal epithelial cell under hyperoxic stress [121]
rs986211	2p12	CTNNA2 (catenin (cadherin-associated protein), alpha 2)	Cadherins, proto-cadherins, catenins and other proteins in the cadherin and Wnt signaling pathway are key for cell adhesion, axonogenesis, and synapse organization	We find several SNPs in genes of the alpha and delta catenin gene families in significant interactions. CTNNA2 is the most prominent
rs2376275 rs13419511 rs3924668	2q12.1	MRPS9 (mitochondrial ribosomal protein S9)	Mitochondrial DNA damage detection and repair	Macular-specific mitochondrial DNA damage and mutations [120, 122–124]
rs774764	3q13.2	TAGLN3 (Transgelin 3/ Neuronal protein22)	Synapse protein binds to membranes containing cholesterol [125–127]	Suggested as part of a neuronal morphology complex which associates with cytoskeletal elements [128, 129]
rs9849379 rs7618451	3q13.31			*Interacting with regions in 10q26.13, 17p12, 18q21.2, 1p36.11 and 1q31.2. Another AMD GWAS has found a weak association for rs12637095, at less than 10kbp [130]*

(continued)

Table 4
(continued)

SNPs	Cytoband	Gene	Function	Role
rs7809670 (+others)	7q35	CNTNAP2 (contactin-associated protein-like 2)		*Interacting with SLC24A2 (9p22-p13).* GWAS associate CNTNAP2 to secondary locus for pseudoexfoliation syndrome/glaucoma [131, 132]. We also observe several interactions for other members of the neurexin family (CNTNAP5, CNTNAP4, NRXN1, NRXN3)
rs7040903	9p22-p13	SLC24A2 (solute carrier family 24 (sodium/potassium/calcium exchanger), member 2)	Retinal cone Na-Ca+K exchanger	*Interacting with CNTNAP2 (7q35).* Variations in SLC24A2 were investigated in relation with retinal diseases (including AMD), but no conclusive evidence was found [133]. We also observe several interactions for other proteins in the calcium ion channel SLC family 24 (SLC24A3, SLC24A6)
rs11789058 rs10983479 (+others)	9q33.1	ASTN2 (Astrotactin 2)	Contains fibronectin III, immunoglobulin-like fold and EGF 3 domains	Involved in development of the laminar architecture of cortical regions of the mammalian brain [134]
rs7932415 rs7105512 rs10891592	11q23.2	ZW10 (kinetochore associated, homolog (Drosophila))	Regulator of cytoskeleton-associated motor protein dynein. Subcellular transport	Also involved with neuron and growth-cone migration [135]
(+others)		CLDN25 (Claudin-25)	Cell junction. Cell adhesion	Four other Claudin family members appear in topmost interactions (CLDN1, CLDN10, CLDN14, CLDN17)
rs8054265	16p13.12	ERCC4 (excision repair cross-complementing rodent repair deficiency, complementation group 4)	DNA repair	*Interacting w/Netrin-1 receptor DCC (18q21.2).* A synergic effect between CFH and another excision repair gene, ERCC6, has been reported as associated to AMD [136]
rs2731783 rs2731780 rs2550333	16q21	CSNK2A2 (casein kinase 2, alpha prime polypeptide)	Cadherin and Wnt signaling pathways	Regulates neovascularization and cell shape in retinal astrocytes [137, 138]

| rs9966348 | 18q21.2 | Netrin-1 receptor DCC (deleted in colorectal carcinoma) | Netrin receptor. Axonal guidance [83], neuronal migration and retinal development [84], steering of axons out of the retina [85], guidance of commissural neurons toward the floor plate [86], maintenance of axo-oligodendroglial paranodal junctions [87, 88], retinal neovascularization [93, 94], neurite growth and in retinal ganglion cell axon pathfinding [89, 90] | Associated to AMD [16]. Netrin-1 receptor DCC (−/−) mice retina exhibit reduced ganglion cell layers [92]. We observe significant interactions w/CFHR4 for three Netrins (NTN1, NTNG1, NTN4) and Netrin-1 receptors (UNC5B, UNC5C, UNC5D) and Netrin-G1 receptor LRRC4C |
| rs6030189 rs2425483 | 20q12 | PTPRT (protein tyrosine phosphatase, receptor type, T) | Exclusively expressed in CNS. Developmental neurite outgrowth [139] | Related to cell adhesion mechanisms in cancer [140, 141]. Regulation of synapse formation through interaction w/cell adhesion molecules [142] |

The most significant regions and associated SNPs observed in interactions in Tables 2 and 3 and expanded list available at the supporting site are provided with their plausible functional relation to AMD. For some of the regions, a notice has been added indicating whenever we observed significant interactions for other members of the same protein family, or its receptors

functions seem to be axon guidance, regulation of Wnt signaling, and regulation of metalloproteases [95, 98], which, in turn, have roles in light-induced phototoxic processes [99].

Six SNPs in three members of the Netrin family (NTN1, NTNG1, and NTN4) exhibit significant interactions with the CFHR4 region. In addition to Netrin-1 receptor DCC, we also find significant interactions of SNPs in three other Netrin-1 receptors (UNC5B, UNC5C, and UNC5D), and the Netrin-G1 receptor LRRC4C (leucine-rich repeat containing 4C) interacts with CFHR4.

3.9.4 Gene–Environment

Our method also highlights gene–environment interactions that are not detectable through usual individual SNP-based analyses. Again, traditional meta-analytic methods are not prepared to search interactions between loci and environment exhibiting a purely epistatic effect. These results are summarized in Table 5.

The strongest interaction found is between plasma fibrinogen level and SNP *rs7529530* in the ST6GALNAC3 neural gene (1p31.1, ST6 (alpha-*N*-acetyl-neura-minyl-2,3-beta-galactosyl-1,3)-*N*-acetylgalactosaminide alpha-2,6-sialyltransferase 3). The interaction P-value is 4×10^{-9}, while the relative P-value is 4.4×10^{-7}. There is another strong interaction with a neighbor SNP, *rs12742405*, with a relative P-value of 1.4×10^{-6}. The product of ST6GALNAC3 catalyzes sialic acid transfer to carbohydrates on glycoproteins and glycolipids [81] and has been linked to subcutaneous fat thickness [81].

The interaction between fibrinogen and *rs363043*, a SNP in the SNAP25 gene (20p12, synaptosomal-associated protein, 25 kDa) is also more than three orders of magnitude more significant than the SNP or fibrinogen itself (interaction relative P-value 3.7×10^{-5}). SNAP25 is predominantly expressed in neural and neuroendocrine cells, and it is a key protein in replenishment of the light sensing machinery in the retina. It is involved in protein trafficking, synaptic vesicle exocytosis, neuro-peptide release, axon outgrowth, and photoreceptor differentiation, and is present in retinal horizontal cells [100–104].

4   Notes

The important facts about the methodology are summarized here:

1. The identification of the "kernel" of highly interconnected SNPs from the exhaustive scan of SNP–SNP Boolean interactions allows us to find the most individually significant SNPs associated with disease and a richer set of highly relevant epistatic and gene-environment interactions, even in modestly sized AMD GWAS. These results point to a handful of functional groups and affected processes, which are new or supported by less confirmed evidence.

Table 5
Significant gene–environment loci identified by our method

Variable	SNP	Cytoband	Function	Role
Fibrinogen	rs7529530 rs12742405	1p31.1	ST6GALNAC3 (ST6 (alpha-N-acetyl-neuraminyl-2,3-beta-galactosyl-1,3)-*N-acetylgalactosaminide* alpha-2,6-sialyltransferase 3)	Transfer of sialic acids to carbohydrate groups on glycoproteins and glycolipids [81]
	rs363043	20p12	SNAP25 (synaptosomal-associated protein, 25 kDa)	Retinal light-sensing machinery replenishment. Protein trafficking, synaptic vesicle exocytosis, neuro-peptide release, axon outgrowth, and photoreceptor differentiation; present in retinal horizontal cells [100–104]
Omega-3	rs10200650 rs7355261	2q23	CRYG (gamma-crystallin gene-cluster)	Major proteins of eye lens. Mutations related to cataract formation [143–145]
	rs3748391 rs12934213 rs7193979 rs2059883 rs1864145	16q24.2		Region weakly associated with AMD [130], study also found association of AMD with a LIPC variant (hepatic lipase)
Smoker	rs1143768	18q21.2	Netrin-1 receptor DCC (deleted in colorectal carcinoma)	DCC associated with AMD. See in-text remarks about DCC

The most significant regions and associated SNPs observed in interactions between clinical/environment variables and SNPs (the full list of topmost SNP-environment interactions is available at the supporting site). Regions are listed with their plausible functional relation to AMD

2. The use of GPUs for computation drastically reduces the computation time to a reasonable interval, even when significance is measured through the exact Fisher statistic. Our implementation can make use of multiple GPUs present in the machine for increased performance, however this is not a requirement and use of inexpensive hardware is possible.

3. Further gains can be expected when using other significance measures, such as those based on Information Entropy, as they are computationally simpler to evaluate.

4. While computing the maximum cliques in a graph is a computationally challenging problem (such combinatorial optimization problems in this class are called NP-hard), the relative

sparsity of the interaction graph allows this computation to be performed in a reasonable amount of time, using open-source software.

5. This work contributes to the possibility of modularity studies of genome-wide analysis of weighted graphs of all SNP–SNP and SNP × clinical/environment variables. This type of Boolean interaction analysis may become a complementary alternative to univariate statistical models to further extract knowledge and help dissect the different components of multifactorial diseases.

Acknowledgements

This work was supported by the National Health and Medical Research Council [Grant 512423 *Genes and environment in the risk of early age-related macular degeneration: a population-based case-control study*, to J.J.W., P.M., and J.A., and fellowship scheme to E.G.H.]. We are grateful to Katrina Bogan for her careful revision of the manuscript.

References

1. Manolio TA, Collins FS, Cox NJ, Goldstein DB, Hindorff LA, Hunter DJ, McCarthy MI, Ramos EM, Cardon LR, Chakravarti A, Cho JH, Guttmacher AE, Kong A, Kruglyak L, Mardis E, Rotimi CN, Slatkin M, Valle D, Whittemore AS, Boehnke M, Clark AG, Eichler EE, Gibson G, Haines JL, Mackay TFC, McCarroll SA, Visscher PM (2009) Finding the missing heritability of complex diseases. Nature 461(7265):747–753

2. Purcell S, Neale B, Todd-Brown K, Thomas L, Ferreira MAR, Bender D, Maller J, Sklar P, de Bakker PIW, Daly MJ, Sham PC (2007) PLINK: A tool set for whole-genome association and population-based linkage analyses. Am J Hum Genet 81(3):559–575

3. Ruczinski I, Kooperberg C, LeBlanc M (2003) Logic regression. J Comput Graph Stat 12(3):475–511

4. Ruczinski I (2004) Exploring interactions in high-dimensional genomic data: an overview of Logic Regression, with applications. J Multivar Anal 90(1):178–195

5. Kooperberg C, Ruczinski I (2005) Identifying interacting SNPs using Monte Carlo logic regression. Genet Epidemiol 28(2):157–170

6. Wan X, Yang C, Yang Q, Xue H, Fan X, Tang NLS, Yu W (2010) BOOST: A fast approach to detecting gene-gene interactions in genome-wide case-control studies. Am J Hum Genet 87(3):325–340

7. Ritchie MD, Hahn LW, Roodi N, Bailey LR, Dupont WD, Parl FF, Moore JH (2001) Multifactor-dimensionality reduction reveals high-order interactions among estrogen-metabolism genes in sporadic breast cancer. Am J Hum Genet 69(1):138–147

8. Hahn LW, Ritchie MD, Moore JH (2003) Multifactor dimensionality reduction software for detecting gene-gene and gene-environment interactions. Bioinformatics 19(3):376–382

9. Sinnott-Armstrong N, Greene C, Cancare F, Moore J (2009) Accelerating epistasis analysis in human genetics with consumer graphics hardware. BMC Res Notes 2(1):149

10. Hu X, Liu Q, Zhang Z, Li Z, Wang S, He L, Shi Y (2010) SHEsisEpi, a GPU-enhanced genome-wide SNP-SNP interaction scanning algorithm, efficiently reveals the risk genetic epistasis in bipolar disorder. Cell Res 20(7):854–857

11. Yung LS, Yang C, Wan X, Yu W (2011) GBOOST: a GPU-based tool for detecting gene-gene interactions in genome-wide case control studies. Bioinformatics 27(9):1309–1310

12. Hageman GS, Anderson DH, Johnson LV, Hancox LS, Taiber AJ, Hardisty LI, Hageman

JL, Stockman HA, Borchardt JD, Gehrs KM, Smith RJH, Silvestri G, Russell SR, Klaver CCW, Barbazetto I, Chang S, Yannuzzi LA, Barile GR, Merriam JC, Smith RT, Olsh AK, Bergeron J, Zernant J, Merriam JE, Gold B, Dean M, Allikmets R (2005) A common haplotype in the complement regulatory gene factor H (HF1/CFH) predisposes individuals to age-related macular degeneration. Proc Natl Acad Sci U S A 102(20):7227–7232

13. Klein RJ, Zeiss C, Chew EY, Tsai J-Y, Sackler RS, Haynes C, Henning AK, SanGiovanni JP, Mane SM, Mayne ST, Bracken MB, Ferris FL, Ott J, Barnstable C, Hoh J (2005) Complement factor H polymorphism in age-related macular degeneration. Science 308(5720):385–389

14. Scholl HPN, Weber BHF, Nöthen MM, Wienker T, Holz FG (2005) Y402H polymorphism in complement factor H and age-related macula degeneration (AMD). Ophthalmologe 102(11):1029–1035

15. Rivera A, Fisher SA, Fritsche LG, Keilhauer CN, Lichtner P, Meitinger T, Weber BHF (2005) Hypothetical LOC387715 is a second major susceptibility gene for age-related macular degeneration, contributing independently of complement factor H to disease risk. Hum Mol Genet 14(21):3227–3236

16. Edwards AO, Fridley BL, James KM, Sharma AK, Sharma AS, Cunningham JM, Tosakulwong N (2008) Evaluation of clustering and genotype distribution for replication in genome wide association studies: the age-related eye disease study. PLoS One 3(11):e3813

17. Janik-Papis K, Mg Z, Skłodowska A, Ulińska M, Borucka AI, Błasiak J (2009) Genetic aspects of age-related macular degeneration. Klinika Oczna 111(4–6):178–182

18. Lin W-Y, Lee W-C (2010) Incorporating prior knowledge to facilitate discoveries in a genome-wide association study on age-related macular degeneration. BMC Res Notes 3:26

19. Wegscheider BJ, Weger M, Renner W, Steinbrugger I, März W, Mossböck G, Temmel W, El-Shabrawi Y, Schmut O, Jahrbacher R, Haas A (2007) Association of complement factor H Y402H gene polymorphism with different subtypes of exudative age-related macular degeneration. Ophthalmology 114(4): 738–742

20. Zhang H, Morrison MA, Dewan A, Adams S, Andreoli M, Huynh N, Regan M, Brown A, Miller JW, Kim IK, Hoh J, Deangelis MM (2008) The NEI/NCBI dbGAP database: genotypes and haplotypes that may specifically predispose to risk of neovascular age-related macular degeneration. BMC Med Genet 9:51

21. Yu Y, Bhangale TR, Fagerness J, Ripke S, Thorleifsson G, Tan PL, Souied EH, Richardson AJ, Merriam JE, Buitendijk GHS, Reynolds R, Raychaudhuri S, Chin KA, Sobrin L, Evangelou E, Lee PH, Lee AY, Leveziel N, Zack DJ, Campochiaro B, Campochiaro P, Smith RT, Barile GR, Guymer RH, Hogg R, Chakravarthy U, Robman LD, Gustafsson O, Sigurdsson H, Ortmann W, Behrens TW, Stefansson K, Uitterlinden AG, van Duijn CM, Vingerling JR, Klaver CCW, Allikmets R, Brantley MA, Baird PN, Katsanis N, Thorsteinsdottir U, Ioannidis JPA, Daly MJ, Graham RR, Seddon JM (2011) Common variants near FRK/COL10A1 and VEGFA are associated with advanced age-related macular degeneration. Hum Mol Genet 20(18): 3699–3709

22. VanderWeele TJ (2010) Epistatic interactions. Stat Appl Genet Mol Biol 9(1):Article 1

23. Ozkiris A (2010) Anti-VEGF agents for age-related macular degeneration. Expert Opin Ther Pat 20(1):103–118

24. Gotoh N, Yamada R, Nakanishi H, Saito M, Iida T, Matsuda F, Yoshimura N (2008) Correlation between CFH Y402H and HTRA1 rs11200638 genotype to typical exudative age-related macular degeneration and polypoidal choroidal vasculopathy phenotype in the Japanese population. Clin Exp Ophthalmol 36(5):437–442

25. Kondo N, Honda S, S-i K, Negi A (2009) Coding variant I62V in the complement factor H gene is strongly associated with polypoidal choroidal vasculopathy. Ophthalmology 116(2): 304–310

26. Lee KY, Vithana EN, Mathur R, Yong VH, Yeo IY, Thalamuthu A, Lee M-W, Koh AH, Lim MC, How AC, Wong DW, Aung T (2008) Association analysis of CFH, C2, BF, and HTRA1 gene polymorphisms in Chinese patients with polypoidal choroidal vasculopathy. Invest Ophthalmol Vis Sci 49(6): 2613–2619

27. Neuner B, Komm A, Wellmann J, Dietzel M, Pauleikhoff D, Walter J, Busch M, Hense H-W (2009) Smoking history and the incidence of age-related macular degeneration–results from the Muenster Aging and Retina Study (MARS) cohort and systematic review and meta-analysis of observational longitudinal studies. Addict Behav 34(11):938–947

28. Yasuda M, Kiyohara Y, Hata Y, Arakawa S, Yonemoto K, Doi Y, Iida M, Ishibashi T (2009) Nine-year incidence and risk factors for age-related macular degeneration in a defined Japanese population the Hisayama study. Ophthalmology 116(11):2135–2140

29. Kalariya NM, Wills NK, Ramana KV, Srivastava SK, van Kuijk FJGM (2009) Cadmium-induced apoptotic death of human retinal pigment epithelial cells is mediated by MAPK pathway. Exp Eye Res 89(4):494–502

30. Wang JJ, Rochtchina E, Smith W, Klein R, Klein BEK, Joshi T, Sivakumaran TA, Iyengar S, Mitchell P (2009) Combined effects of complement factor H genotypes, fish consumption, and inflammatory markers on long-term risk for age-related macular degeneration in a cohort. Am J Epidemiol 169(5):633–641

31. Conley YP, Thalamuthu A, Jakobsdottir J, Weeks DE, Mah T, Ferrell RE, Gorin MB (2005) Candidate gene analysis suggests a role for fatty acid biosynthesis and regulation of the complement system in the etiology of age-related maculopathy. Hum Mol Genet 14(14): 1991–2002

32. Singerman LJ, Brucker AJ, Jampol LM, Lim JI, Rosenfeld P, Schachat AP, Spaide RF (2005) Neovascular age-related macular degeneration: roundtable. Retina 25(7 Suppl):S1–S22

33. Oldmeadow CJ, Riveros C, Holliday EG, Scott R, Moscato P, Wang JJ, Mitchell P, Buitendijk GHS, Vingerling JR, Klaver CCW, Klein R, Attia J (2011) Sifting the wheat from the chaff: prioritizing GWAS results by identifying consistency across analytical methods. Genet Epidemiol 35(8):745–754

34. Mitchell P, Smith W, Attebo K, Wang JJ (1995) Prevalence of age-related maculopathy in Australia. The Blue Mountains Eye Study. Ophthalmology 102(10):1450–1460

35. Foran S, Wang JJ, Mitchell P (2003) Causes of visual impairment in two older population cross-sections: the Blue Mountains Eye Study. Ophthalmic Epidemiol 10(4):215–225

36. Wang JJ, Rochtchina E, Lee AJ, Chia E-M, Smith W, Cumming RG, Mitchell P (2007) Ten-year incidence and progression of age-related maculopathy: the Blue Mountains Eye Study. Ophthalmology 114(1):92–98

37. Tan JSL, Wang JJ, Flood V, Mitchell P (2009) Dietary fatty acids and the 10-year incidence of age-related macular degeneration: the Blue Mountains Eye Study. Arch Ophthalmol 127(5):656–665

38. Tan JSL, Mitchell P, Kifley A, Flood V, Smith W, Wang JJ (2007) Smoking and the long-term incidence of age-related macular degeneration: the Blue Mountains Eye Study. Arch Ophthalmol 125(8):1089–1095

39. Fayyad UM, Irani KB (1993) Multi-interval discretization of continuous-valued attributes for classification learning. Int Joint Conf Artif Intel 13:1022–1027

40. Fisher RA (1932) Statistical methods for research workers. Genesis, New Delhi

41. Agresti A (1992) A survey of exact inference for contingency tables. Stat Sci 7(1):131–153

42. Baldi P, Sr B, Chauvin Y, Andersen CAF, Nielsen H (2000) Assessing the accuracy of prediction algorithms for classification: an overview. Bioinformatics 16(5):412–424

43. Matthews BW (1975) Comparison of the predicted and observed secondary structure of T4 phage lysozyme. Biochim Biophys Acta 405(2):442–451

44. Cotta C, Langston M, Moscato P (2007) Combinatorial and algorithmic issues for microarray data analysis. In: González TF (ed) Handbook of Approximation Algorithms and Metaheuristics. Champman & Hall/CRC, pp 74.1–74.14

45. NVIDIA Corporation (2010) CUDA Home Page. In: Nvidia Developer Zone. https://developer.nvidia.com/category/zone/cuda-zone. Accessed 27 Nov 2013

46. R Core Team (2011) R: A Language and Environment for Statistical Computing. In: R Foundation for Statistical Computing, Vienna, Austria. http://www.R-project.org. Accessed 11 May 2011

47. Csárdi G, Nepusz T (2006) The igraph software package for complex network research. InterJournal Complex Systems:1695

48. Oliveros JC (2007) VENNY: An interactive tool for comparing lists with Venn Diagrams. http://bioinfogp.cnb.csic.es/tools/venny/index.html

49. Krzywinski M, Schein J, Birol I, Connors J, Gascoyne R, Horsman D, Jones SJ, Marra MA (2009) Circos: an information aesthetic for comparative genomics. Genome Res 19(9): 1639–1645

50. Chen J, Bardes EE, Aronow BJ, Jegga AG (2009) ToppGene Suite for gene list enrichment analysis and candidate gene prioritization. Nucleic Acids Res 37:W305–W311, Web server issue

51. Chen W, Stambolian D, Edwards AO, Branham KE, Othman M, Jakobsdottir J, Tosakulwong N, Pericak-Vance MA, Campochiaro PA, Klein ML, Tan PL, Conley YP, Kanda A, Kopplin L, Li Y, Augustaitis KJ, Karoukis AJ, Scott WK, Agarwal A, Kovach JL, Schwartz SG, Postel EA, Brooks M, Baratz KH, Brown WL, Brucker AJ, Orlin A, Brown G, Ho A, Regillo C, Donoso L, Tian L, Kaderli B, Hadley D, Hagstrom SA, Peachey NS, Klein R, Klein BEK, Gotoh N, Yamashiro K, Ferris Iii F, Fagerness JA, Reynolds R, Farrer LA, Kim IK, Miller JW, Cortón M, Carracedo A, Sanchez-Salorio M, Pugh EW, Doheny KF, Brion M,

Deangelis MM, Weeks DE, Zack DJ, Chew EY, Heckenlively JR, Yoshimura N, Iyengar SK, Francis PJ, Katsanis N, Seddon JM, Haines JL, Gorin MB, Abecasis GR, Swaroop A (2010) Genetic variants near TIMP3 and high-density lipoprotein-associated loci influence susceptibility to age-related macular degeneration. Proc Natl Acad Sci U S A 107(16): 7401–7406

52. Pai AS-I, Mitchell P, Rochtchina E, Iyengar S, Wang JJ (2009) Complement factor H and the bilaterality of age-related macular degeneration. Arch Ophthalmol 127(10):1339–1344

53. Tsuchihashi T, Mori K, Horie-Inoue K, Gehlbach PL, Kabasawa S, Takita H, Ueyama K, Okazaki Y, Inoue S, Awata T, Katayama S, Yoneya S (2011) Complement factor H and high-temperature requirement A-1 genotypes and treatment response of age-related macular degeneration. Ophthalmology 118(1):93–100

54. Kloeckener-Gruissem B, Barthelmes D, Labs S, Schindler C, Kurz-Levin M, Michels S, Fleischhauer J, Berger W, Sutter F, Menghini M (2011) Genetic association with response to intravitreal ranibizumab in patients with neovascular AMD. Invest Ophthalmol Vis Sci 52(7):4694–4702

55. Farwick A, Dasch B, Weber BHF, Pauleikhoff D, Stoll M, Hense HW (2009) Variations in five genes and the severity of age-related macular degeneration: results from the Muenster aging and retina study. Eye (Lond) 23(12): 2238–2244

56. Ding X, Patel M, Chan C-C (2009) Molecular pathology of age-related macular degeneration. Prog Retin Eye Res 28(1):1–18

57. Montezuma SR, Sobrin L, Seddon JM (2007) Review of genetics in age related macular degeneration. Semin Ophthalmol 22(4): 229–240

58. Seitsonen S, Järvelä I, Meri S, Tommila P, Ranta P, Immonen I (2008) Complement factor H Y402H polymorphism and characteristics of exudative age-related macular degeneration lesions. Acta Ophthalmol (Copenh) 86(4):390–394

59. Swaroop A, Branham KE, Chen W, Abecasis G (2007) Genetic susceptibility to age-related macular degeneration: a paradigm for dissecting complex disease traits. Human Mol Genet 16 (Spec No. 2):R174–182

60. Chen Y, Zeng J, Zhao C, Wang K, Trood E, Buehler J, Weed M, Kasuga D, Bernstein PS, Hughes G, Fu V, Chin J, Lee C, Crocker M, Bedell M, Salasar F, Yang Z, Goldbaum M, Ferreyra H, Freeman WR, Kozak I, Zhang K (2011) Assessing susceptibility to age-related macular degeneration with genetic markers and environmental factors. Arch Ophthalmol 129(3):344–351

61. Ryu E, Fridley BL, Tosakulwong N, Bailey KR, Edwards AO (2010) Genome-wide association analyses of genetic, phenotypic, and environmental risks in the age-related eye disease study. Mol Vis 16:2811–2821

62. Dong L, Qu Y, Jiang H, Dai H, Zhou F, Xu X, Bi H, Pan X, Dang G (2011) Correlation of complement factor H gene polymorphisms with exudative age-related macular degeneration in a Chinese cohort. Neurosci Lett 488(3):283–287

63. Nakanishi H, Yamashiro K, Yamada R, Gotoh N, Hayashi H, Nakata I, Saito M, Iida T, Oishi A, Kurimoto Y, Matsuo K, Tajima K, Matsuda F, Yoshimura N (2010) Joint effect of cigarette smoking and CFH and LOC387715/HTRA1 polymorphisms on polypoidal choroidal vasculopathy. Invest Ophthalmol Vis Sci 51(12): 6183–6187

64. Teixeira AG, Silva AS, Lin FLH, Velletri R, Bavia L, Belfort R Jr, Isaac L (2010) Association of complement factor H Y402H polymorphism and age-related macular degeneration in Brazilian patients. Acta Ophthalmol (Copenh) 88(5):e165–e169

65. Raychaudhuri S, Ripke S, Li M, Neale BM, Fagerness J, Reynolds R, Sobrin L, Swaroop A, Ga A, Seddon JM, Daly MJ (2010) Associations of CFHR1-CFHR3 deletion and a CFH SNP to age-related macular degeneration are not independent. Nat Genet 42(7):553–555, author reply 555–556–553

66. Hayashi H, Yamashiro K, Gotoh N, Nakanishi H, Nakata I, Tsujikawa A, Otani A, Saito M, Iida T, Matsuo K, Tajima K, Yamada R, Yoshimura N (2010) CFH and ARMS2 variations in age-related macular degeneration, polypoidal choroidal vasculopathy, and retinal angiomatous proliferation. Invest Ophthalmol Vis Sci 51(11):5914–5919

67. Yang X, Hu J, Zhang J, Guan H (2010) Polymorphisms in CFH, HTRA1 and CX3CR1 confer risk to exudative age-related macular degeneration in Han Chinese. Br J Ophthalmol 94(9):1211–1214

68. Liu X, Zhao P, Tang S, Lu F, Hu J, Lei C, Yang X, Lin Y, Ma S, Yang J, Zhang D, Shi Y, Li T, Chen Y, Fan Y, Yang Z (2010) Association study of complement factor H, C2, CFB, and C3 and age-related macular degeneration in a Han Chinese population. Retina 30(8): 1177–1184

69. McKay GJ, Dasari S, Patterson CC, Chakravarthy U, Silvestri G (2010) Complement component 3: an assessment of association with AMD and analysis of gene-gene and

gene-environment interactions in a Northern Irish cohort. Mol Vis 16:194–199

70. Goto A, Akahori M, Okamoto H, Minami M, Terauchi N, Haruhata Y, Obazawa M, Noda T, Honda M, Mizota A, Tanaka M, Hayashi T, Tanito M, Ogata N, Iwata T (2009) Genetic analysis of typical wet-type age-related macular degeneration and polypoidal choroidal vasculopathy in Japanese population. J Ocul Biol Dis Infor 2(4):164–175

71. Mori K, Horie-Inoue K, Gehlbach PL, Takita H, Kabasawa S, Kawasaki I, Ohkubo T, Kurihara S, Iizuka H, Miyashita Y, Katayama S, Awata T, Yoneya S, Inoue S (2010) Phenotype and genotype characteristics of age-related macular degeneration in a Japanese population. Ophthalmology 117(5):928–938

72. Katta S, Kaur I, Chakrabarti S (2009) The molecular genetic basis of age-related macular degeneration: an overview. J Genet 88(4): 425–449

73. Leveziel N, Puche N, Richard F, Somner JEA, Zerbib J, Bastuji-Garin S, Cohen SY, Korobelnik J-F, Sahel J, Soubrane G, Benlian P, Souied EH (2010) Genotypic influences on severity of exudative age-related macular degeneration. Invest Ophthalmol Vis Sci 51(5):2620–2625

74. Losonczy G, Fekete Á, Vokó Z, Takács L, Káldi I, Ajzner É, Kasza M, Vajas A, Berta A, Balogh I (2011) Analysis of complement factor H Y402H, LOC387715, HTRA1 polymorphisms and ApoE alleles with susceptibility to age-related macular degeneration in Hungarian patients. Acta Ophthalmol (Copenh) 89(3):255–262

75. Cui L, Zhou H, Yu J, Sun E, Zhang Y, Jia W, Jiao Y, Snellingen T, Liu X, Lim A, Wang N, Liu N (2010) Noncoding variant in the complement factor H gene and risk of exudative age-related macular degeneration in a Chinese population. Invest Ophthalmol Vis Sci 51(2): 1116–1120

76. Scholl HPN, Fleckenstein M, Fritsche LG, Schmitz-Valckenberg S, Göbel A, Adrion C, Herold C, Keilhauer CN, Mackensen F, Mössner A, Pauleikhoff D, Weinberger AWA, Mansmann U, Holz FG, Becker T, Weber BHF (2009) CFH, C3 and ARMS2 are significant risk loci for susceptibility but not for disease progression of geographic atrophy due to AMD. PLoS One 4(10):e7418

77. Ricci F, Zampatti S, D'Abbruzzi F, Missiroli F, Martone C, Lepre T, Pietrangeli I, Sinibaldi C, Peconi C, Novelli G, Giardina E (2009) Typing of ARMS2 and CFH in age-related macular degeneration: case-control study and assessment of frequency in the Italian population. Arch Ophthalmol 127(10):1368–1372

78. Zerbib J, Seddon JM, Richard F, Reynolds R, Leveziel N, Benlian P, Borel P, Feingold J, Munnich A, Soubrane G, Kaplan J, Rozet J-M, Souied EH (2009) rs5888 variant of SCARB1 gene is a possible susceptibility factor for age-related macular degeneration. PLoS One 4(10):e7341

79. Chen M, Muckersie E, Robertson M, Forrester JV, Xu H (2008) Up-regulation of complement factor B in retinal pigment epithelial cells is accompanied by complement activation in the aged retina. Exp Eye Res 87(6): 543–550

80. Chen M, Forrester JV, Xu H (2007) Synthesis of complement factor H by retinal pigment epithelial cells is down-regulated by oxidized photoreceptor outer segments. Exp Eye Res 84(4):635–645

81. Lee K-T, Byun M-J, Kang K-S, Park E-W, Lee S-H, Cho S, Kim H, Kim K-W, Lee T, Park J-E, Park W, Shin D, Park H-S, Jeon J-T, Choi B-H, Jang G-W, Choi S-H, Kim D-W, Lim D, Park H-S, Park M-R, Ott J, Schook LB, Kim T-H, Kim H (2011) Neuronal genes for subcutaneous fat thickness in human and pig are identified by local genomic sequencing and combined SNP association study. PLoS One 6(2):e16356

82. Rajasekharan S, Kennedy TE (2009) The netrin protein family. Genome Biol 10(9):239

83. Masuda T, Watanabe K, Sakuma C, Ikenaka K, Ono K, Yaginuma H (2008) Netrin-1 acts as a repulsive guidance cue for sensory axonal projections toward the spinal cord. J Neurosci Off J Soc Neurosci 28(41):10380–10385

84. Livesey FJ, Hunt SP (1997) Netrin and netrin receptor expression in the embryonic mammalian nervous system suggests roles in retinal, striatal, nigral, and cerebellar development. Mol Cell Neurosci 8(6):417–429

85. de la Torre JR, Höpker VH, Ming GL, Poo MM, Tessier-Lavigne M, Hemmati-Brivanlou A, Holt CE (1997) Turning of retinal growth cones in a netrin-1 gradient mediated by the netrin receptor DCC. Neuron 19(6): 1211–1224

86. Meriane M, Tcherkezian J, Webber CA, Danek EI, Triki I, McFarlane S, Bloch-Gallego E, Lamarche-Vane N (2004) Phosphorylation of DCC by Fyn mediates Netrin-1 signaling in growth cone guidance. J Cell Biol 167(4): 687–698

87. Jarjour AA, Bull S-J, Almasieh M, Rajasekharan S, Baker KA, Mui J, Antel JP, Di Polo A, Kennedy TE (2008) Maintenance of axo-oligodendroglial paranodal junctions requires DCC and netrin-1. J Neurosci Off J Soc Neurosci 28(43):11003–11014

88. Spassky N, de Castro F, Le Bras B, Heydon K, Quéraud-LeSaux F, Bloch-Gallego E, Chédotal A, Zalc B, Thomas J-L (2002) Directional guidance of oligodendroglial migration by class 3 semaphorins and netrin-1. J Neurosci Off J Soc Neurosci 22(14): 5992–6004

89. Manitt C, Nikolakopoulou AM, Almario DR, Nguyen SA, Cohen-Cory S (2009) Netrin participates in the development of retinotectal synaptic connectivity by modulating axon arborization and synapse formation in the developing brain. J Neurosci Off J Soc Neurosci 29(36):11065–11077

90. Oster SF, Deiner M, Birgbauer E, Sretavan DW (2004) Ganglion cell axon pathfinding in the retina and optic nerve. Semin Cell Dev Biol 15(1):125–136

91. Smith CJ, Watson JD, VanHoven MK, Colón-Ramos DA, Iii DMM (2012) Netrin (UNC-6) mediates dendritic self-avoidance. Nat Neurosci 15(5):731–737

92. Shi M, Zheng M-H, Liu Z-R, Hu Z-L, Huang Y, Chen J-Y, Zhao G, Han H, Ding Y-Q (2010) DCC is specifically required for the survival of retinal ganglion and displaced amacrine cells in the developing mouse retina. Dev Biol 348(1):87–96

93. Xu H, Liu J, Xiong S, Y-z L, Xia X (2012) Suppression of retinal neovascularization by lentivirus-mediated netrin-1 small Hairpin RNA. Ophthalmic Res 47(3):163–169

94. Tian X-F, Xia X-B, Xiong S-Q, Jiang J, Liu D, Liu J-L (2011) Netrin-1 overexpression in oxygen-induced retinopathy correlates with breakdown of the blood-retina barrier and retinal neovascularization. Ophthalmologica 226(2):37–44

95. Bányai L, Patthy L (1999) The NTR module: domains of netrins, secreted frizzled related proteins, and type I procollagen C-proteinase enhancer protein are homologous with tissue inhibitors of metalloproteases. Protein Sci 8(8):1636–1642

96. Kaur I, Rathi S, Chakrabarti S (2010) Variations in TIMP3 are associated with age-related macular degeneration. Proc Natl Acad Sci U S A 107(28):E112–E113

97. Strunnikova NV, Maminishkis A, Barb JJ, Wang F, Zhi C, Sergeev Y, Chen W, Edwards AO, Stambolian D, Abecasis G, Swaroop A, Munson PJ, Miller SS (2010) Transcriptome analysis and molecular signature of human retinal pigment epithelium. Hum Mol Genet 19(12):2468–2486

98. Bekhouche M, Kronenberg D, Vadon-Le Goff S, Bijakowski C, Lim NH, Font B, Kessler E, Colige A, Nagase H, Murphy G, Hulmes DJS, Moali C (2010) Role of the netrin-like domain of procollagen C-proteinase enhancer-1 in the control of metalloproteinase activity. J Biol Chem 285(21):15950–15959

99. Sanchez-Ramos C, Vega JA, Del Valle ME, Fernandez-Balbuena A, Bonnin-Arias C, Benitez-Del Castillo JM (2010) Role of metalloproteases in retinal degeneration induced by violet and blue light. Adv Exp Med Biol 664:159–164

100. Hirano AA, Brandstätter JH, Morgans CW, Brecha NC (2011) SNAP25 expression in mammalian retinal horizontal cells. J Comp Neurol 519(5):972–988

101. Mazelova J, Ransom N, Astuto-Gribble L, Wilson MC, Deretic D (2009) Syntaxin 3 and SNAP-25 pairing, regulated by omega-3 docosahexaenoic acid, controls the delivery of rhodopsin for the biogenesis of cilia-derived sensory organelles, the rod outer segments. J Cell Sci 122(Pt 12):2003–2013

102. Morgans C, Brandstätter JH (2000) SNAP-25 is present on the Golgi apparatus of retinal neurons. Neuroreport 11(1):85–88

103. Yang H, Standifer KM, Sherry DM (2002) Synaptic protein expression by regenerating adult photoreceptors. J Comp Neurol 443(3): 275–288

104. Greenlee MHW, Wilson MC, Sakaguchi DS (2002) Expression of SNAP-25 during mammalian retinal development: thinking outside the synapse. Semin Cell Dev Biol 13(2):99–106

105. Kono N, Inoue T, Yoshida Y, Sato H, Matsusue T, Itabe H, Niki E, Aoki J, Arai H (2008) Protection against oxidative stress-induced hepatic injury by intracellular type II platelet-activating factor acetylhydrolase by metabolism of oxidized phospholipids in vivo. J Biol Chem 283(3):1628–1636

106. Demos C, Bandyopadhyay M, Br R (2008) Identification of candidate genes for human retinal degeneration loci using differentially expressed genes from mouse photoreceptor dystrophy models. Mol Vis 14:1639–1649

107. Burzynski GM, Delalande J-M, Shepherd I (2009) Characterization of spatial and temporal expression pattern of SCG10 during zebrafish development. Gene Expr Patterns 9(4):231–237

108. Liedtke W, Leman EE, Fyffe REW, Raine CS, Schubart UK (2002) Stathmin-deficient mice develop an age-dependent axonopathy of the central and peripheral nervous systems. Am J Pathol 160(2):469–480

109. Hasegawa A, Hisatomi O, Yamamoto S, Ono E, Tokunaga F (2007) Stathmin expression during newt retina regeneration. Exp Eye Res 85(4):518–527

110. Finnegan S, Robson J, Hocking PM, Ali M, Inglehearn CF, Stitt A, Curry WJ (2010) Proteomic profiling of the retinal dysplasia and degeneration chick retina. Mol Vis 16:7–17

111. Kompass KS, Agapova OA, Li W, Kaufman PL, Rasmussen CA, Hernandez MR (2008) Bioinformatic and statistical analysis of the optic nerve head in a primate model of ocular hypertension. BMC Neurosci 9:93

112. Wang A-G, Chen C-H, Yang C-W, Yen M-Y, Hsu W-M, Liu J-H, Fann M-J (2002) Change of gene expression profiles in the retina following optic nerve injury. Brain Res Mol Brain Res 101(1–2):82–92

113. Nakazawa T, Nakano I, Furuyama T, Morii H, Tamai M, Mori N (2000) The SCG10-related gene family in the developing rat retina: persistent expression of SCLIP and stathmin in mature ganglion cell layer. Brain Res 861(2):399–407

114. Yanagisawa H, Komuta Y, Kawano H, Toyoda M, Sango K (2010) Pleiotrophin induces neurite outgrowth and up-regulates growth-associated protein (GAP)-43 mRNA through the ALK/GSK3[beta]/[beta]-catenin signaling in developing mouse neurons. Neurosci Res 66(1):111–116

115. Perez-Pinera P, Zhang W, Chang Y, Vega JA, Deuel TF (2007) Anaplastic Lymphoma Kinase Is Activated Through the Pleiotrophin/Receptor Protein-tyrosine Phosphatase β/ζ Signaling Pathway. Journal of Biological Chemistry 282:28683–28690. doi: 10.1074/jbc.M704505200

116. Chen H, Campbell RA, Chang Y, Li M, Wang CS, Li J, Sanchez E, Share M, Steinberg J, Berenson A, Shalitin D, Zeng Z, Gui D, Perez-Pinera P, Berenson RJ, Said J, Bonavida B, Deuel TF, Berenson JR (2009) Pleiotrophin produced by multiple myeloma induces trans-differentiation of monocytes into vascular endothelial cells: a novel mechanism of tumor-induced vasculogenesis. Blood 113(9): 1992–2002

117. Yeh HJ, He YY, Xu J, Hsu CY, Deuel TF (1998) Upregulation of pleiotrophin gene expression in developing microvasculature, macrophages, and astrocytes after acute ischemic brain injury. J Neurosci Off J Soc Neurosci 18(10):3699–3707

118. Rao KN, Nagireddy S, Chakrabarti S (2011) Complex genetic mechanisms in glaucoma: an overview. Indian J Ophthalmol 59(Suppl): S31–S42

119. Tamm ER (2011) Development of the irido-corneal angle and congenital glaucoma. Ophthalmologe 108(7):610–617

120. Yu-Wai-Man P, Griffiths PG, Chinnery PF (2011) Mitochondrial optic neuropathies—Disease mechanisms and therapeutic strategies. Prog Retin Eye Res 30(2):81–114

121. Tang Y, Scheef EA, Gurel Z, Sorenson CM, Jefcoate CR, Sheibani N (2010) CYP1B1 and endothelial nitric oxide synthase combine to sustain proangiogenic functions of endothelial cells under hyperoxic stress. Am J Physiol Cell Physiol 298(3):C665–C678

122. Lin H, Xu H, Liang F-Q, Liang H, Gupta P, Havey AN, Boulton ME, Godley BF (2011) Mitochondrial DNA damage and repair in RPE associated with aging and age-related macular degeneration. Invest Ophthalmol Vis Sci 52(6):3521–3529

123. Schrier SA, Falk MJ (2011) Mitochondrial disorders and the eye. Curr Opin Ophthalmol 22(5):325–331

124. Blasiak J, Szaflik JP (2011) DNA damage and repair in age-related macular degeneration. Front Biosci 16:1291–1301

125. Epand RF, Maekawa S, Epand RM (2003) Specificity of membrane binding of the neuronal protein NAP-22. J Membr Biol 193(3):171–176

126. Epand RM, Braswell EH, Yip CM, Epand RF, Maekawa S (2003) Quaternary structure of the neuronal protein NAP-22 in aqueous solution. Biochim Biophys Acta 1650(1–2): 50–58

127. Khan TK, Yang B, Thompson NL, Maekawa S, Epand RM, Jacobson K (2003) Binding of NAP-22, a calmodulin-binding neuronal protein, to raft-like domains in model membranes. Biochemistry 42(17):4780–4786

128. de las-Heras R, Depaz I, Jaquet V, Kroon P, Wilce PA (2007) Neuronal protein 22 colocalises with both the microtubule and microfilament cytoskeleton in neurite-like processes. Brain Res 1128(1):12–20

129. Depaz IM, Wilce PA (2006) The novel cytoskeleton-associated protein Neuronal protein 22: elevated expression in the developing rat brain. Brain Res 1081(1):59–64

130. Neale BM, Fagerness J, Reynolds R, Sobrin L, Parker M, Raychaudhuri S, Tan PL, Oh EC, Merriam JE, Souied E, Bernstein PS, Li B, Frederick JM, Zhang K, Brantley MA, Lee AY, Zack DJ, Campochiaro B, Campochiaro P, Ripke S, Smith RT, Barile GR, Katsanis N, Allikmets R, Daly MJ, Seddon JM (2010) Genome-wide association study of advanced age-related macular degeneration identifies a role of the hepatic lipase gene (LIPC). Proc Natl Acad Sci 107(16):7395–7400

131. Krumbiegel M, Pasutto F, Schlötzer-Schrehardt U, Uebe S, Zenkel M, Mardin

CY, Weisschuh N, Paoli D, Gramer E, Becker C, Ekici AB, Weber BHF, Nürnberg P, Kruse FE, Reis A (2011) Genome-wide association study with DNA pooling identifies variants at CNTNAP2 associated with pseudoexfoliation syndrome. Eur J Hum Genet 19(2):186–193

132. Schlötzer-Schrehardt U (2011) Genetics and genomics of pseudoexfoliation syndrome/glaucoma. Middle East Afr J Ophthalmol 18(1):30–36

133. Sharon D, Yamamoto H, McGee TL, Rabe V, Szerencsei RT, Winkfein RJ, Prinsen CFM, Barnes CS, Andreasson S, Fishman GA, Schnetkamp PPM, Berson EL, Dryja TP (2002) Mutated alleles of the rod and cone Na-Ca + K-exchanger genes in patients with retinal diseases. Invest Ophthalmol Vis Sci 43(6):1971–1979

134. Wilson PM, Fryer RH, Fang Y, Hatten ME (2010) ASTN2, a novel member of the astrotactin gene family, regulates the trafficking of ASTN1 during glial-guided neuronal migration. J Neurosci Off J Soc Neurosci 30(25):8529–8540

135. Vallee RB, Seale GE, Tsai J-W (2009) Emerging roles for myosin II and cytoplasmic dynein in migrating neurons and growth cones. Trends Cell Biol 19(7):347–355

136. Tuo J, Ning B, Bojanowski CM, Lin Z-N, Ross RJ, Reed GF, Shen D, Jiao X, Zhou M, Chew EY, Kadlubar FF, Chan C-C (2006) Synergic effect of polymorphisms in ERCC6 5 flanking region and complement factor H on age-related macular degeneration predisposition. Proc Natl Acad Sci 103(24):9256–9261

137. Kramerov AA, Saghizadeh M, Pan H, Kabosova A, Montenarh M, Ahmed K, Penn JS, Chan CK, Hinton DR, Grant MB, Ljubimov AV (2006) Expression of protein kinase CK2 in astroglial cells of normal and neovascularized retina. Am J Pathol 168(5):1722–1736

138. Kramerov AA, Golub AG, Bdzhola VG, Yarmoluk SM, Ahmed K, Bretner M, Ljubimov AV (2010) Treatment of cultured human astrocytes and vascular endothelial cells with protein kinase CK2 inhibitors induces early changes in cell shape and cytoskeleton. Mol Cell Biochem 349(1–2):125–137

139. Xie Y, Yeo TT, Zhang C, Yang T, Tisi MA, Massa SM, Longo FM (2001) The leukocyte common antigen-related protein tyrosine phosphatase receptor regulates regenerative neurite outgrowth in vivo. J Neurosci Off J Soc Neurosci 21(14):5130–5138

140. Yu J, Becka S, Zhang P, Zhang X, Brady-Kalnay SM, Wang Z (2008) Tumor-derived extracellular mutations of PTPRT/PTPrho are defective in cell adhesion. Mol Cancer Res 6(7):1106–1113

141. Zhang P, Becka S, Craig SEL, Lodowski DT, Brady-Kalnay SM, Wang Z (2009) Cancer-derived mutations in the fibronectin III repeats of PTPRT/PTPrho inhibit cell-cell aggregation. Cell Commun Adhes 16(5–6):146–153

142. Lim S-H, Kwon S-K, Lee MK, Moon J, Jeong DG, Park E, Kim SJ, Park BC, Lee SC, Ryu S-E, Yu D-Y, Chung BH, Kim E, Myung P-K, Lee J-R (2009) Synapse formation regulated by protein tyrosine phosphatase receptor T through interaction with cell adhesion molecules and Fyn. EMBO J 28(22):3564–3578

143. Kumar M, Agarwal T, Khokhar S, Kumar M, Kaur P, Roy TS, Dada R (2011) Mutation screening and genotype phenotype correlation of $-crystallin, $-crystallin and GJA8 gene in congenital cataract. Mol Vis 17:693–707

144. Roshan M, Vijaya PH, Lavanya GR, Shama PK, Santhiya ST, Graw J, Gopinath PM, Satyamoorthy K (2010) A novel human CRYGD mutation in a juvenile autosomal dominant cataract. Mol Vis 16:887–896

145. Zhang L-Y, Yam GH-F, Tam PO-S, Lai RY-K, Lam DS-C, Pang C-P, Fan DS-P (2009) An alphaA-crystallin gene mutation, Arg12Cys, causing inherited cataract-microcornea exhibits an altered heat-shock response. Mol Vis 15:1127–1138

Chapter 13

Epistasis Analysis Using Information Theory

Jason H. Moore and Ting Hu

Abstract

Here we introduce entropy-based measures derived from information theory for detecting and characterizing epistasis in genetic association studies. We provide a general overview of the methods and highlight some of the modifications that have greatly improved its power for genetic analysis. We end with a few published studies of complex human diseases that have used these measures.

Key words Epistasis, Information theory, Entropy, Association studies, Genetic analysis, Gene–gene interaction

1 Introduction

Epistasis has been defined in multiple ways [1–3] and is likely to play central roles in the genetic architecture of common human diseases [4–10]. Multiple modeling approaches have been developed to detect epistasis in genetic association studies, including logistic regression methods such as the focused interaction testing framework (FITF) [11] and logic regression [12–14] as well as machine learning approaches such as multifactor dimensionality reduction (MDR) [15–23]. These approaches are particularly well suited for detecting epistasis or nonadditive gene–gene interactions in the absence of significant independent effects. We focus here on the use of entropy-based measures from information theory that are powerful and computationally efficient detectors of epistatic interactions.

As reviewed by Moore [24], information theory was launched as a formal discipline in 1948 with the publication of Shannon's paper on "A Mathematical Theory of Communication" [25]. Shannon, at that time, worked for Bell Labs and was interested in mathematical methods of encoding information for transmission as electronic signals. The basic problem was to encode a message, transmit it as a signal, receive it, and decode it with minimal noise such that the original message was not lost. It was in these seminal

Jason H. Moore and Scott M. Williams (eds.), *Epistasis: Methods and Protocols*, Methods in Molecular Biology, vol. 1253, DOI 10.1007/978-1-4939-2155-3_13, © Springer Science+Business Media New York 2015

papers that Shannon introduced and defined entropy as a measure of uncertainty to help maximize the efficiency and accuracy of encoding, sending, receiving, and decoding a message. Building on Shannon's work, in 1954 McGill introduced the concept of interaction information [26]. This measure has been rediscovered multiple times as an entropy-based measure for nonadditive or synergistic interactions between attributes [27, 28]. We review here interaction information for the detection of epistasis and present some recent extensions. We then provide several examples of its use from real data.

2 Materials

2.1 Getting Started

Interaction information for genetic analysis is implemented in the open-source MDR software package that is freely available from http://sourceforge.net/projects/mdr and the Visualization of Statistical Epistasis Networks (ViSEN) software package [29] that is freely available from http://sourceforge.net/projects/visen. Both are programmed entirely in Java and are compatible with any operating system that has Java installed. We provide elsewhere in this volume an overview of the graphic user interface (GUI) of the MDR software package. Command line options are also available and are described in a document labeled MDR_command_line_help.txt that is distributed with MDR. The interaction information algorithms are available in the Entropy tab and the Network tab of MDR. It is important to note that the base interaction information algorithm is accessible in several other freely available data mining software packages including Orange [30] (*see* **Note 1**).

2.2 Data Format

MDR, Orange, and ViSEN accept a tab-delimited text file as input (*see* **Note 2**). Data are loaded in MDR using the Load Datafile button in the upper right corner of the main screen. The data can also be viewed using the View Datafile button. Sample data files with discrete and continuous outcomes and simulated gene–gene interactions are provided as part of the MDR software package. Briefly, each row is an observation or instance (e.g., a human subject) and each column is a discrete variable or attribute (e.g., a single-nucleotide polymorphism or SNP). Typically, the attributes are polymorphisms with three genotypes coded 0, 1, and 2 or covariates such as gender coded 0 and 1. The last column is saved for the phenotype or outcome, which can be binary (e.g., case-control status) or continuous (e.g., body mass index). MDR requires a complete data set with no missing values. Missing data can be handled in two ways. First, the missing data can be coded with a number not used to represent a genotype (e.g., [9]) and will be treated as an extra level in the analysis (e.g., a fourth genotype). This is acceptable as long as your data are missing at random. As a second option data imputation methods can be used to fill in the missing values (*see* **Note 3**).

3 Methods

3.1 Interaction Information

For a discrete variable X with alphabet X and probability mass function $p(x)$, its entropy $H(X)$ is defined as

$$H(X) = -\sum_{x \in X} p(x) \log p(x).$$

When there is more than one random variable, the definition of entropy can be extended as follows. The *joint entropy* of two discrete random variables X and Y with a joint distribution $p(x,y)$ is defined as

$$H(X,Y) = -\sum_{x \in X} \sum_{y \in Y} p(x,y) \log p(x,y),$$

and the conditional entropy of X, given knowledge of X, can be obtained by the chain rule as

$$H(X \mid Y) = H(X,Y) - H(Y).$$

Dependency between two random variables can be described using *mutual information* [31]. This is a measure of the amount of information that one random variable contains about the other, or can be thought of as the reduction of uncertainly of one random variable given the knowledge of the other. In the context of genetic association studies, mutual information can be very useful to quantify how much of a phenotypic status is explained by genotypic variations. We consider a genetic attribute G_1 and the phenotypic class C, e.g., case or control, are both discrete, random variables. The mutual information $I(G_1; C)$ measures the reduction in the uncertainty of the class C due to knowledge about the genotype of G_1, defined as

$$I(G_1; C) = H(C) - H(C \mid G_1).$$

Intuitively, $I(G_1; C)$ can be used as a measure of the main effect of the genetic attribute G_1 on the class C.

Mutual information can also be extended to measure the interaction effect between two attributes. Given two genetic attributes G_1 and G_2, the mutual information

$$I(G_1, G_2; C) = H(C) - H(C \mid G_1, G_2)$$

measures how much of the phenotypic class can be explained by joining G_1 and G_2 together. By subtracting the individual main effects of G_1 and G_2 from their joint effect $I(G_1, G_2; C)$, i.e.,

$$IG(G_1; G_2; C) = I(G_1, G_2; C) - I(G_1; C) - I(G_2; C),$$

the *information gain* $IG(G_1, G_2; C)$, the gain of mutual information, knowing both G_1 and G_2 with respect to the class C is obtained. A positive value of $IG(G_1; G_2; C)$ indicates *synergy* between G_1 and G_2, while a negative value indicates *redundancy* or correlation between them. As first discussed by Moore et al. [28] synergy can be used to measure epistatic interactions between two genetic attributes. This pairwise information-gain measure has been successfully applied in many epistasis studies thanks to its model-free, non-parametric, and fast implementation [28, 32–43]. Hu et al. have shown that these pairwise measure can also facilitate the search for three-way epistatic interactions [33].

3.2 Three-Way Interaction Information

Extension of the interaction information measure to more than two genetic attributes is important for epistasis studies because many complex human diseases are likely to involve genetic interactions higher than two-way. There is no widely accepted formal definition of information gain including genetic attributes higher than two. In a previous attempt, Anastassiou and Varadan [44, 45] defined three-way information gain by comparing the integrated joint mutual information to the best-achieved subset mutual information after breaking the whole into partitions, mathematically written as

$$IG_{\text{partition}}\left(G_1; G_2; G_3; C\right) = I\left(G_1, G_2, G_3; C\right) - \max \begin{cases} I\left(G_1, G_2; C\right) + I\left(G_3; C\right) \\ I\left(G_1, G_3; C\right) + I\left(G_2; C\right) \\ I\left(G_2, G_3; C\right) + I\left(G_1; C\right) \\ I\left(G_1; C\right) + I\left(G_2; C\right) + I\left(G_3; C\right). \end{cases}$$

The partition of the set $\{G_1, G_2, G_3\}$ chosen in this formula is the one that maximizes the sum of the amounts of mutual information connecting the subsets with the phenotypic class. This is referred as "maximum-information partition" of the set $\{G_1, G_2, G_3\}$ with respect to the class C. This three-way $IG_{\text{partition}}\left(G_1; G_2; G_3; C\right)$ quantifies the information that can be gained by combining G_1, G_2, and G_3 together and comparing to its maximum-information partition. Although technically sound, this formula might include false-positive errors for pure three-way epistasis. For instance, assuming $I(G_1, G_2; C)$ and $I(G_3; C)$ is the maximum-information partition, after combining G_3 with $\{G_1, G_2\}$, the gained information could be the result of either the pure three-way epistasis, or the pairwise epistasis between G_3 and one (or both) of $\{G_1, G_2\}$, or the mixture of all of the above.

A more strict alternative measure [46] was proposed as follows:

$$IG_{\text{alternative}}\left(G_1; G_2; G_3; C\right) = I\left(G_1, G_2, G_3; C\right)$$
$$- IG\left(G_1; G_2; C\right) - IG\left(G_1; G_3; C\right) - IG\left(G; G_3; C\right)$$
$$- I\left(G_1; C\right) - I\left(G_2; C\right) - I\left(G_3; C\right),$$

where all the lower order effects are subtracted. However, as reviewed and pointed out in [44], this formula fails in the extreme redundancy case where G_1, G_2, and G_3 provide the same full amount of information on C, i.e., $G_1 = G_2 = G_3 = C$. In this case, $I(G_i; C) = I(G_i, G_j; C) = I(G_i, G_j, G_k; C) = H(C)$, where i, j, k are different values taken from $\{1, 2, 3\}$. Therefore $IG_{\text{alternative}}(G_1; G_2; G_3; C) = H(C)$, which indicates the contradictory extreme synergy.

We have previously proposed [43] a new strict measure by modifying $IG_{\text{alternative}}$ as

$$IG_{\text{strict}}(G_1; G_2; G_3; C) = I(G_1, G_2, G_3; C)$$
$$-\max\left\{\begin{matrix} IG(G_1; G_2; C) \\ 0 \end{matrix}\right. -\max\left\{\begin{matrix} IG(G_1; G_3; C) \\ 0 \end{matrix}\right. -\max\left\{\begin{matrix} IG(G_2; G_3; C) \\ 0 \end{matrix}\right.$$
$$-I(G_1; C) - I(G_2; C) - I(G_3; C).$$

We only subtract pairwise synergies, i.e., positive information gain, because the failure of $IG_{\text{alternative}}$ is due to the fact that it adds back information by subtracting negative information gain. By subtracting all lower order effects and synergies, IG_{strict} measures the *pure* three-way synergy that is observable only by considering three attributes together (*see* **Note 4**).

3.3 Statistical Epistasis Networks

The methods summarized above provide measures of interaction information for small sets of genetic variants. These measures reflect the degree of nonadditive or synergistic effects on the phenotype. Hu et al. have provided a global approach to summarizing and visualizing these measures across many sets of genetic variants using network science [34]. The working hypothesis is that this statistical epistasis network approach reveals global information about genetic architecture that is not captured in any one set of genetic variants. We briefly review this methodology here and later provide an example application to real data.

Networks are formalized mathematically by graphs. A graph G is composed of a set $V(G)$ of vertices and a set $E(G)$ of edges [47]. In epistasis networks [34], each vertex corresponds to an SNP, and we use v_A to denote the vertex corresponding to SNP A. An edge linking a pair of vertices, for instance v_A and v_B, corresponds to an interaction between SNPs A and B. We first assigned a weight to each SNP and each pair of SNPs to quantify how much of the disease status the corresponding SNP and SNP pair genotypes explain. In analogy to statistical models, those weights correspond to the strength of the main and the interaction effects, and stronger effects translate into higher weights. Here, interaction information as described above is used to assign weights to each pair of SNPs.

An important challenge in building statistical epistasis networks from measures such as interaction information is determining the threshold that is used. A low threshold will result in a very highly connected network while a high threshold results in a very small and potentially uninteresting network. Hu et al. [34] used the percolation threshold but other measures such as statistical significance using permutation testing could be explored. With the percolation threshold method the size of the network is plotted by the threshold. A percolation threshold can be defined as an inflection point in the bivariate relationship such that after a certain threshold the network size changes rapidly. This provides an objective threshold value and can be a nice compromise between a small uninteresting network and a large highly connected network that is hard to interpret.

Once a network is constructed there are a number of different network measures that can be performed [48]. Hu et al. specifically looked at four measures. The first is the number of edges or synergistic interactions exceeding the threshold. The second is the number of vertices or genetic variants that have at least one edge. The third is the size of the largest connected component. This latter measure is important because there are often many self-contained networks. The final measure is the vertex degree distribution. The vertex degree for any single SNP is the number of edges connecting to it. Other measures such as centrality are covered elsewhere in this volume.

3.4 Visualization of Statistical Epistasis Networks

Hu et al. [29] provide a methodology and open-source software package called Visualization of Statistical Epistasis Networks (ViSEN) for visualization of two-way and three-way interactions estimated from the interaction information methods described above. The key to this approach is the ability to visualize three-way epistatic interactions as triangles where the points of the shape connect to individual SNPs and the size of the shape is proportional to the strength of the three-way synergistic effect as defined by Hu et al. [43].

ViSEN has a set of controls to read user data and to save the graph layout. The lists of one to three orders of effects follow the standard tab-delimited network file format with each line of text consisting of an attribute name (two and three attribute names for two-way and three-way files, respectively) followed by the effect strength. The format of the user population SNP data is also tab-delimited plaintext. The first line contains a header row of labels assigned to each column of data, and each subsequent line contains a data row. The last column of the file is the class. This is the same format described above for MDR.

After the initial layout is displayed, the user can reposition the nodes and triangles for fine-tuning. ViSEN also provides the user with a set of controls to turn on and off the labels for the strength of epistatic effects as needed. In addition, the user can control the number of pairwise and three-way interactions being visualized in the GUI. While edges and triangles are inserted into or removed from the layout, ViSEN animates these changes for the user to

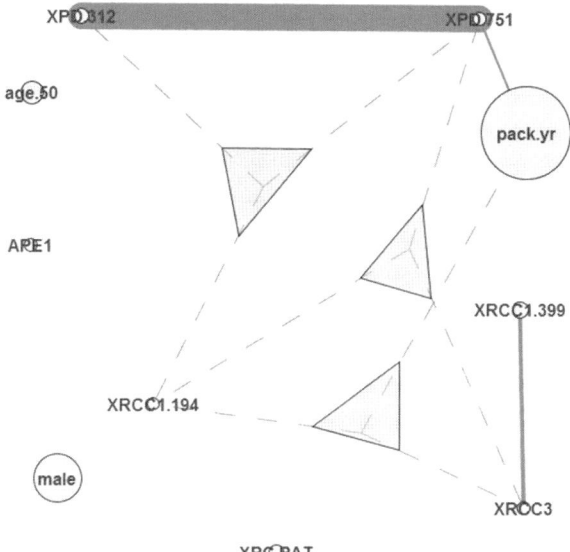

Fig. 1 A ViSEN network of synergistic interactions among a set of SNPs and other factors such as pack-years of smoking (pack.yr), age above and below 50 (age.50), and gender (male). Independent effects of each factor are denoted by the size of the circle. Two-way synergistic interactions are denoted by a line. Three-way synergistic interactions are denoted by a triangle. The strength of the interactions I are proportional to the thickness of the line or size of the triangle. Note that in both case the lower order effects have been subtracted leaving only the synergistic interaction

observe how the network evolves. When a satisfactory layout is achieved, ViSEN can export the visualization to a PNG file. An example of the output is provided in Fig. 1.

3.5 Application of ViSEN to Bladder Cancer

We have selected the following machine learning study to highlight these entropy-based methods. Andrew et al. [49] carried out an epidemiologic study to identify genetic and environmental predictors of bladder cancer susceptibility in a sample of Caucasians ($n = 914$) from New Hampshire. This study focused specifically on genes that play an important role in the repair of DNA sequences that have been damaged by chemical compounds (e.g., carcinogens). Seven SNPs were measured including two from the *X-ray repair cross-complementing group 1* gene (*XRCC1*), one from the *XRCC3* gene, two from the *xeroderma pigmentosum group D* (*XPD*) gene, one from the *nucleotide excision repair* gene (*XPC*), and one from the *AP endonuclease 1* gene (*APE1*). Each of these genes plays an important role in DNA repair. Smoking is a known risk factor for bladder cancer and was included in the analysis along with gender and age for a total of 10 attributes. Age was discretized to $>$ or ≤ 50 years.

A parametric statistical analysis of each attribute individually revealed a significant independent main effect of smoking as expected. However, none of the measured SNPs were significant predictors of bladder cancer individually. Andrew et al. used MDR to exhaustively evaluate all possible two-, three-, and four-way interactions among the genetic and environmental variables. The best MDR model had a testing accuracy of 0.66 ($p < 0.001$) and included two SNPs from the *XPD* gene and smoking. Decomposition of this model using the measures of information gain demonstrated that the effects of the two *XPD* SNPs were non-additive or synergistic suggestive of epistasis or gene–gene interaction. The same analysis indicated that the effect of smoking was mostly independent from the gene–gene interaction effect. Analysis of the same data using a learning classifier system approach confirmed this result, suggesting that smoking and the epistatic interaction may actually be associated with bladder cancer in different subsets of subjects (i.e., heterogeneity) [50].

Application of the ViSEN software (Fig. 1) shows smoking by gender and age as having the strongest independent effects (biggest circles). The strongest pairwise interaction (denoted by the thickest line) is between the two *XPD* polymorphisms previously described as having strong synergistic interactions. A moderate synergistic interaction is shown between the *XRCC1* and *XRCC3* polymorphisms. The triangles indicate the three-way synergistic interactions with the one-way and two-way effects subtracted out as described above in the Methods (*see* **Note 5**). Note the three-way interaction between the polymorphisms in *XPD* and *XRCC1*. A reexamination of the MDR results revealed that this was the fourth best three-way MDR model. When just these three SNPs are considered, the three-way MDR model has a training accuracy of 0.64 and a testing accuracy of 0.60. This was statistically significant but not as good as the two *XPD* SNPs with smoking (*see* **Note 6**).

3.6 Application of Statistical Epistasis Networks to Bladder Cancer

As mentioned above, Hu et al. [34] used statistical epistasis networks to investigate the genetic architecture of approximately 1,500 SNPs across nearly 500 cancer susceptibility genes in a population-based study of bladder cancer. Details of the study and the data have been provided elsewhere [51]. The authors found that resulting networks were much larger and more complex than expected by chance. The largest connected component consisted of 39 SNPs representing an overabundance of genes from the aryl hydrocarbon receptor pathway that plays an important role in responding to toxic chemicals that enter cells as carcinogens such as benzo[a]pyrene from tobacco smoke. This was tested using gene set enrichment methods in a subsequent study [52]. The authors generated a functional annotation of this epistasis network by treating bladder cancer cells with benzo[a]pyrene and then measuring changes in gene expression using DNA microarrays.

The authors found nonrandom sets of differentially expressed genes were enriched in the epistasis network. Although not proof, these results suggest that there may be a biologically important pattern that is captured by statistical epistasis network analysis. This is a plausible finding since benzo[a]pyrene is a major carcinogen in tobacco smoke, and smoking is a known strong risk factor for bladder cancer. If true, this suggests that statistical epistasis networks are able to capture patterns of genetic effects for which there may be biological basis (*see* **Note** 7).

3.7 Conclusions

We have reviewed here the use of information theoretic methods for measuring non-additive interactions between multiple genetic variants in population-based studies of human health. The tools described are available in several open-source software packages including Orange, MDR, and ViSEN. We have also provided several real-world examples where these algorithms have detected networks of interacting genetic variants. More information about how these methods fit into a general bioinformatics strategy for detecting epistasis is provided by Moore et al. [53]. Information on other methods for detecting and characterizing interactions has been provided by Cordell [4].

4 Notes

The following are some important tips for using interaction information algorithms:

1. Orange is a user-friendly machine learning and data mining software package that uses graphical programming. Use the Interaction Graph widget located in the Unsupervised tab of the Orange Canvas. Be sure and install the graphvis package from http://www.graphviz.org/ prior to analysis.

2. A sample data file with an example of the tab-delimited format is provided with the MDR software.

3. A simple frequency-based imputation approach may work well when there are very few missing genotypes. Of course, genetic data is ideally suited to multivariate imputation methods that use linkage disequilibrium to infer missing genotypes. The freely available PLINK software is commonly used to impute genotype data [54].

4. When applying to genetics data, the mutual-information and information-gain measures can be normalized by dividing the class entropy $H(C)$. The normalized measures range from $[-1, 1]$, and provide the percentage of explaining the phenotypic class C by giving the knowledge of one or multiple genetic attributes.

5. It is important to note that entropy-based scores can be used as expert knowledge to help guide a stochastic search algorithm such as genetic programming [55–57]. More details of this approach can be found in the computational evolution chapter in this volume.

6. The bigger the gap between the training and testing accuracies the more likely it is that the model is overfitting the data. The closeness of two accuracies can be used as a measure of model fit or quality.

7. The promise of systems biology is to bring population level statistical results closer to individual-level biology. We have previously written about this challenge specifically for understanding biological and statistical epistasis [7].

Acknowledgements

This work was supported by National Institutes of Health (NIH) grants AI59694, EY022300, GM103534, GM103506, LM009012, LM010098, and LM011360.

References

1. Phillips PC (2008) Epistasis – the essential role of gene interactions in the structure and evolution of genetic systems. Nat Rev Genet 9:855–867

2. Phillips PC (1998) The language of gene interaction. Genetics 149:1167–1171

3. Cordell HJ (2002) Epistasis: what it means, what it doesn't mean, and statistical methods to detect it in humans. Hum Mol Genet 11:2463–2468

4. Cordell HJ (2009) Detecting gene–gene interactions that underlie human diseases. Nat Rev Genet 10:392–404

5. Moore JH (2003) The ubiquitous nature of epistasis in determining susceptibility to common human diseases. Hum Hered 56:73–82

6. Moore JH (2005) A global view of epistasis. Nat Genet 37:13–14

7. Moore JH, Williams SM (2005) Traversing the conceptual divide between biological and statistical epistasis: systems biology and a more modern synthesis. Bioessays 27:637–646

8. Tyler AL, Asselbergs FW, Williams SM et al (2009) Shadows of complexity: what biological networks reveal about epistasis and pleiotropy. Bioessays 31:220–227

9. Cowper-Sal lari R, Cole MD, Karagas MR et al (2011) Layers of epistasis: genome-wide regulatory networks and network approaches to genome-wide association studies. Wiley Interdiscip Rev Syst Biol Med 3:513–526

10. Moore JH, Williams SM (2009) Epistasis and its implications for personal genetics. Am J Hum Genet 85:309–320

11. Millstein J, Conti DV, Gilliland FD et al (2006) A testing framework for identifying susceptibility genes in the presence of epistasis. Am J Hum Genet 78:15–27

12. Kooperberg C, Ruczinski I (2005) Identifying interacting SNPs using Monte Carlo logic regression. Genet Epidemiol 28:157–170

13. Kooperberg C, Ruczinski I, LeBlanc ML et al (2001) Sequence analysis using logic regression. Genet Epidemiol 21(Suppl 1):S626–S631

14. Schwender H, Ruczinski I (2010) Logic regression and its extensions. Adv Genet 72:25–45

15. Hahn LW, Ritchie MD, Moore JH (2003) Multifactor dimensionality reduction software for detecting gene–gene and gene–environment interactions. Bioinformatics 19:376–382

16. Ritchie MD, Hahn LW, Roodi N et al (2001) Multifactor-dimensionality reduction reveals high-order interactions among estrogen-metabolism genes in sporadic breast cancer. Am J Hum Genet 69:138–147

17. Hahn LW, Moore JH (2004) Ideal discrimination of discrete clinical endpoints using multi-locus genotypes. In Silico Biol 4:183–194

18. Ritchie MD, Hahn LW, Moore JH (2003) Power of multifactor dimensionality reduction for detecting gene–gene interactions in the presence of genotyping error, missing data, phenocopy, and genetic heterogeneity. Genet Epidemiol 24:150–157

19. Moore JH (2004) Computational analysis of gene–gene interactions using multifactor dimensionality reduction. Expert Rev Mol Diagn 4:795–803

20. Moore JH (2010) Detecting, characterizing, and interpreting nonlinear gene–gene interactions using multifactor dimensionality reduction. Adv Genet 72:101–116

21. Velez DR, White BC, Motsinger AA et al (2007) A balanced accuracy function for epistasis modeling in imbalanced datasets using multifactor dimensionality reduction. Genet Epidemiol 31:306–315

22. Pattin KA, White BC, Barney N et al (2009) A computationally efficient hypothesis testing method for epistasis analysis using multifactor dimensionality reduction. Genet Epidemiol 33:87–94

23. Moore JH (2007) Genome-wide analysis of epistasis using multifactor dimensionality reduction: feature selection and construction in the domain of human genetics. In: Zhu X, Davidson I (eds) Knowledge discovery and data mining: challenges and realities. IGI Global, Hershey, PA, pp 17–30

24. Moore JH (2008) Bases, bits and disease: a mathematical theory of human genetics. Eur J Hum Genet 16:143–144

25. Shannon CE (1948) A mathematical theory of communication. Bell Syst Tech J 27:379–423

26. McGill WJ (1954) Multivariate information transmission. Psychometrika 19:97–116

27. Jakulin A, Bratko I (2003) Analyzing attribute dependencies. In: Lavrač N, Gamberger D, Todorovski L et al (eds) Knowledge discovery in databases: PKDD 2003. Springer, Berlin, pp 229–240

28. Moore JH, Gilbert JC, Tsai C-T et al (2006) A flexible computational framework for detecting, characterizing, and interpreting statistical patterns of epistasis in genetic studies of human disease susceptibility. J Theor Biol 241:252–261

29. Hu T, Chen Y, Kiralis JW et al (2013) ViSEN: methodology and software for visualization of statistical epistasis networks. Genet Epidemiol 37(3):283–285

30. Demšar J, Curk T, Erjavec A et al (2013) Orange: data mining toolbox in python. J Mach Learn Res 14:2349–2353

31. Cover TM, Thomas JA (2006) Elements of information theory. Wiley-Interscience, Hoboken, NJ

32. Fan R, Zhong M, Wang S et al (2011) Entropy-based information gain approaches to detect and to characterize gene–gene and gene–environment interactions/correlations of complex diseases. Genet Epidemiol 35:706–721

33. Hu T, Andrew AS, Karagas MR et al (2013) Statistical epistasis networks reduce the computational complexity of searching three-locus genetic models. Pac Symp Biocomput 397–408

34. Hu T, Sinnott-Armstrong NA, Kiralis JW et al (2011) Characterizing genetic interactions in human disease association studies using statistical epistasis networks. BMC Bioinformatics 12:364

35. McKinney BA, Reif DM, White BC et al (2007) Evaporative cooling feature selection for genotypic data involving interactions. Bioinformatics 23:2113–2120

36. Dong C, Chu X, Wang Y et al (2008) Exploration of gene–gene interaction effects using entropy-based methods. Eur J Hum Genet 16:229–235

37. Kang G, Yue W, Zhang J et al (2008) An entropy-based approach for testing genetic epistasis underlying complex diseases. J Theor Biol 250:362–374

38. Wu C, Li S, Cui Y (2012) Genetic association studies: an information content perspective. Curr Genomics 13:566–573

39. Chanda P, Zhang A, Brazeau D et al (2007) Information-theoretic metrics for visualizing gene–environment interactions. Am J Hum Genet 81:939–963

40. Sucheston L, Chanda P, Zhang A et al (2010) Comparison of information-theoretic to statistical methods for gene–gene interactions in the presence of genetic heterogeneity. BMC Genomics 11:487

41. Chanda P, Zhang A, Ramanathan M (2011) Modeling of environmental and genetic interactions with AMBROSIA, an information-theoretic model synthesis method. Heredity 107:320–327

42. Tritchler DL, Sucheston L, Chanda P et al (2011) Information metrics in genetic epidemiology. Stat Appl Genet Mol Biol 10, Article 12

43. Hu T, Chen Y, Kiralis JW et al (2013) An information-gain approach to detecting three-way epistatic interactions in genetic association studies. J Am Med Inform Assoc 20:630–636

44. Anastassiou D (2007) Computational analysis of the synergy among multiple interacting genes. Mol Syst Biol 3:83

45. Varadan V, Miller DM 3rd, Anastassiou D (2006) Computational inference of the molecular logic for synaptic connectivity in C. elegans. Bioinformatics 22:e497–e506

46. Chechik G, Globerson A, Tishby N et al (2002) Group redundancy measures reveal redundancy reduction in the auditory pathway. In: Becker S, Ghaharamani Z, Dietterich TG (eds) Advances in neural information processing systems. MIT Press, Cambridge, MA, pp 173–180

47. West D (2007) Introduction to graph theory. Prentice Hall PTR, Upper Saddle River, NJ

48. Newman MEJ (2010) Networks: an introduction. Oxford University Press, Oxford, UK

49. Andrew AS, Nelson HH, Kelsey KT et al (2006) Concordance of multiple analytical approaches demonstrates a complex relationship between DNA repair gene SNPs, smoking and bladder cancer susceptibility. Carcinogenesis 27:1030–1037

50. Urbanowicz RJ, Andrew AS, Karagas MR et al (2013) Role of genetic heterogeneity and epistasis in bladder cancer susceptibility and outcome: a learning classifier system approach. J Am Med Inform Assoc 20:603–612

51. Andrew AS, Gui J, Sanderson AC et al (2009) Bladder cancer SNP panel predicts susceptibility and survival. Hum Genet 125:527–539

52. Hu T, Pan Q, Andrew AS et al (2014) Functional genomics annotation of a statistical epistasis network associated with bladder cancer susceptibility. BioData Min 7:5

53. Moore JH, Asselbergs FW, Williams SM (2010) Bioinformatics challenges for genome-wide association studies. Bioinformatics 26:445–455

54. Purcell S, Neale B, Todd-Brown K et al (2007) PLINK: a tool set for whole-genome association and population-based linkage analyses. Am J Hum Genet 81:559–575

55. Greene CS, Hill DP, Moore JH (2010) Environmental sensing of expert knowledge in a computational evolution system for complex problem solving in human genetics. In: Riolo R, O'Reilly U-M, McConaghy T (eds) Genetic programming theory and practice VII. Springer, New York, USA, pp 19–36

56. Moore JH, Andrews PC, Barney N et al (2008) Development and evaluation of an open-ended computational evolution system for the genetic analysis of susceptibility to common human diseases. In: Marchiori E, Moore JH (eds) Evolutionary computation, machine learning and data mining in bioinformatics. Springer, Berlin, pp 129–140

57. Moore JH, Greene CS, Andrews PC et al (2009) Does complexity matter? artificial evolution, computational evolution and the genetic analysis of epistasis in common human diseases. Genetic programming theory and practice VI. Springer, New York, USA, pp 1–19

Chapter 14

Genome-Wide Epistasis and Pleiotropy Characterized by the Bipartite Human Phenotype Network

Christian Darabos and Jason H. Moore

Abstract

Networks are central to turning the colossal amount of information generated by high-throughput genetic technology into manageable sources of knowledge. They are an intuitive way of representing interaction data, yet they offer a full set of sophisticated quantitative tools to analyze the phenomena they model. When combining genetic information, diseases, and phenotypic traits, networks can reveal and facilitate the analysis of pleiotropic and epistatic effects at the genome-wide scale. Genome-wide association study data is publicly available, and so are gene and pathway databases, and many more, making the global overview next to impossible. Networks allow information from these multiple sources to be encompassed. We use connections between the strata of the network to characterize pleiotropy and epistasis effects taking place between traits and biological pathways. The global graph-theory-based quantitative methods reveal that levels of pleiotropy and epistasis are in-line with theoretical expectations. The results of the magnified "glaucoma" region of the network confirm the existence of well-documented interactions, supported by overlapping genes and biological pathways and more obscure associations. They have the potential to generate new hypotheses for yet uncharacterized interactions. As the amount and complexity of genetic data increase, bipartite and, more generally, multipartite networks that combine human diseases and other physical attributes with layers of genetic information have the potential to become ubiquitous tools in the study of complex genetic, phenotypic interactions, and possibly improve personalized medicine.

Key words Pleiotropy, Epistasis, Eye diseases, Glaucoma, Network, GWAS, Human Phenotype Network, SNPs, Pathways

1 Introduction

Conjectures about the complexity of human genetics arose over a century ago. Since then, geneticists have confirmed that most genetic traits do not obey the simple Mendelian one-gene-one-phenotype paradigm. Because of this inherent complexity, we are still working on developing novel methods to diagnose, treat, cure, and prevent genetic diseases. At the center of prevention lies information on individual patients' personal genetic risk landscape.

Jason H. Moore and Scott M. Williams (eds.), *Epistasis: Methods and Protocols*, Methods in Molecular Biology,
vol. 1253, DOI 10.1007/978-1-4939-2155-3_14, © Springer Science+Business Media New York 2015

Moreover, genetic disorders can no longer be studied in isolation from one another or from external and environmental factors. The cascading effects of genomic mutations can extend to entire organisms, thus having a global understanding of the ramifications of mutations, including all the affected phenotypes and diseases, is crucial. Two phenomena are pivotal to complex genetic variations: pleiotropy, when a single mutation affects several traits, and epistasis, when multiple mutations in distant parts of the genome have synergetic, non-linear effects on a single phenotype. Networks are the preferred quantitative analysis and visualization method for interactions of human diseases and traits. Networks offer intuitive representations of phenotypic and genotypic interactions, and they come with a comprehensive analytical toolset to perform quantitative studies of their intrinsic properties.

The concepts of epistasis and pleiotropy are widely underappreciated due to their perceived rarity. State-of-the-art genome-wide association studies (GWAS) most often look for individual genes with large impacts on a single phenotype. The impacts of genetic mutation, like genetic disease, cannot be studied in isolation, even if attempts are made to bridge the gap between a single gene and a single phenotype. In systems biology, the pervasiveness and strength of biomolecular interactions require a step back from reductionist biology and an acknowledgement of the importance of biological networks and pathways. These biological pathways interact at the systems level, and may include environmental and external interactions and signals.

Using bipartite networks to characterize epistasis and pleiotropy at the systems level, we can go beyond the reductionist definitions and embrace complex genetic interaction phenomena. This method offers a bird's eye view of the effects of genetic mutations on human phenotypes. It is often arduous to distinguish between certain types of pleiotropy and epistasis. The effect of a single mutation rippling though a pathway can be confused with the combined effect of distinct mutations. There is therefore a case to be made about studying these phenomena in unison.

This chapter describes the methodology to build and use bipartite Human Phenotype Networks (HPNs) made of two distinct sets of vertices: phenotypic traits and biological pathways. HPNs are constructed with GWAS data and other publicly available genetic databases. This type of network helps characterize the pleiotropic and epistatic interactions at the systems level. Finally, we magnify a specific phenotypic region of the HPN, the "glaucoma" region, which groups the disease and all its first and second neighbors. We offer a close up view of pleiotropic and epistatic interactions within a specific sub-network.

2 Materials

2.1 Concepts of Pleiotropy and Epistasis

Ludwig Platt and William Bateson first introduced the concepts of pleiotropy and epistasis, respectively, to explain observed inconsistencies in Mendelian inheritance and in the one-gene-one-phenotype paradigms [1, 2]. To adapt with the progress in understanding genetics, the definition of pleiotropy has changed since it was first coined in 1910, and remains somewhat loose. A thorough history of pleiotropy in the past 100 years can be found in Stearns' 2010 review [3]. It refers to the general phenomenon in which a single gene dictates two or more seemingly unrelated phenotypic traits. In some cases, the definition is limited to a single mutation in a locus that affects multiple traits. It is, however, widely accepted that there is more than one type of pleiotropy. Grüneberg [4] in 1938 correctly distinguished between two major types called "genuine" and "spurious" pleiotropy. Genuine pleiotropy refers to a single locus responsible for the production of two distinct gene products, whereas spurious pleiotropy involves a single gene product utilized in two different ways. Furthermore, he distinguished a second form of spurious pleiotropy in which the single primary product initiates a cascade of events with different phenotypic consequences. Spurious pleiotropy can be said to perturb biological pathways. Since then, more refined subdivisions have emerged. To help us navigate the various types of pleiotropy, Hodgking's survey offers classifications, descriptions, and examples of seven types of pleiotropy [5] (Table 1).

Actual genetic mechanisms of pleiotropy are extremely diverse. Genuine pleiotropy encapsulates pleiotropy at the mRNA-processing level, multiple or overlapping loci reading frames,

Table 1
A classification of different types of pleiotropy

Type	Situation
Artefactual	Adjacent but functionally unrelated genes affected by the same mutation
Secondary	Simple primary biochemical disorder leading to complex final phenotype
Adoptive	One gene product used for quite different chemical purposes in different tissues
Parsimonious	One gene product used for identical chemical purposes in multiple pathways
Opportunistic	One gene product playing a secondary role in addition to its main function
Combinatorial	One gene product employed in various ways, and with distinct properties, depending on its different protein partners
Unifying	One gene, or cluster of adjacent genes, encoding multiple chemical activities that support a common biological function

Adapted form Hodgkin's study [13]

alternative splicing, and multifunctional proteins, to mention a few. Spurious pleiotropy covers single loci mutations that produce deviation in the gene product affecting other genes or regulatory elements located further downstream in biological pathways. Indeed, new gene products may promote or repress the expression of other genes. They may initiate alternate gene-gene and protein-protein interactions and alternate mRNA and microRNA production, which may, in turn, affect seemingly, unrelated phenotypes. Pleiotropic genes offer a unique insight into the complexities of biomolecular interaction networks.

In epistasis, on the other hand, the phenotypic contribution of a gene and its gene products depends on the specific genotype of a locus at a different genomic position. From the origin of the word, "standing upon," we derive the modern definition of epistasis, or epistatic gene effects, in which the expression of an allele at one locus masks the expression of an allele at another locus [6]. Epistasis is, therefore, usually the result of multiple genetic mutations at different loci. In this age of genome-wide association studies (GWAS), epistatic studies can be conducted at the genome level, quantitatively studying the masking and combined effects of single nucleotide polymorphisms (SNPs).

Both epistasis and pleiotropy are exceptions to the one-gene-one-phenotype Mendelian rules of genetics. They are, however, far from being rare deviations [7]. Epistasis and pleiotropy are ubiquitous inherent properties of biological systems, and they are necessary byproducts of biomolecular networks [8]. Most phenotypes are the result of interactions between thousands of genes, as well as between genes and their environments. Because of the widespread connectivity within networks, the effects of a single mutation or variation can spread through thousands of gene-gene interactions, resulting in multiple phenotypes, or pleiotropy. The connections through which a variant's effects propagate define the molecular basis for epistatic interactions and how they translate into an observed phenotype. Because of their close relatedness, it is not unreasonable to conclude that a similar set of quantitative tools can be applied to study both phenomena, sometimes simultaneously. In this chapter, these tools are Bipartite Human Phenotype Networks.

2.2   Networks and Graphs

Networks (or graphs, in mathematical terms) provide a means of intuitively visualizing and characterizing complex systems, which have proven to be particularly valuable in modeling biological systems (Fig. 1). The statistical analysis of graph properties offers a quantitative and holistic means of revealing underlying connections among vertices, as well as the emergent global properties. Networks are being used with increasing frequency to analyze large-scale systems. A network can take an extraordinarily complex

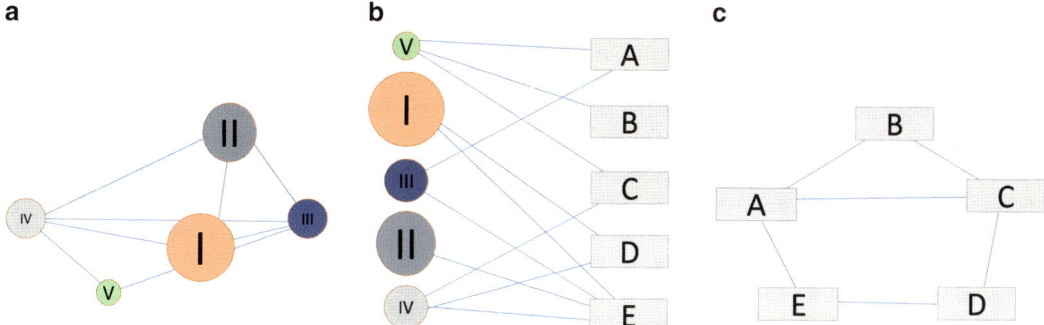

Fig. 1 Bipartite Network schematic. A bipartite network (**b**) made of two data sets the "*circles*," and the "*rectangles*." Projections in the "*circle*" space (**a**) and in "*rectangle*" space (**c**)

system and reduce it to a relatively simple form, revealing underlying connections and important clustering details that would not be evident from studying individual or non-complex relationships among traits [9].

Formally, a network is a collection of nodes (or vertices) and edges (or links) connecting them. The degree, k, of a node is the number of edges incident upon the node, and the degree distribution, $P(k)$ of the network, describes the fraction of nodes in a network with degree k. The degree distribution also characterizes global properties of the graph and how the nodes are connected to one another; for example, if they are connected at random, the nodes' degrees are expected to be homogeneous, and the degree distribution would be uniform binomial. More often in biology, networks are highly heterogeneous, with a "heavy-tailed" degree distribution, placing them in the scale-free family. This means that the degree distribution follows a power law, or exponential decay. Within the network, this translates into the presence of "hubs"—a minority of highly connected nodes. When the degree distribution of a "scale-free" network is plotted on a logarithmic scale, the resulting curve is approximately linear across the top [9]. In the case of relatively small networks, it is impossible to demonstrate the presence of a scale-free network. We can, at best, show the existence of a power-law-type degree distribution and not dismiss the scale-free hypothesis. The clustering coefficient (CC) of a network measures the degree to which nodes tend to form closely knit communities with higher than average connectivity [10]. The CC of networks found in nature, in particular social and biological networks, show higher degrees of clustering than randomized networks of identical size. The average path length (APL) of a network represents the average of the minimum number of edges separating any two vertices.

2.3 Bipartite Networks

Bipartite networks [11] consist of two disjointed sets of nodes. The nodes are connected such that the vertices of one set will have no connections between them, but can only be connected to nodes of the other set. The use of a bipartite network is natural when dealing with two different types of data sets (Fig. 1b), in our case phenotypes and pathways. Two nodes of the same type cannot connect, so one node can only be connected to a node of the other data type. We used a bipartite network to construct the relationships of our data.

From the bipartite network, one can project the data onto either of the data spaces (Fig. 1a, c). In either single dataset space, nodes are connected to one another through a vertex of the other space. By ignoring the different types of data, all network properties described above remain valid in the bipartite network (as a single data set network) and on either projection. This type of network gives us three degree-distributions, one for each projection, and one for the bipartite network. Each degree distribution shows how many links each node has. Nodes in a projection of a bipartite network are connected if they share at least one node in the other group. This gives us the ability to visualize connections within a group.

2.4 Human Disease Networks

In recent years there has been a trend toward studying disease through network based analyses of various systems of connections between diseases. The result was the Human Disease Network (HDN) [12]. The nodes in the HDN represent human genetic disorders, and the edges represent various connections between disorders, including gene-gene or protein-protein interactions. The HDN is helpful in visualizing connections among human disorders on a large scale. The underlying connections of the HDN contribute to the understanding of the basis of disorders, which, in turn, leads to a better comprehension of human diseases.

One study by Goh et al. [12] explored the HDN built on genes shared by different diseases. Another study by Li et al. [13] traced the SNPs connecting disease traits. In 2009, Silpa Suthram et al. [14] found that when diseases were compared by analyses of disease-related mRNA expression data and the human protein interaction network, there were significant similarities between some diseases and between some drug treatments. In 2009, Barrenas et al. [15] further studied the genetic architecture of complex diseases by performing GWAS. They found that complex disease genes are less central than the essential and monogenic disease genes in the human interactome (*see* **Note 1**). In the present work, we expand our study to include not only disease traits, but also behaviors and normal variations in humans, such as hair color, and explore large portions of noncoding variants in the human genome. Links between phenotypes are based on overlapping biological pathways (Subheading 3).

2.5 GWAS Catalog and Pathway Association Data

GWAS identify common genetic variants, such as SNPs, found in the genotype of different individuals in association with phenotypic traits. The catalog of published GWAS maintained by the National Human Genome Research Institute (NHGRI) at the National Institute of Health aggregates studies that report phenotype-to-SNP(s) and phenotype/SNP-to-gene associations (http://www.genome.gov/gwastudies/). The NHGRI catalog, downloaded in March 2013, was the primary source of phenotype-to-gene to association data. It reports over 800 phenotypes associated with approximately 2,300 genes and 6,000 SNPs.

Biological pathways represent elaborate series of cascading biochemical reactions and signals occurring within and without the cell [16]. Pathways govern all major cellular functions, such as cell cycle, cell respiration, or apoptosis (programmed cell death). Biochemical compounds (e.g., nucleic acids, proteins, complexes, and small molecules) participating in reactions form a network of biological processes and are grouped into pathways. Reactome is an open-source, open access, manually curated and peer-reviewed pathway database (http://www.reactome.org). It visually displays structured information about the elements, enzymes, and genes (via their gene products) within many known pathways. The Reactome database was accessed in March 2013.

3 Methods

3.1 The Human Phenotype Network

Using GWAS catalog data, one can extend the HDN to include not only diseases, but also general phenotypes, encompassing behavioral traits and physical attributes, such as hair color, and explore large portions of noncoding variation in the human genome. This more complete representation is called the Human Phenotype Network or HPN [17]. The catalog of published GWAS maintained by the National Human Genome Research Institute (NHGRI) at the National Institute of Health (http://www.genome.gov/gwastudies/) offers the primary source of phenotypic data. It aggregates studies that report SNP(s)-to-phenotype(s) and gene(s)-to-phenotype(s) associations. The NHGR catalog used in this chapter, downloaded in June 2013, reports 646 phenotypes associated with 2,000+ genes and 6,000+ SNPs.

Over 90 % of risk-associated SNPs identified by GWAS fall outside of coding regions [18], stressing the requirement for more global assessment of phenotypic associations. In this work we explore methods of building the HPN that go beyond previously mentioned gene-centric HDN approaches. An interesting side effect of all the methods presented below is that before obtaining an HPN, the algorithm produces a bipartite network (*see* Subheading 2.3), which is the property that allows us to study the pleiotropic and epistatic information in the genetic association data.

The HPN is obtained by projecting the bipartite network onto the phenotype space.

The following sections present our methods for building the HPNs based on the overlap of biological pathways of the associated SNPs and genes. It offers high-density information contained within both the bipartite network and the projected HPN.

3.2 Building the Pathway-Based Human Phenotype Network

Construct a network of human phenotypes based on shared biological pathways of associated genes. This is accomplished by linking genes to phenotypes (or traits) from hundreds of GWAS catalogs at NHGRI. Genes were further linked to pathways using Reactome data. By annotating each trait with its associated SNPs, genes, and pathways, the HPN can now link phenotypes with genes involved in the same pathways. The steps used to build the network are illustrated in Fig. 2 and are performed using the following steps.

1. From the NHGRI catalog, extract all phenotypes and annotate them with their mapped SNPs and genes. Those with no mapped SNPs or genes are omitted.

2. From Reactome, extract all genes in the database and annotate them with their associated pathways.

3. Annotate the genes associated with each phenotype to their associated pathways.

4. Connect phenotypes with overlapping pathways with an undirected edge, setting edge weight as the number of overlapping pathways.

We filter out isolate phenotypes with no connections to the rest of the network. We are only interested in phenotypes that have been associated with a gene, and their possible shared biology.

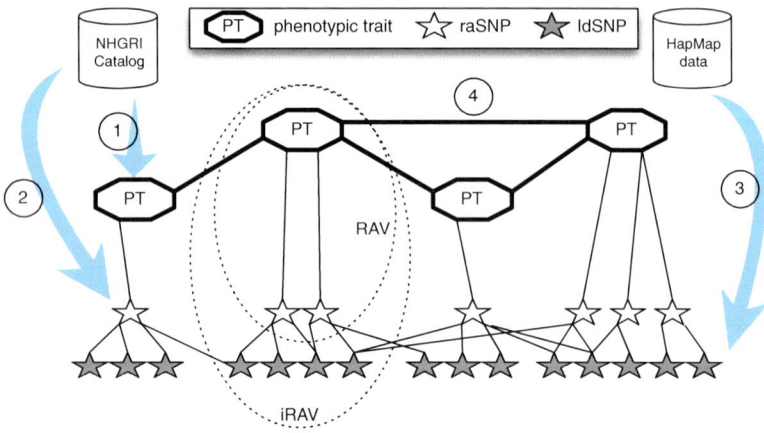

Fig. 2 Step-by-step description of the method to obtain the HPN. The *circled numbers* correspond to the steps described in Subheading 3.1

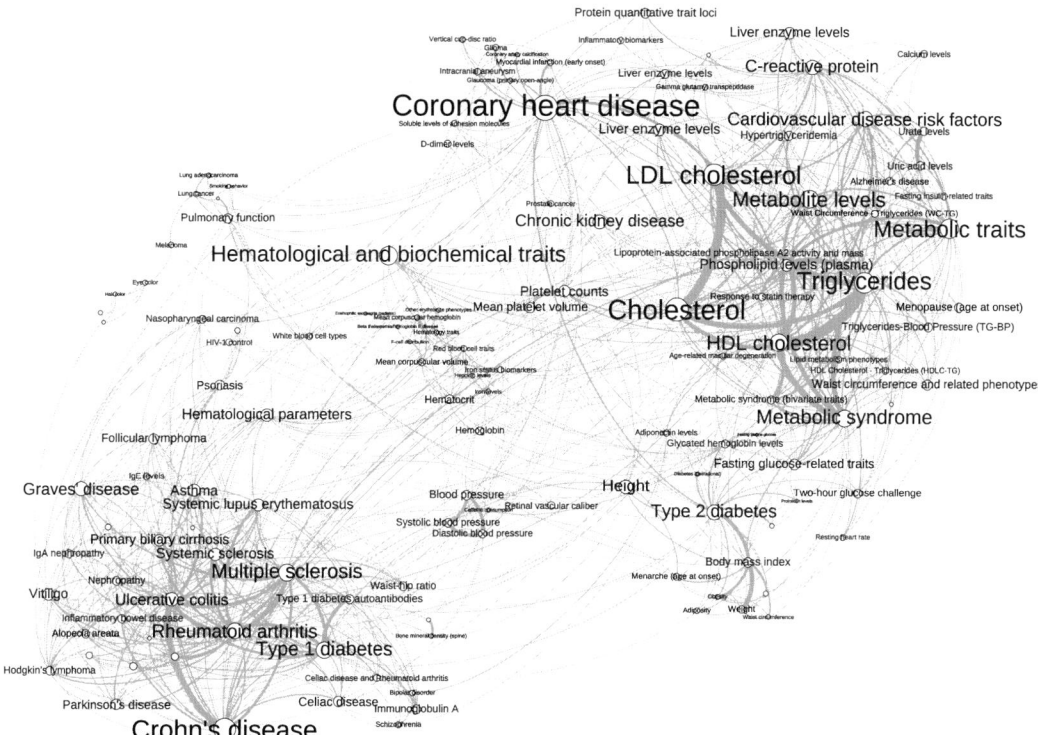

Fig. 3 The Human Phenotype Network. In order to increase the readability, we have filtered out nodes with a degree lower than 5 (i.e. connected to less than 5 other nodes), showing only 137 nodes (about 30 %) and about 45 % of the actual edges. To further facilitate the readability, we have manually merged a number of clearly redundant nodes. The nodes and labels sizes are proportional to the original degree of the phenotype (before filtering). The edge width is proportional to the number of overlapping pathways

The original NHGRI database contained over 800 phenotypes; by removing the isolated nodes, the HPN contains 401 nodes connected to at least one other node (Fig. 3) (*see* **Notes 2** and **3**).

This flexible process of building phenotype-gene-pathway associations also allowed us to examine the network from multiple configurations. Specifically, we were also able to construct a pathway network following the same logic as the HDN (Subheading 2.3): connecting pathways based on shared phenotypes, as well as a bipartite graph with links between phenotypes and pathways (*see* **Note 4**).

The bipartite network shown in Fig. 4 consists of 1,523 vertices (408 phenotypes, 1,115 pathways) and over 10,000 edges, with an average degree $k \approx 7$. We do not show the intermediate stages of genes, as this makes the network difficult to interpret. Indeed, highly connected phenotypes are connected to 40+ pathways, and highly connected pathway, to 100–300 phenotypes. Height is clearly associated with most pathways, forming a major hub. However, it is safe to suppose that the size of the height hub represents a bias, because it is recorded in most studies. It is unclear what the implications of this and other data biases are.

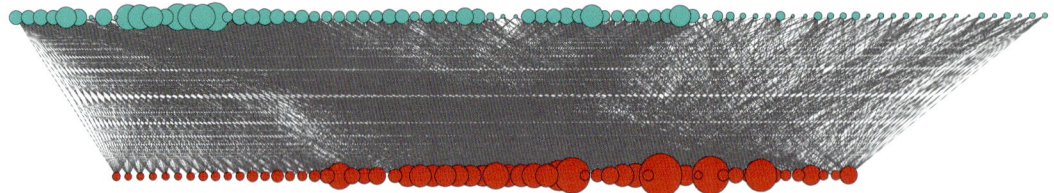

Fig. 4 Bipartite Phenotypes-Pathways Network. The *top* (*blue*) *row* of vertices represents the phenotypes and the *bottom* (*red*) *row* of vertices the pathways. The vertices' sizes are proportionate to its degree. For readability reasons, we omit vertices with degrees <24

3.3 Characterizing and Quantifying Pleiotropy and Epistasis in the HPNs

Early studies have made use of network theory in studying both pleiotropy and epistasis. Global statistical properties of networks, such as the "shape" of the degree distribution and an above average clustering coefficient place gene expression networks in the small-world [10] or scale-free [9] family of networks [19]. This indicates that most of the nodes (genes) in the network are of a low degree k. However, a small minority of the vertices is highly connected (hubs). Put in the context this chapter, it means that a few genes have extensive pleiotropic/epistatic effects, but most genes only affect/are affected by a small number of phenotypes. The quantitative structural analyses of protein interaction networks of model organisms have highlighted the importance of properties such as diameter and average path length. Li et al. [20] determined that the diameter was ~4–5 edges, meaning that each gene in the genomes studied affected on average four or five proteins. This finding also corroborates the conjecture that pleiotropy and epistasis are confined to genomic modules, and cannot generally affect any pairs/sets of loci in the genome [21].

But why limit ourselves to the projected HPN, which does not contain actual interactions between phenotypes and pathways. Instead, we will analyze interactions between the two layers of the bipartite networks. Because of the density and complexity of the HPNs, the following section presents the results of a quantitative overview of pleiotropy and epistasis as properties of the entire network. In addition, Subheading 3.4 reports the clinical implications and the specific effects observed in a region of the HPNs centered on the "glaucoma" vertex and its neighboring phenotypes. The use of pathways as overlapping units necessitates a slightly amended definition of pleiotropy, usually at the gene level, and epistasis, usually at the SNP or gene level. Admittedly, this interpretation of pleiotropy and epistasis may somewhat stray from the commonly accepted definitions, but it is in line with the loose nature of the phenomena, where both have sub-types that relate to multiple levels of granularity.

Table 2
Quantitative properties of the biological pathway-based HPN

	Pathway-based HPN
LCC size (%nodes)	396 (61 %)
#Edges	40K
Avg. degree/weighted	204.1/497.6
APL/diameter	1.48/3
Avg. CC	0.79
Modularity/communities	0.10/4
Isolate vertices	250
Avg. pleiotropy (#phenotype/#pathways)	27.06
Avg. epistasis (#pathways/#genotypes)	5.69

Relying on data in the bipartite HPNs, we calculate the number of phenotypes connected to each phenotype. We use the average connectivity $<k>$ of pathway vertices as a proxy measuring global pleiotropy (Table 2). We also measure the degree distribution of the pathway subset, showing the effect of pleiotropy at each pathway. Inversely, the average epistatic effect of pathways on phenotypes can be calculated as the average degree of the phenotype subset in the bipartite HPN (Table 2). The degree distribution of the phenotype subset conveys the distribution of epistatic effects that different pathways have on phenotypes.

Both pleiotropy and epistasis distributions are "right-skewed" with heavy tails, which denotes the presence of hubs. Pleiotropy distributions show that most pathways only influence a few phenotypes; however, a small minority of pathways influences a large number (50+) of phenotypes. Similarly, the epistasis distribution hints to the fact that most phenotypes can be associated with only a few pathways. However, a small number of phenotypes rely on the signaling of a large number of pathways. Although somewhat simplistic, these results are in line with the findings of Featherstone et al. [19]. We acknowledge that the manner in which the average effects are computed may capture more than just pleiotropic and epistatic effects. However, these results, due to the their ubiquity, reflect a biologically plausible property of the system. We run a full array of quantitative statistical analysis of the different HPNs, including the average pleiotropic and epistatic effects (Table 2).

The analysis of the bipartite HPN's global statistical properties agrees with the findings of Li et al. [20]. The above average CC and the shape of the heavy-tailed degree distributions put the HPN in the scale-free region of the network topology spectrum. The number of isolated nodes provides insight into how many phenotypes have no detected genetic connections to any other phenotype. The average epistasis value shows that each pathway affects about five to six phenotypes. On the other hand, average pleiotropy is cooperatively high, showing that each pathway tends to affect ~27 phenotypes, on average. This denotes the necessity to apply a biologically relevant filter to the HPN in order to extract the "backbone" of the network, containing the most relevant genetic influences [22].

3.4 Clinical Implications: The Example of Glaucoma

The HPN can help to explain the shared etiology of glaucoma and other diseases by revealing a substantial number of interactions unseen in the Human Disease Network. Ultimately, studying different genetic predictive elements, such SNPs or, at the other end of the spectrum, pathways, from a global perspective, using networks, could contribute to novel discoveries of pleiotropic drug therapies.

The HPN confirms well-known interactions, such as those between glaucoma and blood pressure. Studies have linked the two for years, and drugs used to treat glaucoma, such as beta-blockers and alpha-adrenergic agonists [23], are known to affect blood pressure. In fact, patients with cardiovascular problems are advised against taking beta-blockers, a treatment for high intraocular pressure (IOP) associated with glaucoma, because of effects on heart rate and blood pressure [24]. Moreover, many studies have shown that blood pressure and ocular perfusion are important factors in the pathogenesis of glaucoma. For example, studies have linked increases in blood pressure to slight increases in IOP. Going further, the "Blue Mountains Eye Study" found that systemic hypertension was significantly associated with an increased risk of primary open-angle glaucoma (POAG), independent of the effect of blood pressure on IOP. Systemic hypertension was the greatest risk factor for POAG [24]. Blood pressure is a first neighbor to glaucoma in the HPN, suggesting the validity of the model. They are linked by the umbrella "pathways in cancer." Diabetes mellitus is another well-documented disease known to interact with glaucoma [25]. Type 1 diabetes is a direct neighbor, and type 2 diabetes is a second (indirect) neighbor. Type 1 diabetes and glaucoma are linked by the cell cycle and human T-lymphotropic virus 1 (HTLV-I) infection pathways. Type 2 diabetes and glaucoma share common associated gene: cyclin-dependent kinase inhibitor 2B antisense RNA 1 (CDKN2B-AS1). Additionally, the HPN allows us to see connections that are not included in the HDN, which could lead to new advances in treatments for the linked diseases. For instance, Alzheimer's disease is a second neighbor of glaucoma. Both are neurodegenerative diseases and their similarities have recently begun to receive significant attention. Inoue et al. maintain that elevated levels of biomarkers for Alzheimer's are more often found in patients with open-angle glaucoma (OAG) than in patients with cataracts [26]. In addition, Alzheimer's and OAG share pathways such as cell death mechanisms (apoptosis), reactive oxygen species (ROS) production, mitochondrial dysfunction, and vascular abnormalities [27]. Apoptosis of the neural ganglia cells is a major issue in glaucoma. In the HPN, the link between Glaucoma and Alzheimer's disease is not readily apparent by looking at the graph—it becomes a third neighbor. Another interesting link is to the "smoking behavior" phenotype, although this is only readily apparent in the HPN, where it is a first neighbor to glaucoma. The two share the umbrella pathways in cancer. Association studies have shown that smoking behavior is correlated with central corneal thickness in OAG and might also be a risk factor for POAG [28].

3.5 Conclusions

The study of genetic diseases is progressing at an unprecedented pace, thanks to modern high-throughput sequencing technology and to the development of modeling techniques at the crossroads of bioinformatics and mathematics. Bipartite HPN modeling is capable of leveraging the massive amount of GWAS and other readily accessible genetic data, and collapsing the information into a single, manageable source. The projection of the HPN helps analyze phenotypic interactions [17, 22]. The overall structure of the connections between layers of the bipartite HPN, on the other hand, allows us to estimate in a quantitative manner the pleiotropic and epistatic effects at a global scale. Finally, by magnifying regions of the HPN, we are able to highlight previously documented phenotypic interactions, supported by genes and biological pathways evidence as a proof of concept. The bipartite HPN is a flexible, scalable, and intuitive model. The HPN is potentially useful to study phenotypic links, as well as uncover novel pleiotropy and epistasis effects at the single variation level, at the gene level, and all the way to the biological pathway.

4   Notes

The following are some important tips when using bipartite networks of phenotype-to-pathway interactions:

1. As demonstrated by Goh's Human Disease Network, bipartite interaction networks can be constructed at any level of genetic overlap. Work beyond the scope of this chapter discusses HPNs constructed around SNPs, SNPs in linkage disequilibrium, genes, and of course biological pathways. Insight can be gained at any level of granularity [22].

2. There exist numerous software packages and libraries that can help layout, visualize, and run comprehensive quantitative analyses of networks. Amongst those, we can cite Cytoscape,

Gephi, iGraph, JUNG, NetworkX, and Network Workbench. Most allow importing simple "edge list" formatted files and can manage bipartite as well as simple graphs.

3. Achieving a meaningful layout of network models offering intuitive insights of the intrinsic properties of the system might be impractical due to the sheer size or complexity of the data at hand. Quantitative statistical analyses base on graph theory can offer significant information on the global properties of the modeled system without the need for graphical qualitative representation.

4. Most "simple" interaction networks involving a single type of node are, in fact, projections of multilayered networks involving several different types of vertices. Remember that, although potentially more accessible, all projections do lose information that cannot be retrieved.

5. Most networks found in nature are too dense to offer any quantitative or qualitative insight into the modeled system. These types of network necessitate the application of a filter prior to the analysis. Network/graph filters vary immensely in their complexity, efficiency, and applicability, depending on the model at hand.

6. Most domain-specific algorithms applied to networks are actually transferable to other domains. Community detection algorithms derived from social sciences have successfully been applied to Human Phenotype Networks to identify modules of traits with underlying shared biology [17].

Acknowledgements

This work was supported by National Institutes of Health (NIH) grants R01 EY022300, LM009012, LM010098, AI59694.

References

1. Bateson W (1907) Facts limiting the theory of heredity. Science 26(672):649–660
2. Plate L (1910) Vererbungslehre und deszendenztheorie. Festschrift für R Hertwig II:537–610
3. Stearns FW (2010) One hundred years of pleiotropy: a retrospective. Genetics 186(3):767–773
4. Gruneberg H (1938) An analysis of the "pleiotropic" effects of a new lethal mutation in the rat (Mus norvegicus). Proc R Soc Lond B Biol Sci 125(838):123–144
5. Hodgkin J (1998) Seven types of pleiotropy. Int J Dev Biol 42(3):501–505
6. Anthony JF Griffiths, Jeffrey H Miller, David T Suzuki, Richard C Lewontin, and William M Gelbart. Introduction to Genetic Analysis. 7th edition. W. H. Freeman, 2000
7. Moore JH (2003) The ubiquitous nature of epistasis in determining susceptibility to common human diseases. Hum Hered 56(1–3):73–82
8. Tyler AL, Asselbergs FW, Williams SM, Moore JH (2009) Shadows of complexity: what biological networks reveal about epistasis and pleiotropy. Bioessays 31(2):220–227
9. Newman M (2010) Networks: an introduction. Oxford University Press, Inc., New York, NY

10. Watts DJ, Strogatz SH (1998) Collective dynamics of "small-world" networks. Nature 393:440–442

11. Zhou T, Ren J, Medo M, Zhang Y-C (2007) Bipartite network projection and personal recommendation. Phys Rev E 76:046115

12. Goh K-I, Cusick ME, Valle D, Childs B, Vidal M, Barabasi A-L (2007) The human disease network. Proc Natl Acad Sci 104(21):8685–8690

13. Li H, Lee Y, Chen JL, Rebman E, Li J, Lussier YA (2012) Complex-disease networks of trait-associated single-nucleotide polymorphisms (SNPs) unveiled by information theory. J Am Med Inform Assoc 19(2):295–305

14. Suthram S, Dudley JT, Chiang AP, Chen R, Hastie TJ, Butte AJ (2010) Network-based elucidation of human disease similarities reveals common functional modules enriched for pluripotent drug targets. PLoS Comput Biol 6(2):e1000662

15. Barrenas F, Chavali S, Holme P, Mobini R, Benson M (2009) Network properties of complex human disease genes identified through genome-wide association studies. PLoS One 4(11):e8090

16. Schilling CH, Schuster S, Palsson BO, Heinrich R (1999) Metabolic pathway analysis: basic concepts and scientific applications in the post-genomic era. Biotechnol Prog 15(3):296–303

17. Darabos C, Desai K, Cowper-Sal-lari R, Giacobini M, Lupien M, Moore JH. Inferring human phenotype networks from genome-wide genetic associations. In Giacobini M, Vanneschi L, Bush WS (eds.), Evolutionary computation, machine learning and data mining in bioinformatics—11th European Conference, EvoBIO 2013, Vienna, Austria, April 3–5, 2013. Proceedings, Lecture Notes in Computer Science. Springer, to appear, 2013

18. Hindorff LA, Sethupathy P, Junkins HA, Ramos EM, Mehta JP, Collins FS, Manolio TA (2009) Potential etiologic and functional implications of genome-wide association loci for human diseases and traits. Proc Natl Acad Sci U S A 106(23):9362–9367

19. Featherstone DE, Broadie K (2002) Wrestling with pleiotropy: genomic and topological analysis of the yeast gene expression network. Bioessays 24(3):267–274

20. Li R, Tsaih S-W, Shockley K, Ioannis M, Stylianou JW, Paigen B, Churchill GA (2006) Structural model analysis of multiple quantitative traits. PLoS Genet 2(7):e114

21. Welch JJ, Waxman D (2003) Modularity and the cost of complexity. Evolution 57(8):1723–1734

22. Darabos C, White MJ, Graham BE, Leung DN, Williams S, Moore JH (2014) The multiscale backbone of the human phenotype network based on biological pathways. BioData Min 7(1):1

23. Chae B, Cakiner-Egilmez T, Desai M (2013) Glaucoma medications. Insight 38(1):5–9

24. He Z, Vingrys AJ, Armitage JA, Bui BV (2011) The role of blood pressure in glaucoma. Clin Exp Optom 94(2):133–149

25. Oswal KS, Sivaraj RR, Murray PI, Stavrou P (2013) Clinical course and visual outcome in patients with diabetes mellitus and uveitis. BMC Res Notes 6(1):167

26. Inoue T, Kawaji T, Tanihara H (2013) Elevated levels of multiple biomarkers of Alzheimer's disease in the aqueous humor of eyes with open-angle glaucoma. Invest Ophthalmol Vis Sci 54(8):5353–8

27. Wang D, Huang Y, Huang C, Pengfei W, Lin J, Zheng Y, Peng Y, Liang Y, Chen J-H, Zhang M (2012) Association analysis of cigarette smoking with onset of primary open-angle glaucoma and glaucoma-related biometric parameters. BMC Ophthalmol 12:59

28. Ghiso JA (2013) Alzheimer's disease and glaucoma: mechanistic similarities and differences. J Glaucoma 22(Suppl 5):S36–S38

Chapter 15

Network Theory for Data-Driven Epistasis Networks

Caleb A. Lareau and Brett A. McKinney

Abstract

One of the challenges of understanding the genetic basis of complex phenotypes is explaining variability not attributable to individual genes. While most existing methods that investigate variant mutations or differential gene expression focus on individual effects, a complex system of gene interactions (epistasis) and pathways is likely needed to explain phenotypic variation. Herein, we examine methods for treating the interactions in these biological data sets as edges in a network model of the phenotype and review relevant network theory methods for analyzing network structure and identifying important genes. In particular, we review methods for detecting community structure, describing the statistical properties of networks, and computing network centrality of genes that may reveal insights missed by individual genetic effects. We also discuss available tools to facilitate the construction and visualization of epistasis networks of GWAS data.

Key words Epistasis, Network theory, GWAS, SNPs, Biological pathways

1 Introduction

Biological datasets from the cellular and genomic levels have uncovered thousands of associations between genes and complex traits, like disease. Most analyses performed on microarray, genotyping, and sequencing data have yielded insights based on single marker or single-variant associations. However, genes do not often operate as sole effectors. Rather, genes, proteins, and RNA each play a part in a complex interactive system that can ultimately lead to disease. As many previous association models lack the sensitivity that interaction models can bring to biological data, we argue for the increased usage of networks to more properly characterize the genetic architecture of complex diseases. This chapter contains an overview of the theory, construction, and evaluation of epistasis networks, a powerful tool for uncovering hidden biological effects.

1.1 Epistasis and Disease

In the 1850s and 1860s, Gregor Mendel tested thousands of pea plants to establish the notions of dominant and recessive alleles. Mendel demonstrated that physical characteristics of parents could

Jason H. Moore and Scott M. Williams (eds.), *Epistasis: Methods and Protocols*, Methods in Molecular Biology, vol. 1253, DOI 10.1007/978-1-4939-2155-3_15, © Springer Science+Business Media New York 2015

be predicted to appear in clear statistical distributions in later off-spring. Fifty years later, William Bateson coined the term "epista-sis" to describe the perceived genetic interaction between two traits that caused deviation from Mendelian ratios. The canonical work of these two scientists established two biological axioms that we recognize here: (1) traits are heritable, and (2) gene interac-tions affect the heritability of these traits. While these two rules can be easily stated, uncovering their effects in cell cultures, statistical models, and pedigrees has proven considerably more difficult for human disease.

To observe epistasis in a cell system, one must first identify a candidate interaction then execute a series of cumbersome molecu-lar biology techniques to test the interaction hypothesis. While the results of these techniques are invaluable for our knowledge of pathways and cellular processes, the recent boom in genomic data enables fast, efficient discovery of gene interactions using data-driven model inference. For example, one experiment utilized radi-ation hybrid genotyping data to identify nearly seven million potential interactions in the mammalian genome [1]. We also note that dozens of studies have implicated novel epistatic partners in disease-specific systems (such as [2–5]). The complication of disease-specific epistasis renders traditional methods of uncovering interac-tions impractical. As these studies demonstrate that epistasis helps explain the genetic architecture of human disease, statistical methods are needed to uncover the basis of complex traits.

While statistical methods enable more exhaustive interrogation of gene interactions, we note that biological epistasis and statistical epistasis have fundamentally different meanings. Biological epista-sis can be defined as the physical interaction of two or more mol-ecules at the cellular level, whereas statistical epistasis represents a significant deviation from the null model at the population level. One of the challenges of epistasis is to understand the relationship between statistical epistasis and the many potential underlying molecular mechanisms. Though these terms are not interchange-able, both forms of epistasis must be examined to elucidate the mechanism of disease onset. In this chapter, we discuss theoretical and methodological approaches to computing statistical epistasis networks.

1.2 Networks

The recent explosion of social media has resulted in unprecedented interest in social network analyses. Billions of people worldwide are connected via Facebook friendships, Tweets, and LinkedIn con-nections, to name a few. Your personal social network likely con-sists of a conglomeration of people ranging from close friends to professional associates to high school acquaintances. Organizing social relationships in a network structure provides a powerful means to analyze the dissemination of information, such as fashion trends or celebrity gossip. Likewise, hundreds of studies have demonstrated the power of arranging biological relationships in networks. Network theory has demonstrated its utility in modeling biological phenomena ranging from macroscopic structures such as ecosystems and evolution [6] to intracellular interactions to uncover novel pathways [7]. While the genetics underlying complex traits like disease remain elusive, genomic information modeled in a network structure has proven a powerful method to uncover effects. Here, we discuss the functional roles of different elements of a network to establish an informative model of a complex trait at the genomic level.

1.2.1   Nodes

In network theory, nodes or vertices are the fundamental units that are connected by edges as defined by some relationship. As nodes often represent a more concrete entity such as people in a social network or webpages in the Internet, nodes are often the most intuitive component of any network. For epistasis networks, a node usually represents a variant's gene or protein. However, nodes can represent other genomic features. For example, the recently completed ENCODE project emphasized the presence of various RNAs, such as long noncoding RNAs and micro RNAs, that are not translated into amino acids. These nucleic acid residues can be visualized in an interactive network in the same manner as genes, enabling us to investigate the influence that interactions between RNA and protein-coding sequences can have on disease. Furthermore, nodes can represent individual variants that are contained in a gene or linked to a gene. While mapping individual variants to their corresponding genes is a reasonable strategy, some highly functional variants can affect multiple genes by altering expression, protein binding, or other genomic interactions, sometimes warranting the depiction of the variants themselves as independent functional units.

1.2.2   Edges

The other fundamental component of a network is the edge, which connects pairs of nodes. For epistasis networks, we assume that the networks are simple graphs, which have at most one edge between two nodes. Thus, there are no multi-edge connections between a pair of nodes, but edges may be weighted. The edge represents the relationship between two nodes. For biological networks, the possible relationships are numerous. Edges may be computed by correlation, as in co-expression networks, homology, protein-protein interactions, and transcription regulation, to name a few. For epistasis networks, edges represent theoretical interactions between two biological entities computed from a statistical model. While networks like the Internet or a gene regulatory network are comprised of directed edges where information flows from one node to another but not necessarily in reverse, epistasis networks consist of undirected edges.

Several methods have been proposed to compute epistatic interactions ranging from information theory [3, 4], machine learning [8], and multifactor dimensionality reduction [9] to Bayesian methods [10]. Recently, we created an algorithm called regression Genetic Association Interaction Network (reGAIN) that employs a regression-based approach to compute interactions and build a network from genotyping and sequencing data [2]. A GAIN is an extension of an epistasis network, because a GAIN also includes marginal main effect information in the network. One aspect that distinguishes the edges in epistasis networks from many other data-driven networks is the conditional dependence of the connection on the phenotype. In this way, edges are context sensitive, or in other words, disease specific [11]. For microarray and other forms of expression data, the primary method of defining edges has been the correlation between probes, though we note that the reGAIN approach can be applied to expression data. Later in this chapter, we discuss the theory and the advantages and disadvantages of a regression-based approach for constructing the edges in a GAIN.

1.2.3 Prioritizing Nodes

When visualizing or analyzing a network, the sheer number of nodes and edges can obscure important features. Imagine a biological network of 20,000+ human genes and the hundreds of thousands of interactions that connect them. Determining the most important genes in such a network would be daunting without utilizing an algorithm to prioritize nodes in a network. One of the more intuitive methods is to count the number of edges or sum the edge weights of a given node with other nodes in the network. This process, known as *degree centrality*, assumes that the nodes with the most edges are the most important features of a given network.

To represent networks in a computational framework, matrices are a convenient data structure. In particular, we define an *adjacency matrix* for a finite network with n nodes as the $n \times n$ matrix, where non-diagonal entries correspond to the edge between node i and node j. For a weighted adjacency matrix, the elements correspond to the strength of the edge between two nodes. An *incidence matrix* represents the number of edges between two nodes. It is more common to work in the adjacency matrix framework, where the degree centrality (C_D) of a node i, n_i, can be written in terms of the adjacency matrix, A, as

$$C_D\left(n_i\right) = \sum_{i \neq j} A_{i,j} \tag{1}$$

While degree centrality is a simple and intuitive method for ranking the importance of nodes, it lacks sensitivity as this measure treats each connection as equally important. Consider nodes A and

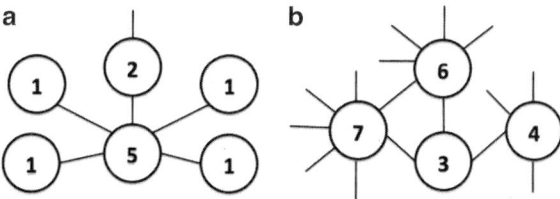

Fig. 1 Two theoretical subgraphs with highlighted nodes A and B. Each node is labeled with its degree. While A has a greater connectivity, node B is likely more important in the network as a whole due to its connectivity with other influential nodes with high degree

B in Fig. 1. Each node has its degree value written in the center, where node A has a degree of 5 and node B has a degree of 3. Thus, ranking by degree centrality would quickly prioritize node A over node B. However, one could argue that node B is actually more influential in this hypothetical network due to its connectivity with other important nodes. While A is connected to nodes with degrees 1,1,1,1, and 2, B is connected to nodes with degree 4, 6, and 7.

Imagine that these sets of nodes were connected in the network of human genes. Despite having fewer connections than A, a mutation in node/gene B would cause more disruption throughout the network due to its connections to other highly connected nodes. In contrast, mutations in gene A would be limited to the five genes that it interacts with directly (radius = 1). This simple, hypothetical example demonstrates the need for a more sensitive algorithm that captures the difference in importance of connected nodes as well as the effect of distant nodes (radius > 1).

Suppose we knew the importance of each node directly connected to A and B. We could then improve on centrality estimates by including a recursive factor $C_E(n_j)$ in Eq. 1 that gives a weight to each connection based on the connection's importance, yielding

$$C_E(n_i) = \sum_{i \neq j} A_{ij} C_E(n_j). \tag{2}$$

In this example, the weighted centrality of node A ($C_E(A)$) from Eq. 2 is 6, while for node B $C_E(B) = 17$. This notion of weighting the connections more accurately prioritizes node B over node A. In general, we do not know the weights of the other nodes ahead of time. However, this problem turns out to be a simple and efficient eigenvector calculation, hence the name of this centrality measure, *eigenvector centrality*.

Eigenvector centrality measures the influence of each node on the network as a whole, an important extension of degree centrality. In our example above, eigenvector centrality would rank node

B over node A due to its connectivity with other important nodes. Though eigenvector centrality has been implemented in dozens of algorithms, one related application is easily the most famous: Google's PageRank. The mathematical form of PageRank looks similar to eigenvector centrality with the addition of a constant term that represents random jumps to other regions of the network. PageRank was designed to prioritize nodes (webpages) based on their connectivity and return a ranked list based on a search query. Many other methods for calculating centrality have been established for different types of networks. Later in this chapter, we focus on SNPrank centrality for biological and epistasis networks. While the World Wide Web is characterized by directed, non-weighted edges, epistasis networks derived from algorithms like reGAIN, produce weighted, undirected edges. This difference plus the inclusion by reGAIN of diagonal elements that represent the main effects of genes requires modification to existing eigenvector centrality algorithms. We devised an algorithm called SNPrank that utilizes eigenvector centrality to prioritize nodes in epistatic networks [4]. We discuss the usage and implementation of SNPrank later in this chapter.

1.2.4 Subgraphs and Community Detection

Further insights can be generated from epistasis and other biological networks by considering smaller units of a network, called subgraphs or communities, which provide an intermediate picture of the network that lies between the individual node level and the global network level. For example, consider a tweet by one of your friends that you find to be humorous or original. Though this tweet is unlikely to trend on Twitter as a whole, it could collect several retweets or favorites from your other friends who likely share common interests with you. Likewise, mutations in a gene may have a minimal effect on all genes in an epistasis network but may have significant impacts on genes that share a pathway with the mutated gene. This mutation can become significant if paired with other mutations in the same pathway, and the aggregation of such effects could significantly influence the phenotype.

In network theory, a *community* is a clustering measure of the structure of networks based on the connectivity of subnetworks. In epistasis networks, the most intuitive interpretation of a module is that the cluster of genes participates in the disruption of the same biological or regulatory pathway. Another interpretation is that modules could form based on genomic relation to the same transcription factor or through evolutionarily conserved functions. A recent study proposed that modularity in biological networks might have increased evolutionary fitness of organisms [12]. Consequently, analyses of the localization of interactions in modules can lead to valuable insights about the evolutionary emergence of disease.

While several algorithms have been devised to separate nodes into communities, *modularity* is an efficient spectral algorithm that has been successful at predicting node groups for a variety of networks [13]. Briefly, modularity finds the class membership (c) for nodes by optimizing the difference between the observed connections (A_{ij}) in the adjacency matrix and the connections that would be expected by chance, where chance is based on the product of the degree (k) for pairs of nodes connected by an edge, where there are m edges in the network. In other words, modularity finds communities whose member nodes are more strongly connected than by random chance. This is quantified by the objective function Q,

$$Q = \frac{1}{2m} \sum_{ij} \left(A_{ij} - \frac{k_i k_j}{2m} \right) \delta\left(c_i, c_j\right), \qquad (3)$$

which can be optimized efficiently using eigenvectors. For modularity optimization, the initial nodes in a network are transformed into points whose coordinates are elements of an eigenvector. This algorithm finds a binary split of the network, but the base modularity algorithm can be used recursively to split a network into two communities repeatedly (recursive binary splits) until there is minimal improvement in modularity.

Hierarchical clustering is another method that detects communities of genes with similar functions in epistasis networks. This method is popular in co-expression network analysis of microarray data [14]. Hierarchical clustering uses a metric or measure of dissimilarity to create a dendrogram for the genes, and communities are determined from the different levels of similarity structure in the dendrogram. For example, in a network of undergraduate students at a particular university, students can be classified according to college (arts and sciences, business, etc.) as well as further grouped based on major (accounting, mathematics, etc.). Given an adjacency matrix of interaction or correlation values, this type of clustering can easily be adapted using dist() and hclust() in the open-source language R. Hierarchies are typically viewed in a diagram similar to the one shown in Fig. 2 where the dotted line distinguishes the communities (in this case, five distinct groups would result).

Once communities have been identified, the genes that comprise subnetworks can be examined for biological relevance. Traditionally, most gene network analyses look for gene enrichment of particular pathways to help understand the function of particular modules. While some gene set databases consider protein-protein interactions in cell systems, examining the similarities of communities of genes at the genomic level can be revealing. Transcription factor enrichment and chromatin structure similarities can theoretically lead to a set of genes appearing in a community in an epistasis network.

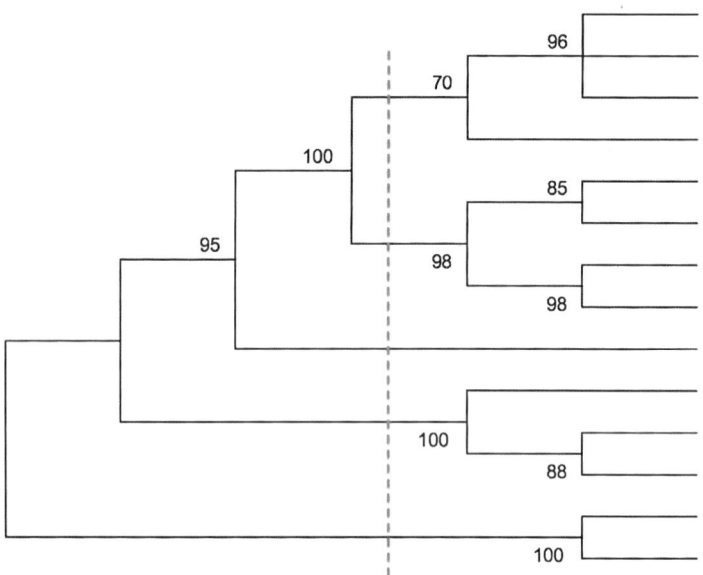

Fig. 2 A theoretical graph from a hierarchical clustering algorithm. The *dotted red line* represents a "cut" that results in five different modules or community substructures (color figure online)

1.2.5 Hubs and Scale-Free Networks

Another important property of networks is their degree distribution. Knowing the underlying degree distribution of epistatic networks may lead to a better understanding of the evolution of genetic networks and the expected connectivity of nodes. For example, do we expect the number of gene interactions to be centered around some mean value (the scale) like a network that is organized randomly or do we expect the vast majority to have a small number of interactions with a few hubs in what is called a scale-free network? In epistasis networks, we consider prominent nodes to be *epistatic hubs*. We define an epistatic hub as a gene with numerous epistatic interactions in which minor mutations lead to significant effects on the module (pathway) or network (phenotype) itself. A study by Hu et al. demonstrated that the topology of epistasis networks may be scale free [3]. Other examples of scale-free topologies include the Internet, most social media, and several biological networks [3]. Thus, this feature of epistasis networks is consistent with other important networks. A network whose distribution of node degrees (k) obeys a power law equation is defined as having scale-free topology:

$$p(k) = Ck^{-\gamma}, 2 \leq \gamma \leq 3. \qquad (4)$$

This equation stipulates that the probability of a node having degree k is proportional to a power of the degree, where the exponent lies between 2 and 3. Figure 3 demonstrates an example of a simulated degree distribution that follows a scale-free network.

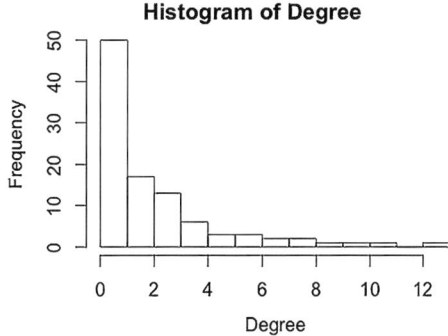

Fig. 3 A histogram of degree distribution in a scale-free network. The distribution follows the negative power law, resulting in few nodes with high connectivity and many nodes with low connectivity. Some epistasis networks have been characterized with this scale-free property

This distribution dictates that scale-free networks have many nodes with low degrees and very few nodes with much higher degrees (hubs). Hu et al. also noted the existence of hubs that did not correlate with main effects, suggesting that genes identified through individual marker or variant methods were no more likely to function as hubs in an epistasis network than a gene with a low main effect [3]. Consequently, epistasis networks are capable of uncovering effects that are normally missed by statistical models that do not consider interactions.

One of the challenges of quantifying whether a weighted network is scale free, which seems to be the case for many biological networks, is the need to specify a threshold to define edges. In order to address whether a weighted network is scale-free, one must employ an independent criterion for choosing the edge threshold to form a binary adjacency matrix from the weighted adjacency matrix. Otherwise, it is possible to choose a threshold that forces a weighted network to appear scale free.

2 Materials

2.1 Software

Here, we describe steps to construct a data-driven epistasis network with centrality gene prioritization and modularity using a software package called Encore [2] and our online visualization software called Enviz. Encore is freely available at http://insilico.utulsa.edu/encore.php and is available for installation on Windows, Mac, and Linux platforms. While several types of datasets can be employed to construct data-driven epistasis networks, Encore specializes in data comprised of single nucleotide polymorphisms (SNPs) or other variants found in Exome or RNA-seq experiments. Encore works flexibly with both binary (case/control) and continuous phenotypes.

The input files for Encore are consistent with those of the popular variant analysis tool PLINK [15]. When handling large SNP datasets in the PLINK format, we recommend using the linkage disequilibrium (LD) pruning option to remove highly correlated variants and reduce network construction time. The following command executes the default LD-prune:

```
encore -i data.bed--d-prune -o LD_data
```

The only input files needed for execution are the standard PLINK binary .bed/.bim/.fam files, and the -i and -o flags are standard input/output indicators. Note that the full array of command line options available in Encore are available by executing the--elp flag. Even after pruning the initial dataset by removing variants in strong LD, many datasets will still have prohibitive numbers of SNPs, requiring further filters. For the integrated network analysis that Encore provides, we recommend utilizing the evaporative cooling (EC) algorithm [EC]. EC ranks all variants in the dataset by combing main effect scores determined by Random Jungle with interactive effects computed by Relief-F [2]. Ranking variants using EC provides a subset enriched with aggregated main effects and interactive effects, likely providing strong candidate effectors in an epistasis network. To employ EC on an LD-pruned dataset, simply execute the following command:

```
encore  -i  data.bed--xtract  LD_data.prune.
in--c -o EC_LD_data
```

The output of EC is a ranked list of the variants in the dataset, which allows one to quickly choose the number of variants used in the demanding step of computational pairwise interaction determination. Encore supports the use of reGAIN, a generalized linear model to compute interactions. For most systems, we recommend filtering down to 10,000 or fewer SNPs from LD and EC filtering, but we note the effectiveness of using as few as 2,000 SNPs [2]. After using the output of EC to extract an appropriate number of SNPs for analysis to create the filtered.bed and associated files, the following command executes reGAIN with the minimum extra options:

```
encore  -i  filtered.bed--egain  -o  regain_
filtered
```

We note that a key advantage of reGAIN over information theoretic itGAIN (employed in Davis et al. [4]) is reGAIN's flexibility in handling covariates and alternate phenotypes. However, itGAIN can be executed in the existing Encore framework by replacing the--egain flag with--ain. Additional commands and their usage can be found on the Encore help screen.

3   Methods

3.1   Epistasis Network Modeling and SNPrank Centrality

While regression models are used ubiquitously in genetic associations, reGAIN extends the normal model for single-variant effects by looking at the effects of combinations of SNPs in a network model of interactions and main effects. We note that the reGAIN regression model is nearly synonymous with an epistatic network model, but this method further generalizes the network approach to include single-gene effects on the diagonal of the weighted matrix. For a phenotype Y, the standard single-locus regression model is defined as

$$Y \sim \beta_0 + \beta_1 \text{SNP}_1, \tag{5}$$

where the standardized coefficient β_1 goes on the reGAIN diagonal. The off-diagonal elements of reGAIN come from the standardized coefficient β_3 in the following interaction model between two SNPs:

$$Y \sim \beta_0 + \beta_1 \text{SNP}_1 + \beta_2 \text{SNP}_2 + \beta_3 \text{SNP}_1 \text{SNP}_2. \tag{6}$$

The output of reGAIN includes several files, notably the .regain file, a matrix of the standardized beta coefficient values computed from the two regression models above. While the diagonal of the .regain file represents the individual effect strength of SNPs on the phenotype, the off-diagonal (standardized β_3 coefficients) represents the interaction strength of pairs of SNPs on the phenotype, analogous to the strength of a biological interaction. The .pvals. regain matrix is similar to the .regain file, but the entries are *p-values* associated with the beta coefficients. This file can be further analyzed to reduce false positive interactions using methods like false discovery rate (FDR) pruning.

The matrix contained in the .regain file is representative of a weighted complete graph. In other words, each variant in the filtered dataset is connected with every other variant by an interaction of varying strength as determined by reGAIN. This file represents an ideal input for prioritization using eigenvector centrality, a feature Encore supports via the tool SNPrank [4]. SNPrank represents a modified version of PageRank designed specifically for epistasis networks. After running reGAIN, the following command can be used to prioritize the features in the epistasis network using eigenvector centrality:

```
encore    -i    regain_filtered--nprank    -o
ranked_regain
```

The theory behind the SNPrank computation involves transforming the reGAIN matrix, B of regression coefficients, into a Markov matrix, T:

$$T_{ij} = \begin{cases} \gamma \dfrac{B_{ij}}{k_j} + \dfrac{1-\gamma}{\mathrm{Tr}(G)}\mathrm{diag}(B)_i \, \delta_{ij}, & k \neq 0 \\[2ex] \dfrac{\mathrm{diag}(B)_i \, \delta_{ij}}{\mathrm{Tr}(B)}, & k_j = 0 \end{cases}, \qquad (7)$$

where $\mathrm{diag}(B)$ and $\mathrm{Tr}(B)$ are the diagonal and trace of the B matrix, k is the weighted degree vector of B, and γ is the so-called damping factor, which is usually given the value 0.85 based on simulation studies. Finally, the SNPrank scores are determined by solving the following equation for r, where r is a vector with elements corresponding to the SNPrank score of each gene or variant:

$$r = Tr. \qquad (8)$$

3.2 Epistasis Network Visualizer (Enviz)

While the prioritization afforded by SNPrank is a valuable tool for identifying key nodes in an epistasis network, a variant or gene ranking has limitations in terms of finding new relationships among genes that affect disease susceptibility. Consequently, we have developed an online visualization tool called Enviz to provide an interactive network-visualization experience. Enviz can be accessed at http://insilico.utulsa.edu/enviz.php. A screenshot of the setup page (Fig. 4) shows some of the available visualization options, such as scaling thickness of edges based on interaction strength, scaling the node radii based on the sum of weighted connections (degree), and detection and coloring of spectral modules. The current version uses a version of sif (simple interaction format) for the input network file format, and allows the node labels in the file to contain a delimiter.

The Enviz input is a .sif file, which may be produced by reGAIN. A .sif file contains three tab-separated columns, where the first and third columns are the node names in a particular net-

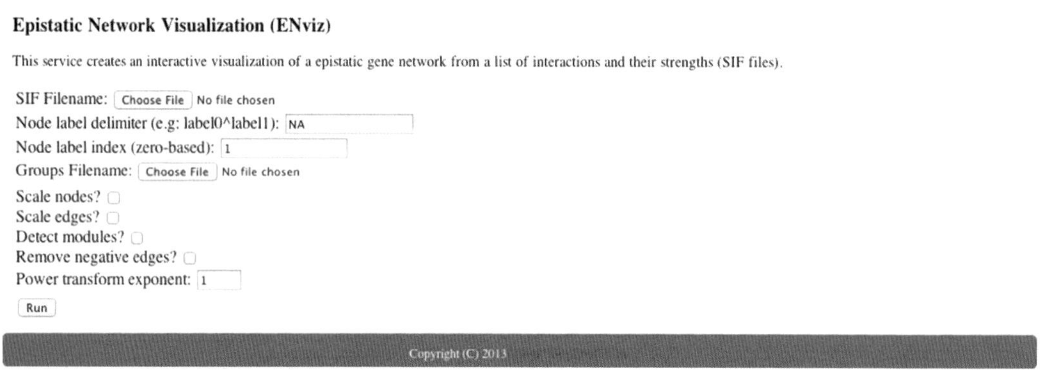

Epistatic Network Visualization (ENviz)

This service creates an interactive visualization of a epistatic gene network from a list of interactions and their strengths (SIF files).

SIF Filename: [Choose File] No file chosen
Node label delimiter (e.g: label0^label1): [NA]
Node label index (zero-based): [1]
Groups Filename: [Choose File] No file chosen
Scale nodes? ☐
Scale edges? ☐
Detect modules? ☐
Remove negative edges? ☐
Power transform exponent: [1]

[Run]

Fig. 4 Screenshot of the Enviz user interface. By uploading files produced by Encore [2], users can visualize and interact with epistasis networks using this online server

work, and the second column is a numeric value representing the strength of an edge between two nodes. As most SNP datasets are not annotated with proper gene names, we recommend using databases like SCANdb to convert SNPs to gene names. We note that Enviz can access both gene and SNP annotations given sufficient information and a proper delimiter in the .sif file. The delimiter and desired label options are in the boxes that follow the .sif file upload. An optional file upload can specify the color of a particular set of nodes, like the blue and orange nodes in Fig. 5. A group number delimited by tab should follow the node name indicated in the .sif file. Other options include scaling nodes based on degree, scaling edges based on weight of interaction, taking the absolute value of edges to ensure that all are positive, and transforming the edge weights using exponentiation to different powers. These options should be utilized as needed to create useful visualization effects.

In particular, we highlight the option to detect spectral modularity in the epistasis network, a key component for visualizing important subgraphs that could have biological meaning, such as pathway enrichment. Once a user uploads the .sif file and specifies the desired options, an interface similar to the one in Fig. 5 will appear. Within the interface, a user can filter the strength of an

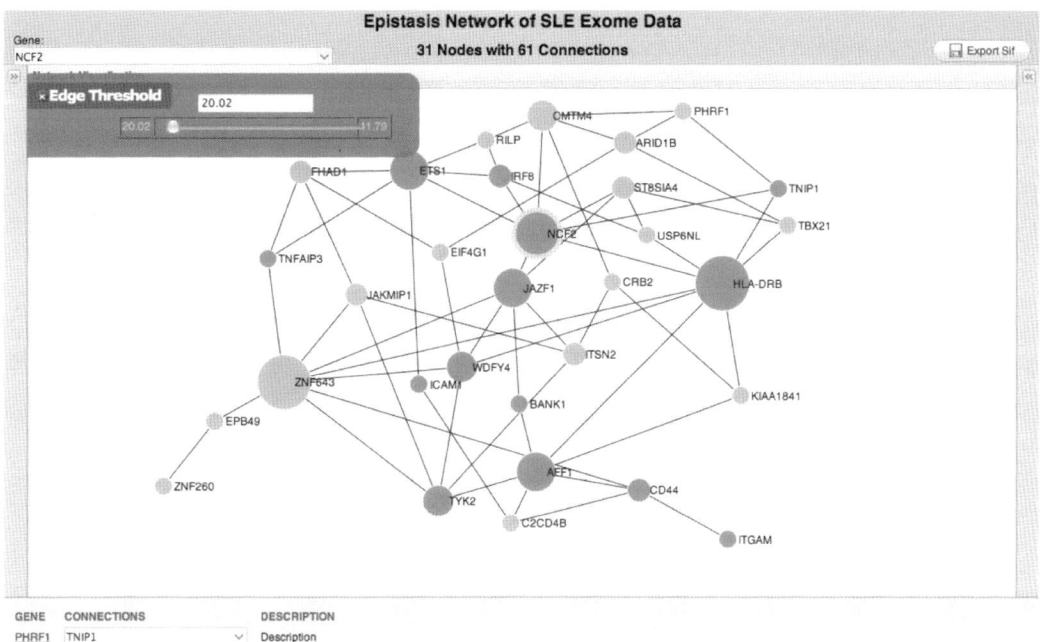

Fig. 5 A screenshot of the Enviz user interface for Epstasis networks. This particular network represents the SLE Exome analysis where known SLE genes are colored in *orange* and *blue* genes represent novel candidates uncovered by a network analysis. Node size is scaled to degree, indicating the importance of genes like HLA-DRB and NCF2 as effectors in this network (color figure online)

interaction using a sliding bar as shown below. Stringent filtering is often needed in epistasis networks to alleviate the "hairball" appearance.

Visualizing epistasis networks using our online tool Enviz is a powerful method to discover novel genes related to a phenotype as well as discover new effects. For example, the epistasis network in Fig. 5 demonstrates the effectiveness of running the Encore pipeline on exome sequencing data from a case/control study of systemic lupus erythematosus (SLE). While orange nodes replicated genes previously associated with SLE via univariate measures, blue genes provide novel effects. This network suggests the importance of interactions such as the one between NCF2 and HLA-DRB, as well as novel effects like CMTM4. The visualization of statistical results provides a powerful means for users to better interpret the results of their statistical analyses. Additional support for Encore and its usage can be found at http://insilico.utulsa.edu/encore-tutorial.php.

3.3 Summary

Network theory has emerged as a vital tool utilized in a myriad of disciplines to characterize various phenomena. Here, we have discussed the merits of analyzing statistical relationships in a biological and epistasis network framework. Several studies using disease-specific datasets have utilized these tools to create more sensitive models of biological activity to gain insights into the genetic basis of complex disease.

For example, Pandey et al. [5] used epistasis network centrality coupled with gene enrichment tools to identify cellular pathways that played roles in bipolar disorder (BD). A highlight of this study was the identification of ankryin-3, a known effector in BD, using a significantly smaller sample size than would be required for a univariate approach to detect an effect [5, 16]. This study demonstrates the potential for epistasis networks to identify effects with smaller sample sizes than traditional statistical methods. Other studies have uncovered novel effects in immune reactions to the smallpox vaccine [4] and helped explain interactions between several genes associated with SLE [2].

In addition to epistatic networks generated from Encore and related tools, webservers such as the integrated multi-species platform [17] and GeneMANIA [18] use an assortment of protein-protein interactions, co-expression samples, protein homology, and several other factors to explore the topology of a gene or set of genes in a network structure. These webservers provide a picture of interactions in a broader scope and can be useful for interpreting results from other statistical and biological methods.

Finally, we note that epistasis networks can be employed for understanding the evolution of genes and gene regulatory systems. As scale-free networks are ubiquitous in biology, one could argue that the resulting pattern of gene interactions may be an emergent property of the evolution of gene networks. Understanding the structure and statistical distributions of gene networks may reveal new evolutionary insights into the etiology of human disease.

4 Notes

Rather than treating statistical associations and epistatic interactions as isolated units of analysis, epistasis networks and Genetic Association Interaction Networks (GAIN) integrate these effects into a global model of disease. These models can then be interrogated with network theory techniques to identify emergent properties. This chapter reviewed some of the relevant network theory techniques, including community structure and module detection, centrality and hubs, and degree distributions. Epistasis networks are a new application domain for network theory, which opens up new opportunities and questions for understanding complex disease genetics. How do these networks relate to other biological networks, such as protein-protein interaction networks, how do epistasis networks help or harm the multiple hypothesis testing burden, what are the functional effects of epistatic hubs, and what role does the degree distribution play in disease susceptibility?

References

1. Lin A, Wang RT et al (2010) A genome-wide map of human genetic interactions inferred from radiation hybrid genotypes. Genome Res 20(8):1122–1132

2. Davis NA, Lareau CA et al (2013) Encore: Genetic Association Interaction Network centrality pipeline and application to SLE exome data. Genet Epidemiol 37(6):614–621

3. Hu T, Sinnott-Armstrong NA et al (2011) Characterizing genetic interactions in human disease association studies using statistical epistasis networks. BMC Bioinform 12:364

4. Davis NA, Crowe JE Jr et al (2010) Surfing a genetic association interaction network to identify modulators of antibody response to smallpox vaccine. Genes Immun 11(8):630–636

5. Pandey A, Davis NA et al (2012) Epistasis network centrality analysis yields pathway replication across two GWAS cohorts for bipolar disorder. Trans Psychiat 2:e154

6. Proulx SR, Promislow DE et al (2005) Network thinking in ecology and evolution. Trends Ecol Evol 20(6):345–353

7. Kaleta C, de Figueiredo LF et al (2011) In silico evidence for gluconeogenesis from fatty acids in humans. PLoS Comput Biol 7(7): e1002116

8. McKinney BA, Reif DM et al (2006) Machine learning for detecting gene-gene interactions: a review. Appl Bioinformatics 5(2):77–88

9. Ritchie MD, Hahn LW et al (2003) Power of multifactor dimensionality reduction for detecting gene-gene interactions in the presence of genotyping error, missing data, phenocopy, and genetic heterogeneity. Genet Epidemiol 24(2):150–157

10. Jiang X, Neapolitan RE et al (2011) Learning genetic epistasis using Bayesian network scoring criteria. BMC Bioinform 12:89

11. McKinney BA, Pajewski NM (2011) Six degrees of epistasis: statistical network models for GWAS. Front Genet 2:109

12. Clune J, Mouret JB et al (2013) The evolutionary origins of modularity. Proc Roy Soc B 280(1755):20122863

13. Newman MEJ (2010) Networks: an introduction. Oxford University Press, Oxford

14. Zhang B, Horvath S (2005) A general framework for weighted gene co-expression network analysis. Stat Appl Gene Mol Biol 4: Article17

15. Purcell S, Neale B et al (2007) PLINK: a tool set for whole-genome association and population-based linkage analyses. Am J Hum Genet 81(3):559–575

16. Schulze TG, Detera-Wadleigh SD et al (2009) Two variants in Ankyrin 3 (ANK3) are independent genetic risk factors for bipolar disorder. Mol Psychiatry 14(5):487–491

17. Wong AK, Park CY et al (2012) IMP: a multi-species functional genomics portal for integration, visualization and prediction of protein functions and networks. Nucleic Acids Res 40(Web Server issue): W484–W490

18. Mostafavi S, Ray D et al (2008) GeneMANIA: a real-time multiple association network integration algorithm for predicting gene function. Genome Biol 9(Suppl 1):S4

Chapter 16

Epistasis Analysis Using Multifactor Dimensionality Reduction

Jason H. Moore and Peter C. Andrews

Abstract

Here we introduce the multifactor dimensionality reduction (MDR) methodology and software package for detecting and characterizing epistasis in genetic association studies. We provide a general overview of the method and then highlight some of the key functions of the open-source MDR software package that is freely distributed. We end with a few examples of published studies of complex human diseases that have used MDR.

Key words Epistasis, Machine learning, Association studies, Genetic analysis, Gene–gene interaction

1 Introduction

Multifactor dimensionality reduction (MDR) was the first machine learning approach designed specifically for detecting, characterizing, and interpreting nonadditive gene–gene interactions in the absence of statistically detectable independent effects [1–7]. MDR is model-free in the sense that no genetic models are assumed, and it is nonparametric in the sense that no parameters are estimated. At the heart of the MDR approach is a feature or attribute construction algorithm that creates a new variable or attribute by pooling, for example, genotypes from multiple SNPs [8]. The process of defining a new attribute as a function of two or more other attributes is referred to as constructive induction or attribute construction and was first developed by Michalski [9]. Constructive induction using the MDR kernel is accomplished in the following way. A multilocus genotype combination is considered high risk if the ratio of cases (subjects with disease) to controls (healthy subjects) in that group exceeds the overall ratio of cases to controls in the data set. Otherwise, it is considered low risk. Genotype combinations considered to be high risk are labeled G_1 while those considered low risk are labeled G_0. This process constructs a new

Jason H. Moore and Scott M. Williams (eds.), *Epistasis: Methods and Protocols*, Methods in Molecular Biology, vol. 1253, DOI 10.1007/978-1-4939-2155-3_16, © Springer Science+Business Media New York 2015

one-dimensional attribute with levels G_0 and G_1. It is this new single variable that is assessed using any classification method. A common measure of model fit is the accuracy of the classifier. A balanced accuracy function is used when the ratio of cases to controls is different than one [10]. The MDR method is based on the idea that changing the representation space of the data will make it easier for a classifier such as a decision tree, logistic regression, or a naive Bayes learner to detect nonadditive gene–gene interactions. The ability to combine MDR with any classification method makes it extremely flexible. This approach has been evaluated using multiple simulation studies (e.g., [2, 6, 7]) and has previously been referred to as a gold standard in the field [12]. We and others have reviewed the importance of epistasis in human disease that motivates methods such as MDR [12–22]. We refer readers to the other chapters in this volume for background on what epistasis is and how it is studied across evolution and genetics.

Like many machine learning methods, MDR is usually implemented within the context of cross-validation to assess the predictive accuracy of models [23]. This allows the generalizability of MDR models to be assessed using a single data set divided into multiple training and testing pieces, which helps confront issues such as overfitting that are common when looking at many high-dimensional models. Statistical significance is assessed using permutation testing [11, 24]. Entropy-based methods have been developed to provide statistical interpretation of MDR models [8].

Numerous extensions and modifications to the MDR method have been developed. These include generalized linear models [25], model-based methods [26], robust methods [27], covariate adjustment [28], risk models [29], family-based methods [30, 31], survival methods [32, 33], high-performance computing [34, 35], analysis of quantitative traits [36], and pathway analysis [37]. MDR and its many extensions have been applied to a wide range of different phenotypes, including pharmacogenetics [29, 38, 39]. We highlight here some of our previous work with the genetic analysis of bladder cancer susceptibility to illustrate the MDR method and software [40, 41].

This chapter provides a basic overview of how to carry out a genetic analysis using the open-source MDR software package. We illustrate the file format, and provide a step-by-step tutorial for data analysis that includes an overview of the parameter settings and the output and interpretation.

2 Materials

2.1 Getting Started

An open-source MDR software package is freely available from http://sourceforge.net/projects/mdr. MDR is programmed entirely in Java and is compatible with any operating system that has Java installed. Here, we provide an overview of using the graphic user interface (GUI) for this MDR software package. Command line options are also available and are described in a document labeled MDR_command_line_help.txt that is distributed with MDR. It is important to note that several extensions to open-source MDR have been developed and distributed by other groups [25]. In addition, there is now an R package for MDR [42]. These additional MDR software solutions will not be discussed here.

2.2 Data Format

MDR accepts a tab-delimited text file as input. Data are loaded using the *Load Datafile* button in the upper right corner of the main screen. The data can also be viewed using the *View Datafile* button. Sample data files with discrete and continuous outcomes and simulated gene–gene interactions are provided as part of the MDR software package. Briefly, each row is an observation or instance (e.g., a human subject) and each column is a discrete variable or attribute (e.g., a single-nucleotide polymorphism or SNP). Typically, the attributes are polymorphisms with three genotypes coded 0, 1, and 2 or covariates such as gender coded 0 and 1. The last column is saved for the phenotype or outcome, which can be binary (e.g., case–control status) or continuous (e.g., body mass index). MDR requires a complete data set with no missing values. Missing data can be handled in two different ways. First, the missing data can be coded with a number not used to represent a genotype (e.g., [9]) and will be treated as an extra level in the analysis (e.g., a fourth genotype). This is acceptable as long as your data are missing at random. As a second option data imputation methods can be used to fill in the missing values. This latter method is preferred by the authors and has been explored specifically for MDR [25].

3 Methods

3.1 Running an Analysis

Once the data set is loaded, MDR can be run using the default settings by pressing the *Run Analysis* button on the main screen (we assume for now that a discrete endpoint is being analyzed) (*see* **Note 1**). Analyses can be saved and loaded for future viewing. A progress bar estimates the time remaining for the run (*see* **Note 2**), and the best results are displayed in the Summary Table of the main screen. With the default settings, MDR completes an exhaustive analysis of all single attributes and all two- to three-way combinations. The best model for each order is shown along with the training and testing accuracy estimated from a tenfold cross-validation (CV) that assesses internal replication. As mentioned above, the accuracy presented is a balanced accuracy that accounts for an imbalance in the numbers of cases and controls [10] (*see* **Note 3**). It is important to note that this accuracy is averaged across all CV intervals regardless of whether the same model was found or not. We call this the CV accuracy (*see* **Note 4**). We also provide in the summary table a measure called CV consistency (CVC) that shows how often the best model was found across the different CV intervals [43]. A higher CVC indicates a more consistent result. A low CVC could be due to a weak signal or complexity in the data such as locus heterogeneity. Other internal validation methods for MDR have been explored [44].

Once the data in summary table have been populated an overall best model can be selected. One approach is to select the mode with the highest testing accuracy (*see* **Note 5**). This represents the model most likely to generalize to independent data. Ideally, the testing accuracy increases with model order until the right model is selected and then decreases as noisy attributes are added leading to overfitting and loss of generalizability. Note that the training accuracy will continue to rise as new attributes are added. This is analogous to the R^2 of a linear regression, which increases with each new attribute added to the model. In general, the closer the training and testing accuracies are to each other the less overfitting there is. It is sometimes necessary to choose among a set of models that have similar testing accuracies. We typically use the CVC to break ties. The size of the model can also be used with smaller models being more parsimonious and thus more likely to generalize to independent data. For the sample data provided with the MDR software, it is clear that attributes X1, X6, and X8 form the overall best model with a training accuracy of 0.87, a testing accuracy of 0.86, and a CVC of 10/10. Alternatives to accuracy have been studied but are not implemented here [45].

In addition to the results provided in the Summary Table, we make available a number of other more detailed results to assist with evaluation and interpretation. For example, in the *Graphical Model* tab we provide a publication-quality plot of the distribution of cases and controls for each genotype combination. High-risk genotypes are shaded dark grey while low risk are shaded light grey. In the *Best Model* tab, the details of the best models found in each of the CV intervals and other statistics such as an odds ratio are shown. The *CV Results* tab shows all the models found in each CV interval. MDR provides only the overall best models in the Summary Table. The *Top Models* tab allows the user to track the second-best, third-best, etc. Finally, in the *Entropy* tab we provide methods based on information theory for statistical interpretation of the nature of the genetic effects. More details on the methods and interpretation of the results have been described previously [8, 46, 47]. Briefly, the amount of information about case–control status that is provided by each pair of attributes is estimated using measures of entropy. A positive information gain indicates a synergistic or nonadditive effect (e.g., epistasis) while a negative value indicates redundancy or correlation (e.g., due to linkage disequilibrium). A value close to zero indicates independence or a balance of synergy and redundancy. In addition to the raw data, we provide publication-quality graphical output in the form of dendrograms and graphs. These relationships can be explored in more detail using the Network tab. Methodological details for the entropy-based network analysis have been previously provided [47–49].

3.2 Parameter Settings

We provide several parameter settings for the MDR analysis. These can be found in the *Configuration* tab. The first option is the random seed that can be set to ensure reproducibility of results. This is primarily used to determine how the data will be divided for CV. The next setting is the Attribute Count Range that controls the order of the analysis (*see* **Note 1**). Next is the CV Count that has a default of 10. Tenfold CV is commonly used in data mining and machine learning. Next is a Compute Fitness Landscape check box that allows the user to view the distribution of all MDR models evaluated during a run. This is off by default to save memory for large runs. Next is the Track Top Models option that indicates how many top models should be saved for further inspection. Also provided is a check box for Paired Analysis. Checking this box will keep paired samples together during CV. Also provided are some options for how to handle tie cells (e.g., genotypes with equal numbers of cases and controls). The default is to label tie cells as cases but this can be changed to include a Fisher's exact test for more robust declaration according to Gui et al. [27]. Finally, there is an option for Search Method Configuration. Briefly, this provides options for stochastic search when an exhaustive search is not computationally feasible. We provide more details about this option in a later section.

3.3 Determining Statistical Significance

Given the nonparametric nature of MDR, statistical significance is usually determined using permutation testing although other methods such as bootstrapping may be used. The goal of permutation testing is to repeat the entire MDR analysis on a large number of data sets with the phenotype column randomized to simulate results under the null hypothesis of no association. One thousand or more permutations can be used to generate a null distribution for testing the accuracy of the best MDR models. This in turn can be used to estimate a p-value. We have provided a separate permutation testing module that will soon be integrated into the MDR software. Due to the computational expense of permutation testing, we recommend writing a script that randomizes the data and calls MDR from the command line to distribute the evaluations across many CPUs on a parallel computer cluster. An example script is available from the authors upon request. We also recommend taking advantage of the extreme value distribution that can cut the number of permutations needed from 1,000 to 20, for an example *see* ref. [11].

It is important to note that the permutation test described above is a test of the null hypothesis of no association. This null could be rejected due to additive effects, nonadditive effects, or a combination of the two. Those interested in epistasis are likely to be more interested in knowing whether the null can be rejected solely due to the nonadditive effects. To address this issue, Greene et al. developed an explicit test of epistasis that sorts the data into cases and controls and then randomizes the genotypes for each SNP within those classes [24]. This preserves the genotype frequency differences between cases and controls (i.e., the additive effects) and randomizes everything else. This allows the user to test the null hypothesis that the only effects in the data are additive. Rejection of this null is evidence of nonadditive effects, which are not confounded with the additive effects. This test is supported in the permutation testing feature of MDR (*see* **Note 6**).

3.4 Determining Biological Significance

The biological significance of MDR models has been approached in several different ways. One approach is to map MDR results to biological pathways to look for enrichment of genes in specific functional categories. Kim et al. [37] provide a flowchart for pathway analysis of MDR results that first assigns the accuracy from the best pairwise MDR model to each SNP. SNPs are then mapped to genes and tested for enrichment at the gene level. Each gene is assigned a *p*-value and then mapped to a functional group such as those from Gene Ontology. A gene set enrichment analysis is then performed to determine whether particular pathways have more significant genes than would be expected by chance given their size. This approach discovered a replicable pathway that was not detected in other studies [37]. Another approach is to carry out a functional genomics analysis using a bioinformatics tool called Integrative Multi-species Prediction (IMP) [50]. IMP integrates thousands of gene expression data sets and other sources of information such as protein–protein interactions and then provides Bayesian confidence scores that represent the likelihood that any two genes are functionally connected. IMP is implemented in a very simple web interface at http://imp.princeton.edu and can be used to interpret lists of genes from best MDR models. Experimental studies are also possible [51, 52], but are beyond the scope of this chapter.

3.5 Interleaving

As described in the introduction, MDR is a constructive induction algorithm that combines two or more attributes to create a new single attribute that captures interaction effects. MDR provides an Attribute Construction tab that allows the user to select two or more attributes from a list and then combine them using the MDR model. A new attribute is produced that is interleaved back into the data set (*see* **Note 5**). MDR can now be rerun to include that new variable in the modeling process. This makes it possible to produce hierarchical models as suggested previously [8]. An example of this in practice is provided by Lee et al. [53] (*see* **Note 7**).

3.6 Covariate Adjustment

It is often of interest to adjust out the effects of one or more covariate before performing a genetic association analysis. For example, you might want to know whether a particular interaction between two SNPs is still associated with the outcome after removing the effects of a third SNP with a significant main effect. In MDR, we provide a very simple nonparametric method for doing this. As described by Gui et al. [28], MDR uses a simple resampling approach to remove covariate effects. This method simply resamples the data to provide equal numbers of cases and controls within each genotype of the attribute being adjusted out. A new data set is created that is then analyzed. For example, X1 in the MDR sample dataset has the strongest main effect and shows up as the best one-way MDR model in the Summary Table. After adjusting for the effects of X1, it no longer appears to have a main effect. However, its interactions with X6 and X8 are preserved. Multiple factors can be adjusted sequentially. However, sequential adjustment has not been extensively tested and should be used with caution. The advantage of this approach is that it is simple and does not require the overhead of a parametric linear model.

3.7 Application to Quantitative Traits

MDR now has the capability of modeling quantitative traits such as body mass index, using the methods of Gui et al. [36]. Here, the means of each genotype combination are compared to the global mean to assign each cell to one of two groups. The means of the two groups are compared using a t-test that can be evaluated using permutation testing. This simple quantitative MDR (QMDR) approach is advantageous over MDR methods that use linear models [25], because it is more computationally efficient and there is no loss of power. This is important when applying MDR to the genome-wide analysis of quantitative traits.

3.8 Genome-Wide Association Studies: Parallel Computing

Epistasis analysis in genome-wide association data is challenging due to the combinatorial nature of the analysis. Parallel computing is a brute-force way to increase the number of gene–gene interactions that can be exhaustively analyzed [54]. The open-source MDR software described here automatically detects whether the computer it is being run on has multiple CPUs and caries out a threaded analysis. However, parallel computing requires a script to parse out the attribute combinations across many CPUs distributed on a network. The MDR software can be scripted directly to allow parallel computing. We also make available from the same website a C library for fast MDR analysis with discrete traits. A version of MDR for discrete traits that can be run on a graphics processing unit (GPU) is also available (MDR-GPU) [35]. This latter approach has been used previously to carry out complete two-way MDR analyses with hundreds of thousands of SNPs and 1,000-fold permutation testing in a matter of hours [34]. More information about GPU computing is provided by Payne and Moore [55]. Although powerful, these high-performance computing approaches still have limitations when moving beyond two-way epistasis analyses. We describe below filtering and stochastic searching as alternatives for MDR.

3.9 Genome-Wide Association Studies: Filters

One approach to dealing with the computational limitations of epistasis analysis is to first filter your list of attributes to a more manageable size that can be exhaustively evaluated [17]. This can be done prior to loading the data into MDR or it can be done using several statistical filters from within the MDR software in the *Filter* tab. Most of the filters provided are based on the ReliefF algorithm [56–58] that has been recently reviewed in the context of MDR [59]. ReliefF is a machine learning algorithm that is capable of detecting complex patterns in data without the need for enumerating combinations of attributes. ReliefF uses a nearest-neighbor approach to assign weights to each attribute that can reflect its additive and/or nonadditive effects. For these reasons it is ideal for pre-processing genome-wide data. We provide several variations on the ReliefF algorithm that we have shown are much more powerful for detecting interaction effects. Each has its own set of options and can be used to filter the attribute list down to a desired size. MDR will then use that reduced set in its analysis. Several visualization tools are provided for viewing the distribution of ReliefF scores. An advantage of filtering is that it greatly reduces the number of combinations evaluated by MDR thus reducing the chances of overfitting. We have found this to be true even for smaller sets of SNPs. One disadvantage is that the cutoffs that are used can be arbitrary. Significance tests to address this issue have been proposed [60]. ReliefF-based filter methods are presented in detail in a separate chapter in this volume. In addition to the statistical filters, we recommend other tools that can be used to filter the attributes before loading the data into MDR. For example, BioFilter is a powerful tool that uses biological knowledge to prioritize a list of SNPs [61].

3.10 Genome-Wide Association Studies: Stochastic Search

An alternative to the filter approach described above is a stochastic search. MDR provides several options for search methods in the Search Method Configuration part of the *Configuration* tab. The default setting is an Exhaustive Search. We also provide a Forced Analysis where the user can specify a set of attributes to evaluate (*see* **Note 7**). The Random Search option will perform a random search for a total number of evaluations or a total length of time, which is specified by the user. The Estimation of Distribution Algorithm (EDA) search provides an agent-based stochastic search algorithm that can use expert knowledge such as ReliefF to help guide the search. Additional details of this search method have been provided previously [62, 63]. Briefly, each agent has a uniform probability of selecting each attribute to include in an MDR model. These probabilities are updated in proportion to the quality of the

MDR model evaluated and can be weighted according to expert knowledge loaded into MDR. The advantage of this approach over filtering is that the complete list of attributes is accessible to the algorithm. These methods are still experimental and have not been widely applied to real data.

3.11 Bladder Cancer Example

We have selected the following case study to highlight the MDR method. Andrew et al. [40] carried out an epidemiologic study to identify genetic and environmental predictors of bladder cancer susceptibility in a sample of Caucasians ($n=914$) from New Hampshire. This study focused specifically on genes that play important roles in the repair of DNA sequences that have been damaged by chemical compounds (e.g., carcinogens). Seven SNPs were measured including two from the *X-ray repair cross-complementing group 1* gene (*XRCC1*), one from the *XRCC3* gene, two from the *xeroderma pigmentosum group D (XPD)* gene, one from the *nucleotide excision repair* gene (*XPC*), and one from the *AP endonuclease 1* gene (*APE1*). Each of these genes plays an important role in DNA repair. Smoking is a known risk factor for bladder cancer and was included in the analysis along with gender and age for a total of ten attributes. Age was discretized to > or ≤50 years.

A parametric statistical analysis of each attribute individually revealed a significant independent main effect of smoking, as expected. However, none of the measured SNPs were significant predictors of bladder cancer individually. Andrew et al. used MDR to exhaustively evaluate all possible two-, three-, and four-way interactions among the genetic and environmental variables. Training and testing accuracies were estimated using tenfold CV. A best model was selected that maximized the testing accuracy. The best model had a testing accuracy of 0.66 and included two SNPs from the *XPD* gene and smoking. The distribution of cases and controls with each genotype-smoking combination is illustrated in Fig. 1. Here, the dark cells are high risk for disease and the left bars represent the number of cases. The *p*-value for this model was less than 0.001, suggesting that a testing accuracy of 0.66 or greater is unlikely under the null hypothesis of no association as assessed using a 1,000-fold permutation test.

Decomposition of this model using the measures of information gain described above demonstrated that the effects of the two *XPD* SNPs were nonadditive or synergistic, which was suggestive of epistasis or a gene–gene interaction. The same analysis indicated that the effect of smoking was mostly independent of the gene–gene interaction effect. Figure 2 illustrates an interaction graph summarizing these measures of information gain. Attributes connected by red lines have stronger synergistic interactions than those connected by yellow lines. Green indicates redundancy. Note that the red line connecting the two *XPD* polymorphisms indicates a

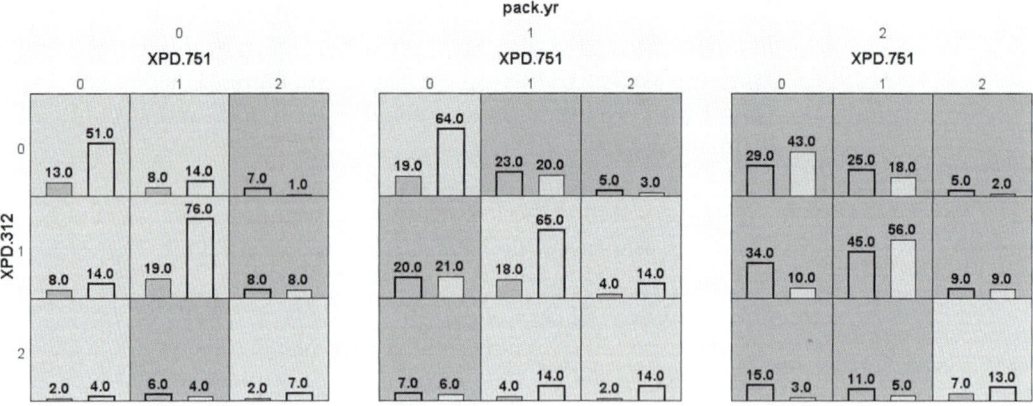

Fig. 1 Illustrates the distribution of cases and controls with each genotype-smoking combination is nonlinear

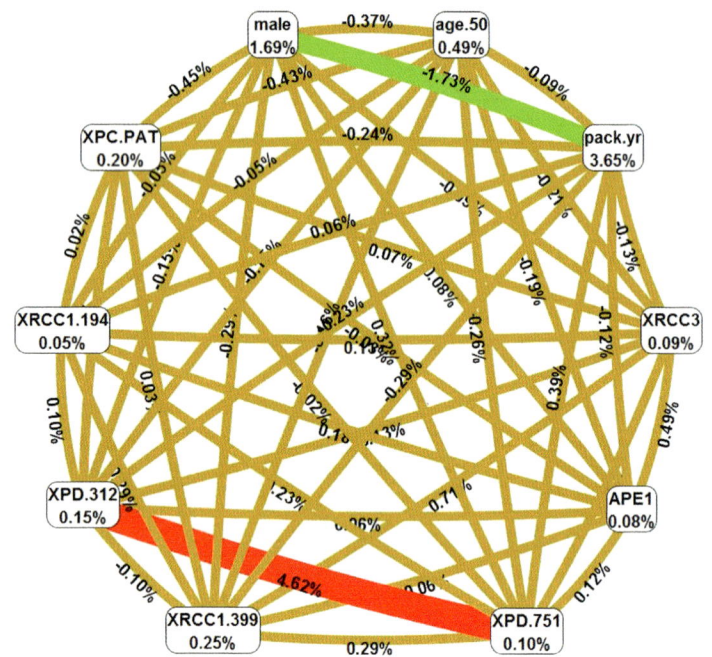

Fig. 2 Illustrates an interaction graph summarizing these measures of information gain

strong synergistic interaction. In other words, combining these two polymorphisms into a single variable using MDR provides more information about case–control status than considering them additively. The green line connecting pack years of smoking to gender suggests that these two variables are correlated or redundant. In other words, information about case–control status is lost when putting these two attributes together. The yellow lines between smoking and the two *XPD* polymorphisms indicate independence.

It is important to note that parametric logistic regression was unable to model this three-attribute interaction due to lack of convergence. It is important to note that this general result was replicated in an independent study [41].

3.12 Conclusions

We have provided a brief overview of the MDR method and the open-source MDR software package that is freely available from sourceforge.net. MDR is a comprehensive software package that provides both analysis and interpretation tools. The software scales to genome-wide association data and has been applied to a wide range of different human diseases. More information about how MDR fits into a general bioinformatics strategy for detecting interactions is provided by Moore et al. [17]. Information on other methods for detecting and characterizing interactions has been provided by Cordell [12].

4 Notes

The following are some important tips for using MDR:

1. It is not advisable to construct MDR models with more than five attributes. Unless you have a very large dataset, each observation will have a unique genotype combination in six or seven dimensions. Unique genotypes will only be present in the training data or the test data during cross-validation, but not both.

2. Running MDR in exhaustive mode on a desktop computer with more than 100 attributes in the data set may require many hours of computing time.

3. As a rule of thumb, MDR models with testing accuracies less than 0.55 are rarely statistically significant with permutation testing.

4. The Forced Analysis option is sometimes useful for getting a more accurate estimate of the training and testing accuracy when the CVC is less than the number of CV intervals (e.g., CVC < 10 for tenfold CV). This option forces the model to be considered in each CV interval. Otherwise, the accuracy estimate may partially come from other models. It should be noted in publications if the Forced Analysis accuracy is used for calculating p-values from permutation tests, since these may be higher than the result in the summary table.

5. MDR can be combined with other methods, such as logistic regression by first running MDR and saving the constructed attributes to a file. That file may then be loaded into a statistical analysis package such as R.

6. Although replication across independent data sets is a gold standard in genetic association studies, it is important to note

that epistasis models might not replicate due to shifts in allele frequencies in different data sets. Greene et al. demonstrated this phenomenon using simulated data [64].

7. A simple way to achieve this uses Forced Analysis with wild-cards (*). To see all three-way models that contain x1 and x2 use x1, x2, *.

Acknowledgements

This work was supported by National Institutes of Health (NIH) grants AI59694, EY022300, GM103534, GM103506, LM009012, LM010098, and LM011360.

References

1. Ritchie MD, Hahn LW, Roodi N et al (2001) Multifactor-dimensionality reduction reveals high-order interactions among estrogen-metabolism genes in sporadic breast cancer. Am J Hum Genet 69:138–147

2. Ritchie MD, Hahn LW, Moore JH (2003) Power of multifactor dimensionality reduction for detecting gene-gene interactions in the presence of genotyping error, missing data, phenocopy, and genetic heterogeneity. Genet Epidemiol 24:150–157

3. Hahn LW, Ritchie MD, Moore JH (2003) Multifactor dimensionality reduction software for detecting gene-gene and gene-environment interactions. Bioinformatics (Oxford, England) 19:376–382

4. Hahn LW, Moore JH (2004) Ideal discrimination of discrete clinical endpoints using multi-locus genotypes. In Silico Biol 4:183–194

5. Moore JH (2010) Detecting, characterizing, and interpreting nonlinear gene-gene interactions using multifactor dimensionality reduction. Adv Genet 72:101–116

6. Moore JH (2004) Computational analysis of gene-gene interactions using multifactor dimensionality reduction. Expert Rev Mol Diagn 4:795–803

7. Moore JH (2007) Genome-wide analysis of epistasis using multifactor dimensionality reduction: feature selection and construction in the domain of human genetics. In: Zhu X, Davidson I (eds) Knowledge discovery and data mining: challenges and realities. IGI Global, Hershey, PA, pp 17–30

8. Moore JH, Gilbert JC, Tsai C-T et al (2006) A flexible computational framework for detecting, characterizing, and interpreting statistical patterns of epistasis in genetic studies of human disease susceptibility. J Theor Biol 241:252–261

9. Michalski RS (1983) A theory and methodology of inductive learning. Artif Intel 20:111–161

10. Velez DR, White BC, Motsinger AA et al (2007) A balanced accuracy function for epistasis modeling in imbalanced datasets using multifactor dimensionality reduction. Genet Epidemiol 31:306–315

11. Pattin KA, White BC, Barney N et al (2009) A computationally efficient hypothesis testing method for epistasis analysis using multifactor dimensionality reduction. Genet Epidemiol 33:87–94

12. Cordell HJ (2009) Detecting gene-gene interactions that underlie human diseases. Nat Rev Genet 10:392–404

13. Moore JH (2003) The ubiquitous nature of epistasis in determining susceptibility to common human diseases. Hum Hered 56:73–82

14. Moore JH (2005) A global view of epistasis. Nat Genet 37:13–14

15. Moore JH, Williams SM (2005) Traversing the conceptual divide between biological and statistical epistasis: systems biology and a more modern synthesis. Bioessays 27:637–646

16. Moore JH, Williams SM (2009) Epistasis and its implications for personal genetics. Am J Hum Genet 85:309–320

17. Moore JH, Asselbergs FW, Williams SM (2010) Bioinformatics challenges for genome-wide association studies. Bioinformatics (Oxford, England) 26:445–455

18. Cordell HJ (2002) Epistasis: what it means, what it doesn't mean, and statistical methods

to detect it in humans. Hum Mol Genet 11:2463–2468

19. Cowper-Sal lari R, Cole MD, Karagas MR et al (2011) Layers of epistasis: genome-wide regulatory networks and network approaches to genome-wide association studies. Wiley Interdiscip Rev Syst Biol Med 3:513–526

20. Tyler AL, Asselbergs FW, Williams SM et al (2009) Shadows of complexity: what biological networks reveal about epistasis and pleiotropy. Bioessays 31:220–227

21. Phillips PC (2008) Epistasis–the essential role of gene interactions in the structure and evolution of genetic systems. Nat Rev Genet 9:855–867

22. Phillips PC (1998) The language of gene interaction. Genetics 149:1167–1171

23. Coffey CS, Hebert PR, Ritchie MD et al (2004) An application of conditional logistic regression and multifactor dimensionality reduction for detecting gene-gene interactions on risk of myocardial infarction: the importance of model validation. BMC Bioinformatics 5:49

24. Greene CS, Himmelstein DS, Nelson HH et al (2010) Enabling personal genomics with an explicit test of epistasis, Pacific Symposium on Biocomputing. Pac Symp Biocomput:327–336

25. Lou X-Y, Chen G-B, Yan L et al (2007) A generalized combinatorial approach for detecting gene-by-gene and gene-by-environment interactions with application to nicotine dependence. Am J Hum Genet 80:1125–1137

26. Calle ML, Urrea V, Malats N et al (2010) mbmdr: an R package for exploring gene-gene interactions associated with binary or quantitative traits. Bioinformatics (Oxford, England) 26:2198–2199

27. Gui J, Andrew AS, Andrews P et al (2011) A robust multifactor dimensionality reduction method for detecting gene-gene interactions with application to the genetic analysis of bladder cancer susceptibility. Ann Hum Genet 75:20–28

28. Gui J, Andrew AS, Andrews P et al (2010) A simple and computationally efficient sampling approach to covariate adjustment for multifactor dimensionality reduction analysis of epistasis. Hum Hered 70:219–225

29. Dai H, Charnigo RJ, Becker ML et al (2013) Risk score modeling of multiple gene to gene interactions using aggregated-multifactor dimensionality reduction. BioData Min 6:1

30. Martin ER, Ritchie MD, Hahn L et al (2006) A novel method to identify gene-gene effects in nuclear families: the MDR-PDT. Genet Epidemiol 30:111–123

31. Cattaert T, Urrea V, Naj AC et al (2010) FAM-MDR: a flexible family-based multifactor dimensionality reduction technique to detect epistasis using related individuals. PLoS One 5:e10304

32. Gui J, Moore JH, Kelsey KT et al (2011) A novel survival multifactor dimensionality reduction method for detecting gene-gene interactions with application to bladder cancer prognosis. Hum Genet 129:101–110

33. Beretta L, Santaniello A, van Riel PLCM et al (2010) Survival dimensionality reduction (SDR): development and clinical application of an innovative approach to detect epistasis in presence of right-censored data. BMC Bioinformatics 11:416

34. Greene CS, Sinnott-Armstrong NA, Himmelstein DS et al (2010) Multifactor dimensionality reduction for graphics processing units enables genome-wide testing of epistasis in sporadic ALS. Bioinformatics (Oxford, England) 26:694–695

35. Sinnott-Armstrong NA, Greene CS, Cancare F et al (2009) Accelerating epistasis analysis in human genetics with consumer graphics hardware. BMC Res Notes 2:149

36. Gui J, Moore JH, Williams SM et al (2013) A simple and computationally efficient approach to multifactor dimensionality reduction analysis of gene-gene interactions for quantitative traits. PLoS One 8:e66545

37. Kim NC, Andrews PC, Asselbergs FW et al (2012) Gene ontology analysis of pairwise genetic associations in two genome-wide studies of sporadic ALS. BioData Min 5:9

38. Wilke RA, Reif DM, Moore JH (2005) Combinatorial pharmacogenetics. Nat Rev Drug Discov 4:911–918

39. Wilke RA, Mareedu RK, Moore JH (2008) The pathway less traveled: moving from candidate genes to candidate pathways in the analysis of genome-wide data from large scale pharmacogenetic association studies. Curr Pharmacogenomics PersonMed 6:150–159

40. Andrew AS, Nelson HH, Kelsey KT et al (2006) Concordance of multiple analytical approaches demonstrates a complex relationship between DNA repair gene SNPs, smoking and bladder cancer susceptibility. Carcinogenesis 27:1030–1037

41. Andrew AS, Karagas MR, Nelson HH et al (2008) DNA repair polymorphisms modify bladder cancer risk: a multi-factor analytic strategy. Hum Hered 65:105–118

42. Winham SJ, Motsinger-Reif AA (2011) An R package implementation of multifactor dimensionality reduction. BioData Min 4:24

43. Moore JH (2003) Cross validation consistency for the assessment of genetic programming results in microarray studies. In: Cagnoni S,

Johnson CG, Cardalda JJR et al (eds) Applications of evolutionary computing. Springer, Berlin, pp 99–106

44. Winham SJ, Slater AJ, Motsinger-Reif AA (2010) A comparison of internal validation techniques for multifactor dimensionality reduction. BMC Bioinformatics 11:394

45. Bush WS, Edwards TL, Dudek SM et al (2008) Alternative contingency table measures improve the power and detection of multifactor dimensionality reduction. BMC Bioinformatics 9:238

46. Fan R, Zhong M, Wang S et al (2011) Entropy-based information gain approaches to detect and to characterize gene-gene and gene-environment interactions/correlations of complex diseases. Genet Epidemiol 35:706–721

47. Hu T, Sinnott-Armstrong NA, Kiralis JW et al (2011) Characterizing genetic interactions in human disease association studies using statistical epistasis networks. BMC Bioinformatics 12:364

48. Hu T, Andrew AS, Karagas MR et al (2013) Statistical epistasis networks reduce the computational complexity of searching three-locus genetic models, Pacific Symposium on Biocomputing. Pac Symp Biocomput:397–408

49. Hu T, Chen Y, Kiralis JW et al (2013) ViSEN: methodology and software for visualization of statistical epistasis networks. Genet Epidemiol 37:283–285

50. Wong AK, Park CY, Greene CS et al (2012) IMP: a multi-species functional genomics portal for integration, visualization and prediction of protein functions and networks. Nucleic Acids Res 40:W484–W490

51. Gaertner BE, Parmenter MD, Rockman MV et al (2012) More than the sum of its parts: a complex epistatic network underlies natural variation in thermal preference behavior in *Caenorhabditis elegans*. Genetics 192:1533–1542

52. Huang W, Richards S, Carbone MA et al (2012) Epistasis dominates the genetic architecture of Drosophila quantitative traits. Proc Natl Acad Sci U S A 109:15553–15559

53. Lee J-H, Moore JH, Park S-W et al (2008) Genetic interactions model among Eotaxin gene polymorphisms in asthma. J Hum Genet 53:867–875

54. Bush WS, Dudek SM, Ritchie MD (2006) Parallel multifactor dimensionality reduction: a tool for the large-scale analysis of gene-gene interactions. Bioinformatics (Oxford, England) 22:2173–2174

55. Payne JL, Sinnott-Armstrong NA, Moore JH (2010) Exploiting graphics processing units for computational biology and bioinformatics. Interdiscip Sci 2:213–220

56. Moore JH, White BC (2007) Tuning ReliefF for genome-wide genetic analysis. In: Marchiori E, Moore JH, Rajapakse JC (eds) Evolutionary computation, machine learning and data mining in bioinformatics. Springer, Berlin, pp 166–175

57. Greene CS, Penrod NM, Kiralis J et al (2009) Spatially uniform relieff (SURF) for computationally-efficient filtering of gene-gene interactions. BioData Min 2:5

58. Greene CS, Himmelstein DS, Kiralis J et al (2010) The informative extremes: using both nearest and farthest individuals can improve relief algorithms in the domain of human genetics. In: Pizzuti C, Ritchie MD, Giacobini M (eds) Evolutionary computation, machine learning and data mining in bioinformatics. Springer, Berlin, pp 182–193

59. Pan Q, Hu T, Moore JH (2013) Epistasis, complexity, and multifactor dimensionality reduction. Methods Mol Biol (Clifton, NJ) 1019:465–477

60. Dai H, Bhandary M, Becker M et al (2012) Global tests of P-values for multifactor dimensionality reduction models in selection of optimal number of target genes. BioData Min 5:3

61. Bush WS, Dudek SM, Ritchie MD (2009) Biofilter: a knowledge-integration system for the multi-locus analysis of genome-wide association studies, Pacific Symposium on Biocomputing. Pac Symp Biocomput:368–379

62. Greene CS, White BC, Moore JH (2008) Ant colony optimization for genome-wide genetic analysis. In: Dorigo M, Birattari M, Blum C et al (eds) Ant colony optimization and swarm intelligence. Springer, Berlin, pp 37–47

63. Gilmore JM, Greene CS, Andrews PC et al (2011) An analysis of new expert knowledge scaling methods for biologically inspired computing. In: Kampis G, Karsai I, Szathmáry E (eds) Advances in artificial life. Darwin meets von Neumann. Springer, Berlin, pp 286–293

64. Greene CS, Penrod NM, Williams SM et al (2009) Failure to replicate a genetic association may provide important clues about genetic architecture. PLoS One 4:e5639

Chapter 17

Epistasis Analysis Using ReliefF

Jason H. Moore

Abstract

Here we introduce the ReliefF machine learning algorithm and some of its extensions for detecting and characterizing epistasis in genetic association studies. We provide a general overview of the method and then highlight some of the modifications that have greatly improved its power for genetic analysis. We end with a few examples of published studies of complex human diseases that have used ReliefF.

Key words Epistasis, Machine learning, Association studies, Genetic analysis, Gene–gene interaction

1 Introduction

Machine learning methods provide an alternative to parametric statistical methods such as logistic regression for detecting and modeling combinations of attributes that are predictive of an outcome, such as case–control status in a population-based study of common human disease [1–3]. Methods such as multifactor dimensionality reduction or MDR [4–12] are particularly well suited for detecting epistasis or nonadditive gene–gene interactions [3, 13–21], but can be too computationally intense to implement in genome-wide studies with hundreds of thousands or millions of measured genetic variants such as single-nucleotide polymorphisms or SNPs [1]. This is primarily because of the effectively infinite number of variant combinations that need to be evaluated, which can result in a needle in the haystack problem. One approach is to apply a biological or statistical filter to generate a more informative subset of genetic variants that is more computationally tractable. Biological filters have been presented elsewhere [22–24]. We focus here on the ReliefF machine learning algorithm as a statistical filter for rapid assessment of genetic variant quality that takes epistasis and other types of genetic effects into consideration.

The Relief algorithm was developed by Kira and Rendell [25] as a method for estimating the qualities of attributes for the purpose of selecting subsets for further analysis. This approach was

Jason H. Moore and Scott M. Williams (eds.), *Epistasis: Methods and Protocols*, Methods in Molecular Biology, vol. 1253, DOI 10.1007/978-1-4939-2155-3_17, © Springer Science+Business Media New York 2015

later extended for robustness and called ReliefF [26, 27]. Numerous extensions and modifications of ReliefF have been proposed, including RReliefF for quantitative outcomes. The algorithm was comprehensively reviewed in 2003 [26]. Since that time, ReliefF has been explored as a rapid quality assessment and filtering algorithm for genetic studies. This chapter provides a basic overview of the ReliefF algorithm and its extensions specific for genetic studies. We provide several real data analysis results to illustrate the approach.

2   Materials

2.1   Getting Started

ReliefF and its extensions for genetic analysis are implemented in the open-source MDR software package that is freely available from http://sourceforge.net/projects/mdr. MDR is programmed entirely in Java and is compatible with any operating system that has Java installed. We provide elsewhere in this volume an overview of using the graphic user interface (GUI) for the MDR software package. Command line options are also available and are described in a document labeled MDR_command_line_help.txt that is distributed with MDR. The ReliefF algorithms are available in the Filter tab of MDR. It is important to note that the base ReliefF algorithm is accessible in several other freely available data mining software packages, including Orange [28] and Weka [29].

2.2   Data Format

MDR accepts a tab-delimited text file as input. Data are loaded using the Load Datafile button in the upper right corner of the main screen. The data can also be viewed using the View Datafile button. Sample data files with discrete and continuous outcomes and simulated gene–gene interactions are provided as part of the MDR software package. Briefly, each row is an observation or instance (e.g., a human subject) and each column is a discrete variable or attribute (e.g., a single-nucleotide polymorphism or SNP). Typically, the attributes are polymorphisms with three genotypes coded 0, 1, and 2 or covariates such as gender coded 0 and 1. The last column is saved for the phenotype or outcome, which can be binary (e.g., case–control status) or continuous (e.g., body mass index). MDR requires a complete dataset with no missing values. Missing data can be handled in two different ways. First, the missing data can be coded with a number not used to represent a genotype (e.g., 9) and will be treated as an extra level in the analysis (e.g., a fourth genotype). This is acceptable as long as your data are missing at random. As a second option, data imputation methods can be used to fill in the missing values (*see* **Note 1**). This latter method is preferred by the authors and has been explored specifically for MDR [25].

3 Methods

3.1 The Relief and ReliefF Algorithms

Kira and Rendell [25] developed an algorithm called Relief that is capable of detecting attribute dependencies. Relief estimates the quality of attributes through a type of nearest neighbor algorithm that selects neighbors (instances) from the same class and from the different class based on the vector of values across attributes. Weights (W) or quality estimates for each attribute (A) are estimated based on whether the nearest neighbor (nearest hit, H) of a randomly selected instance (R) from the same class and the nearest neighbor from the other class (nearest miss, M) have the same or different values. This process of adjusting weights is repeated for m instances. The algorithm produces weights for each attribute ranging from –1 (worst) to +1 (best). The Relief pseudocode is outlined below:

```
set all weights W[A] = 0
for i = 1 to m do begin
        randomly select an instance Ri
        find nearest hit H and nearest miss M
        for A = 1 to a do
            W[A] = W[A] - diff(A,   Ri,   H)/m +
            diff(A, Ri, M)/m
        end
```

The function diff(A, I_1, I_2) calculates the difference between the values of the attribute A for two instances I_1 and I_2. For nominal attributes such as SNPs it is defined as:

```
diff(A,   I1,   I2) = 0   if   genotype(A,   I1)
= genotype(A,   I2),
    1 otherwise
```

The time complexity of Relief is $O(m*n*a)$ where m is the number of instances randomly sampled from a dataset with n total instances and a attributes. Kononenko [27] improved upon Relief by choosing n nearest neighbors instead of just one. This new ReliefF algorithm has been shown to be more robust to noisy attributes [26, 27] and is widely used in data mining applications.

Moore and White [30] carried out a simulation study to evaluate the power of ReliefF compared to a naïve chi-square test of independence for selecting functional attributes with epistasis effects in a filtered subset. Five genetic models in the forms of penetrance functions were generated. Each model consisted of two SNPs that defined nonlinear relationships with disease susceptibility. The heritability of each model was 0.1, which reflected the moderate genetic effect size. Each of the five models was used to generate 100 replicate datasets with sample sizes of 200, 400, 800, 1,600, 3,200, and 6,400 (*see* **Note 2**). This range of sample sizes represents a spectrum that is consistent with small to medium size genetic studies. Each dataset consisted of an equal number of case (disease) and control (no disease) subjects. Each pair of functional SNPs was combined within a genome-wide set of 998 randomly generated SNPs for a total of 1,000 attributes. A total of 600 datasets were generated and analyzed.

ReliefF and the univariate chi-square test of independence were applied to each of the datasets. The 1,000 SNPs were sorted according to their qualities using each method and the top 50, 100, 150, 200, 250, 300, 350, 400, 450, and 500 SNPs out of 1,000 were selected. From each subset the authors counted the number of times the two functional SNPs were selected out of each set of 100 replicates. This proportion is an estimate of the power or how likely it is to find the true SNPs if they exist in the dataset. The number of times each method found the correct two SNPs was statistically compared. A difference in counts (i.e., power) was considered statistically significant at a type I error rate of 0.05. Moore and White [30] found that the power of ReliefF to pick (filter) the correct two functional attributes was consistently better ($P \le 0.05$) than a naïve chi-square test of independence across subset sizes and models when the sample size was 800 or larger. These results suggest that ReliefF is capable of identifying interacting SNPs with a moderate genetic effect size (heritability = 0.1) in moderate sample sizes. This result provides a baseline for the extensions presented below.

3.2 Tuned ReliefF (TuRF)

ReliefF is able to capture attribute interactions, because it selects nearest neighbors using the entire vector of genotypic values across all attributes. However, this advantage is also a disadvantage because the presence of many noisy attributes can reduce the signal that the algorithm is trying to capture. Moore and White [30] proposed a "tuned" ReliefF algorithm (TuRF) that systematically removes attributes that have low-quality estimates. After removal, the ReliefF values of the remaining attributes are re-estimated. The pseudocode for TuRF is outlined below:

```
let a be the number of attributes
for i=1 to n do begin
        estimate ReliefF
        sort attributes
        remove worst n/a attributes
    end
return last ReliefF estimate for each attribute
```

The motivation behind this algorithm is to improve the ReliefF estimates of the true functional attributes by removing the noisy attributes from the dataset.

Moore and White [30] compared the power of TuRF to the power of ReliefF using the same simulation study described above. They found that the TuRF algorithm was consistently better ($P \le 0.05$) than ReliefF across small SNP subset sizes (50, 100, and

150) and across all five models when the sample size was 1,600 or larger. These initial simulation results suggested that algorithms based on ReliefF show promise for filtering interacting attributes in this domain and that there is room for improvement.

3.3 Evaporative Cooling ReliefF (ECRF)

Evaporative cooling (EC) was first proposed as an experimental technique for cooling a collection of atoms [31]. McKinney et al. [32] adapted this idea to the ReliefF algorithm to help address the problem of noisy attributes in genome-wide data. Here, the noisy attributes are evaporated away leaving the signal in place. Simulation studies showed that this new Evaporative Cooling ReliefF (ECRF) algorithm significantly outperformed ReliefF for identifying epistatic effects. When applied to a study of adverse reactions to a smallpox vaccine, ECRF identified several SNPs out of 1,442 total that matched genes identified in previous proteomic analyses [33]. This study demonstrated how ReliefF could be combined with other algorithms such as those from information theory to produce hybrid approaches that have much greater power than either approach alone. This study also reinforced the idea discussed above with respect to TuRF by showing that noisy attributes reduce the power of ReliefF to identify important interacting attributes.

3.4 Random Chemistry ReliefF (RCRF)

Kauffman introduced the idea of "random chemistry" to identify small sets of L auto-catalytic molecules from a large number N of candidates. As described by Eppstein et al. [34], consider putting a random half of the molecules in a test tube. In this simple example, any given molecule has a 0.5 probability of ending up in the tube, and the probability that all the L desired molecules are in the tube is nearly 0.5^L. Thus, it is expected that one out of almost every 2^L such tubes will contain all L interacting molecules. By doubling the number of tubes to 2^{L+1}, the chance that at least one tube will be "positive" for all L molecules is also doubled. Now, assuming that the L molecules interact in some detectable way (e.g., by forming a product), one could simply screen for a "positive" tube, and repeat the process on the molecules in that tube, until only the correct L molecules remain. Note that this process will require only $2^{L+1}\log_2 N/L$ tests to pick out the L interacting molecules. This process can be viewed as feature selection, where the features Kauffman sought were nonlinearly interacting molecules. Eppstein et al. [34] applied this idea to ReliefF creating the Random Chemistry ReliefF (RCRF) algorithm and showed, using simulated data, that it was able to identify epistatic pairs of SNPs from thousands of candidates (*see* **Note 3**). This approach has not been implemented in an open-source software package.

3.5 Spatially Uniform ReliefF (SURF)

Central to the ReliefF algorithm is the parameter that specifies the number of nearest neighbors to select. This has traditionally been set to 10 in much of the machine learning literature and is very often the default setting in software packages. Greene et al. [35] challenged this assumption by developing the Spatially Uniform ReliefF or SURF algorithm that picks all neighbors within some distance or radius from each instance (*see* **Note 4**). The theoretical and empirical results suggest that a distance should be selected that includes between 25 % and 50 % of the instances as neighbors. For a typical dataset with a sample size in the thousands this would mean hundreds of nearest neighbors, which is significantly more than the default of 10 used in most ReliefF applications. Using the same simulated data presented above, Greene et al. [35] showed that SURF dramatically outperforms ReliefF and has even better performance when combined with TuRF. This approach was improved even further by considering information contained in the furthest neighbors. Greene et al. called this extension SURF* [36]. Small improvements in the SURF* algorithm have been made with the addition of multiple individual-level thresholds (multiSURF*) [37] and sigmoid weighting (SWRF*) [38]. SURF, SURF*, and multiSURF* are all implemented in the open-source MDR software package.

3.6 Application to Bladder Cancer

We have selected the following machine learning study to highlight the ReliefF method. Andrew et al. [39] carried out an epidemiologic study to identify genetic and environmental predictors of bladder cancer susceptibility in a sample of Caucasians ($n = 914$) from New Hampshire. This study focused specifically on genes that play an important role in the repair of DNA sequences that have been damaged by chemical compounds (e.g., carcinogens). Seven SNPs were measured including two from the *X-ray repair cross-complementing group 1* gene (*XRCC1*), one from the *XRCC3* gene, two from the *xeroderma pigmentosum group D (XPD)* gene, one from the *nucleotide excision repair* gene (*XPC*), and one from the *AP endonuclease 1* gene (*APE1*). Each of these genes plays an important role in DNA repair. Smoking is a known risk factor for bladder cancer and was included in the analysis along with gender and age for a total of 10 attributes. Age was discretized to > or ≤ 50 years.

A parametric statistical analysis of each attribute individually revealed a significant independent main effect of smoking as expected. However, none of the measured SNPs were significant predictors of bladder cancer individually. Andrew et al. used MDR to exhaustively evaluate all possible two-, three-, and four-way interactions among the genetic and environmental variables. The best MDR model had a testing accuracy of 0.66 ($p < 0.001$) and included two SNPs from the *XPD* gene and smoking. Decomposition of this model using the measures of information gain demonstrated that the effects of the two *XPD* SNPs were nonadditive or synergistic suggestive of epistasis or gene–gene interaction. The same analysis indicated that the effect of smoking was mostly independent from the gene–gene interaction effect. Analysis of the same data by a Learning Classifier System (LCS) approach confirmed this result, and additional analyses suggested that smoking and the epistatic interaction may actually be associated with bladder cancer in different subsets of subjects (i.e., heterogeneity) [40]. Application of the ReliefF algorithm gives smoking the highest weight followed by the two epistatic *XPD* SNPs. Thus, using ReliefF as a filter in this dataset yielded the three strongest signals according to both MDR and LCS machine learning analyses.

3.7 Application to Pulmonary Tuberculosis

Collins et al. used ReliefF as a filter to reduce the total number of SNPs used in an MDR analysis of pulmonary tuberculosis [41]. The dataset used in this study was originally analyzed by Olesen et al. and includes 321 pulmonary TB cases and 347 healthy controls genotyped as part of the Bandim Health Project in Guinea Bissau [42]. Each individual was genotyped for 19 SNPs from immunological candidate genes VDR, DC-SIGN, PTX3, TLR2, TLR4, and TLR9. These genes were all selected as strong biological candidates. Missing data were imputed using a frequency-based imputation.

The authors first applied ReliefF using 100 nearest neighbors and filtered the SNPs to the top five. This yielded SNPs rs187084 (TLR9), rs4986790 (TLR4), rs11465421 (DC-SIGN), rs2305619 (PTX3), rs1840680 (PTX3), and rs2287886 (DC-SIGN). It is important to note that none of these five SNPs alone had statistically significant main effects. The authors then carried out an exhaustive MDR analysis on all possible combinations of these five SNPs to test for epistasis or nonadditive gene–gene interactions. The overall best model consisted of SNPs rs2305619, rs187084, and rs1145421. These three SNPs had a training accuracy of 0.6115 and a testing accuracy of 0.5878. The cross-validation consistency of this model was 10/10. Permutation testing confirmed the statistical significance of the model ($p = 0.008$). Additional permutation testing using the explicit test of epistasis [43] revealed that the nonadditive effects in the model were also statistically significant ($p = 0.013$). Taken together, these results suggest a role for high-order, nonadditive epistatic effects. In this study ReliefF effectively reduced the total number of MDR models that were considered (*see* **Note 5**). This is important given the combinatorial nature of methods such as MDR and the concern about the discovery of false-positives due to multiple testing. Subsequent analyses of these data using measures of interaction information provided additional evidence for a three-way epistatic interaction among these SNPs that was independent of any two-way or one-way effects [44].

3.8 Conclusions

We have provided a brief overview of the ReliefF method and some of its extensions that have been developed specifically for genome-wide genetic analyses. ReliefF is able to detect epistasis in a computationally efficient manner and is thus a good candidate for statistical filtering of large lists of SNPs prior to machine learning analysis with approaches such as MDR (*see* **Note 6**). Some of the ReliefF algorithms have been implemented in the open-source MDR software package and are thus readily available. More information about how ReliefF fits into a general bioinformatics strategy for detecting epistasis is provided by Moore et al. [1]. Information on other methods for detecting and characterizing interactions has been provided by Cordell [3].

4 Notes

The following are some important tips for using ReliefF algorithms:

1. A simple frequency-based imputation approach may work well when there are very few missing genotypes. Of course, genetic data is ideally suited to multivariate imputation methods that use linkage disequilibrium to infer missing genotypes. The freely available PLINK software is commonly used to impute genotype data [45].

2. The epistasis models and simulation data have been previously described and are freely available from the authors [10]. The development of new epistasis models and their use for simulation can be quickly accomplished using the GAMETES software package that is freely available from sourceforge.net. The method behind GAMETES has been described by Urbanowicz et al. [46, 47]. These epistasis models can also be used to simulate population-level data with realistic linkage disequilibrium patterns using the GenomeSIMLA software package [48–50]. Other methods for developing and discovering epistasis models for simulation or the simulation of complex data have been described [51, 52].

3. The random chemistry ReliefF approach was published using a different set of simulated data and is thus not directly comparable to ReliefF, SURF, and SURF*. However, unpublished work has shown that this general approach may be more powerful. This is intuitive given the general findings with TuRF and ECRF that are designed to remove noisy SNPs.

4. It is important to note that Relief and ReliefF produce weights on the scale of –1 to 1. SURF and SURF* scores exceed this range.

5. Although not implemented here, permutation testing methods for ReliefF have been developed [53]. These approaches

allow for filtering cutoffs that are based on statistical significance rather than an arbitrary value.

6. Although the focus of this chapter is on filtering a list of genetic variants, it is important to note that ReliefF scores can be used as expert knowledge to help guide a stochastic search algorithm like genetic programming [54–56]. More details on this approach can be found in the computational evolution chapter in this volume.

Acknowledgements

This work was supported by National Institutes of Health (NIH) grants AI59694, EY022300, GM103534, GM103506, LM 009012, LM010098, and LM011360.

References

1. Moore JH, Asselbergs FW, Williams SM (2010) Bioinformatics challenges for genome-wide association studies. Bioinformatics 26:445–455

2. McKinney BA, Reif DM, Ritchie MD et al (2006) Machine learning for detecting gene-gene interactions: a review. Appl Bioinforma 5:77–88

3. Cordell HJ (2009) Detecting gene-gene interactions that underlie human diseases. Nat Rev Genet 10:392–404

4. Hahn LW, Ritchie MD, Moore JH (2003) Multifactor dimensionality reduction software for detecting gene-gene and gene-environment interactions. Bioinformatics 19:376–382

5. Ritchie MD, Hahn LW, Roodi N et al (2001) Multifactor-dimensionality reduction reveals high-order interactions among estrogen-metabolism genes in sporadic breast cancer. Am J Hum Genet 69:138–147

6. Hahn LW, Moore JH (2004) Ideal discrimination of discrete clinical endpoints using multi-locus genotypes. In Silico Biol 4:183–194

7. Ritchie MD, Hahn LW, Moore JH (2003) Power of multifactor dimensionality reduction for detecting gene-gene interactions in the presence of genotyping error, missing data, phenocopy, and genetic heterogeneity. Genet Epidemiol 24:150–157

8. Moore JH (2004) Computational analysis of gene-gene interactions using multifactor dimensionality reduction. Expert Rev Mol Diagn 4:795–803

9. Moore JH (2010) Detecting, characterizing, and interpreting nonlinear gene-gene interactions using multifactor dimensionality reduction. Adv Genet 72:101–116

10. Velez DR, White BC, Motsinger AA et al (2007) A balanced accuracy function for epistasis modeling in imbalanced datasets using multifactor dimensionality reduction. Genet Epidemiol 31:306–315

11. Pattin KA, White BC, Barney N et al (2009) A computationally efficient hypothesis testing method for epistasis analysis using multifactor dimensionality reduction. Genet Epidemiol 33:87–94

12. Moore JH (2007) Genome-wide analysis of epistasis using multifactor dimensionality reduction: feature selection and construction in the domain of human genetics. In: Zhu X, Davidson I (eds) Knowledge discovery and data mining: challenges and realities. IGI Global, Hershey, PA, pp 17–30

13. Cordell HJ (2002) Epistasis: what it means, what it doesn't mean, and statistical methods to detect it in humans. Hum Mol Genet 11:2463–2468

14. Phillips PC (2008) Epistasis—the essential role of gene interactions in the structure and evolution of genetic systems. Nat Rev Genet 9:855–867

15. Phillips PC (1998) The language of gene interaction. Genetics 149:1167–1171

16. Moore JH (2003) The ubiquitous nature of epistasis in determining susceptibility to common human diseases. Hum Hered 56:73–82

17. Moore JH, Williams SM (2005) Traversing the conceptual divide between biological and statistical epistasis: systems biology and a more modern synthesis. Bioessays 27:637–646

18. Moore JH (2005) A global view of epistasis. Nat Genet 37:13–14

19. Tyler AL, Asselbergs FW, Williams SM et al (2009) Shadows of complexity: what biological networks reveal about epistasis and pleiotropy. Bioessays 31:220–227

20. Cowper-Sal lari R, Cole MD, Karagas MR et al (2011) Layers of epistasis: genome-wide regulatory networks and network approaches to genome-wide association studies, Wiley interdisciplinary reviews. Syst Biol Med 3:513–526

21. Moore JH, Williams SM (2009) Epistasis and its implications for personal genetics. Am J Hum Genet 85:309–320

22. Bush WS, Dudek SM, Ritchie MD (2009) Biofilter: a knowledge-integration system for the multi-locus analysis of genome-wide association studies. Pacific symposium on biocomputing, pp 368–379

23. Pendergrass SA, Verma SS, Holzinger ER et al (2013) Next-generation analysis of cataracts: determining knowledge driven gene-gene interactions using Biofilter, and gene-environment interactions using the PhenX Toolkit. Pacific symposium on biocomputing, pp 147–158

24. Grady BJ, Torstenson ES, McLaren PJ et al (2011) Use of biological knowledge to inform the analysis of gene-gene interactions involved in modulating virologic failure with efavirenz-containing treatment regimens in ART-naïve ACTG clinical trials participants. Pacific symposium on biocomputing, pp 253–264

25. Kira K, Rendell LA (1992) A practical approach to feature selection. In: Proceedings of the ninth international workshop on machine learning, Morgan Kaufmann Publishers, San Francisco, CA, pp 249–256

26. Robnik-Šikonja M, Kononenko I (2003) Theoretical and empirical analysis of ReliefF and RReliefF. Mach Learn 53:23–69

27. Kononenko I (1994) Estimating attributes: analysis and extensions of RELIEF. Lecture Notes in Computer Science 784:171–182

28. Demšar J, Curk T, Erjavec A et al (2013) Orange: data mining toolbox in python. J Mach Learn Res 14:2349–2353

29. Hall M, Frank E, Holmes G et al (2009) The WEKA data mining software: an update. SIGKDD Explor Newsl 11:10–18

30. Moore JH, White BC (2007) Tuning ReliefF for genome-wide genetic analysis. In: Marchiori E, Moore JH, Rajapakse JC (eds) Evolutionary computation, machine learning and data mining in bioinformatics. Springer, Berlin, pp 166–175

31. Hess HF (1986) Evaporative cooling of magnetically trapped and compressed spin-polarized hydrogen. Phys Rev B Condens Matter 34:3476–3479

32. McKinney BA, Reif DM, White BC et al (2007) Evaporative cooling feature selection for genotypic data involving interactions. Bioinformatics 23:2113–2120

33. McKinney BA, Reif DM, Rock MT et al (2006) Cytokine expression patterns associated with systemic adverse events following smallpox immunization. J Infect Dis 194:444–453

34. Eppstein MJ, Payne JL, White BC et al (2007) Genomic mining for complex disease traits with "random chemistry". Genet Program Evolvable Mach 8:395–411

35. Greene CS, Penrod NM, Kiralis J et al (2009) Spatially uniform relieff (SURF) for computationally-efficient filtering of gene-gene interactions. BioData Min 2:5

36. Greene CS, Himmelstein DS, Kiralis J et al (2010) The informative extremes: using both nearest and farthest individuals can improve relief algorithms in the domain of human genetics. In: Pizzuti C, Ritchie MD, Giacobini M (eds) Evolutionary computation, machine learning and data mining in bioinformatics. Springer, Berlin, pp 182–193

37. Granizo-Mackenzie D, Moore JH (2013) Multiple threshold spatially uniform ReliefF for the genetic analysis of complex human diseases. In: Vanneschi L, Bush WS, Giacobini M (eds) Evolutionary computation, machine learning and data mining in bioinformatics. Springer, Berlin, pp 1–10

38. Stokes ME, Visweswaran S (2012) Application of a spatially-weighted Relief algorithm for ranking genetic predictors of disease. BioData Min 5:20

39. Andrew AS, Nelson HH, Kelsey KT et al (2006) Concordance of multiple analytical approaches demonstrates a complex relationship between DNA repair gene SNPs, smoking and bladder cancer susceptibility. Carcinogenesis 27:1030–1037

40. Urbanowicz RJ, Andrew AS, Karagas MR et al (2013) Role of genetic heterogeneity and epistasis in bladder cancer susceptibility and outcome: a learning classifier system approach. J Am Med Inform Assoc 20:603–612

41. Collins RL, Hu T, Wejse C et al (2013) Multifactor dimensionality reduction reveals a three-locus epistatic interaction associated with susceptibility to pulmonary tuberculosis. BioData Min 6:4

42. Olesen R, Wejse C, Velez DR et al (2007) DC-SIGN (CD209), pentraxin 3 and vitamin D receptor gene variants associate with pulmonary tuberculosis risk in West Africans. Genes Immun 8:456–467

43. Greene CS, Himmelstein DS, Nelson HH et al (2010) Enabling personal genomics with an explicit test of epistasis. Pacific symposium on biocomputing, pp 327–336

44. Hu T, Chen Y, Kiralis JW et al (2013) An information-gain approach to detecting three-way epistatic interactions in genetic association studies. J Am Med Inform Assoc 20:630–636

45. Purcell S, Neale B, Todd-Brown K et al (2007) PLINK: a tool set for whole-genome association and population-based linkage analyses. Am J Hum Genet 81:559–575

46. Urbanowicz RJ, Kiralis J, Sinnott-Armstrong NA et al (2012) GAMETES: a fast, direct algorithm for generating pure, strict, epistatic models with random architectures. BioData Min 5:16

47. Urbanowicz RJ, Kiralis J, Fisher JM et al (2012) Predicting the difficulty of pure, strict, epistatic models: metrics for simulated model selection. BioData Min 5:15

48. Edwards TL, Bush WS, Turner SD et al (2008) Generating linkage disequilibrium patterns in data simulations using genomeSIMLA. In: Proceedings of the 6th European conference on evolutionary computation, machine learning and data mining in bioinformatics, Springer, Berlin, pp 24–35

49. Ritchie MD, Bush WS (2010) Genome simulation approaches for synthesizing in silico datasets for human genomics. Adv Genet 72:1–24

50. Dudek SM, Motsinger AA, Velez DR et al (2006) Data simulation software for whole-genome association and other studies in human genetics. Pacific symposium on biocomputing, pp 499–510

51. Moore JH, Hahn LW, Ritchie MD et al (2004) Routine discovery of complex genetic models using genetic algorithms. Appl Soft Comput 4:79–86

52. Himmelstein DS, Greene CS, Moore JH (2011) Evolving hard problems: generating human genetics datasets with a complex etiology. BioData Min 4:21

53. Dai H, Bhandary M, Becker M et al (2012) Global tests of P-values for multifactor dimensionality reduction models in selection of optimal number of target genes. BioData Min 5:3

54. Greene CS, Hill DP, Moore JH (2010) Environmental sensing of expert knowledge in a computational evolution system for complex problem solving in human genetics. In: Riolo R, O'Reilly U-M, McConaghy T (eds) Genetic programming theory and practice, vol VII. Springer, Ann Arbor, MI, pp 19–36

55. Moore JH, Andrews PC, Barney N et al (2008) Development and evaluation of an open-ended computational evolution system for the genetic analysis of susceptibility to common human diseases. In: Marchiori E, Moore JH (eds) Evolutionary computation, machine learning and data mining in bioinformatics. Springer, Berlin, pp 129–140

56. Moore JH, Greene CS, Andrews PC et al (2009) Does complexity matter? artificial evolution, computational evolution and the genetic analysis of epistasis in common human diseases. In: Genetic programming theory and practice, vol VI. Springer, pp 1–19

Chapter 18

Epistasis Analysis Using Artificial Intelligence

Jason H. Moore and Doug P. Hill

Abstract

Here we introduce artificial intelligence (AI) methodology for detecting and characterizing epistasis in genetic association studies. The ultimate goal of our AI strategy is to analyze genome-wide genetics data as a human would using sources of expert knowledge as a guide. The methodology presented here is based on computational evolution, which is a type of genetic programming. The ability to generate interesting solutions while at the same time learning how to solve the problem at hand distinguishes computational evolution from other genetic programming approaches. We provide a general overview of this approach and then present a few examples of its application to real data.

Key words Epistasis, Machine learning, Artificial intelligence, Association studies, Genetic analysis, Gene–gene interaction, Genetic programming

1 Introduction

A hallmark of common human diseases is complexity in the genotype–phenotype mapping relationship due to phenomena such as epistasis or nonadditive gene–gene interactions [1–10]. Machine learning methods such as multifactor dimensionality reduction (MDR) [11–15] have been developed as alternatives to parametric methods such as logistic regression, because they are better at modeling complex patterns in high-dimensional data [5, 16]. Artificial intelligence (AI) takes these approaches several steps beyond model generation and discovery by incorporating elements of human intelligence into the algorithms.

Our general AI approach involves pre-processing genome-wide genetic data to identify useful information, implementing a machine learning algorithm that is able to model nonlinear genetic effects, implementing a stochastic search algorithm that is able to exploit expert knowledge, and implementing post-processing methods that are able to enhance statistical and biological interpretations of the results. Effective integration of these pieces yields a system that can solve a complex problem, and, importantly, learn *how* to solve the

Jason H. Moore and Scott M. Williams (eds.), *Epistasis: Methods and Protocols*, Methods in Molecular Biology, vol. 1253, DOI 10.1007/978-1-4939-2155-3_18, © Springer Science+Business Media New York 2015

problem. The overarching goal is to generate genetic models that are interesting to the user and that are unexpected given the current knowledge base. This will be key for revealing the so-called missing heritability for common diseases [17, 18].

We have previously developed AI methods for genetic analysis that are based on genetic programming (GP). GP is an automated computational discovery tool that is inspired by Darwinian evolution and natural selection [19, 20]. Accessible introductions to the basics of GP are available [20, 21]. The goal of GP is to "evolve" computer programs to solve complex problems, which is accomplished by generating or initializing a population of random computer programs that are composed of the basic building blocks needed to solve or approximate a solution to the problem. Genetic programming and its many variations have been applied successfully in a wide range of different problem domains, including bioinformatics [22] and genetic analysis [5, 23–27]. GP is an attractive approach to the genetic analysis problem, because it is inherently flexible in its representation of solutions and it is stochastic, parallel, and easily adapted to exploit expert knowledge [25].

The early GP algorithms were black box approaches that relied heavily on the stochastic search component of the algorithm and high-performance computing to search through many solutions [19]. As we have previously pointed out, GP and other stochastic search algorithms need additional layers of computation to solve complex real-world problems like those in human genetics and epidemiology [25]. This is consistent with others who have called for additional layers of complexity. For example, Banzhaf et al. [28] suggested that we transform GP to a computational evolution system (CES) that mimics the complexity of real biological systems. Evolution by natural selection solves problems by building complexity. As such, it is hypothesized that computational systems inspired by evolution should do the same. We took this on as a research question and were able to show that by adding complexity to GP as part of CES, we were able to more effectively model gene–gene interactions in simulated data [29, 30]. A variety of extensions have been developed to tailor this approach to the analysis of genome-wide genetic data from population-based studies of common human diseases [31–36]. Here we describe a CES for epistasis analysis in genetic association studies and then provide several real-world examples.

2   Materials

2.1   *Getting Started*

The single most important part of using CES, GP, or any other evolutionary computing method is to know what your building blocks are [37]. That is, you need to be able to identify the fundamental units that are needed to build good genetic models.

This is particularly critical for detecting epistasis in genome-wide data, because individual polymorphisms without main effects will not produce good models unless their interacting partners are also present. This effectively reduces GP to a random search looking for a needle in a haystack. The key to success is to weight important polymorphisms a priori and then provide the GP mechanisms for exploiting this expert knowledge. This is what Goldberg has called a competent genetic algorithm [38]. In practice this means looking at available knowledge about a set of polymorphisms and their corresponding genes or genomic regions and using it to assign weights based on their biological or statistical importance. For example, assigning higher weights to glucose metabolism genes would be reasonable for the genetic analysis of type II diabetes [39]. Other forms of knowledge such as protein–protein interactions can also be used [31, 40, 41] (*see* **Note 1**). Alternatively, prior statistical results from genetic association or linkage studies could be used to weight polymorphisms (*see* **Note 2**). Without this knowledge the algorithm is carrying out an effectively infinite random search for an epistatic needle in a genomic haystack. In our first two studies with CES we pre-processed the data using the ReliefF machine learning algorithm [29, 30]. ReliefF is able to estimate weights for each polymorphism in a way that takes into account nonlinear interactions in addition to any independent main effects (*see* **Note 3**). These ReliefF scores are then provided to CES as expert knowledge that the algorithm can use to bias the search to that portion of the solution space. We also provided an example of using protein–protein interactions to guide the search [31]. Of course, the result of the search is only as good as the expert knowledge that is provided.

The CES software is programmed in C++ and is currently available as a command-line tool by request from the authors as open source. A single run of the system with a population of 576 solutions on a 36×36 grid for 1,000 generations takes approximately 10 min on a single 3.0 GHz AMD Opteron processor. General guidelines for carrying out a CES analysis are provided below. A user-friendly Java version of CES is under development. Contact the authors for more information about available CES software.

2.2 Data Format

CES accepts a tab-delimited text file as input. Briefly, each row is an observation or instance (e.g., a human subject) and each column is a discrete variable or attribute (e.g., a single-nucleotide polymorphism or SNP). Typically, the attributes are polymorphisms with three genotypes coded 0, 1, and 2 or covariates such as gender coded 0 and 1. The last column is saved for the phenotype or outcome, which can be binary (e.g., case–control status) or continuous (e.g., body mass index). CES requires a complete data set with no missing values. Missing data can be handled in two different ways. First, the missing data can be coded with a number not used to represent a genotype (e.g., 9) and will be treated as an extra level in the analysis (e.g., a fourth genotype). This is acceptable as long as your data are missing at random. As a second option data imputation methods can be used to fill in the missing values (*see* **Note 4**).

3 Materials

3.1 Overview of a Computational Evolution System

Figure 1 provides a graphical overview of the base CES, which is both hierarchically organized and spatially explicit. The bottom level of the hierarchy consists of a lattice of solutions (Fig. 1d), which compete with one another within spatially localized, overlapping neighborhoods. The second layer of the hierarchy contains

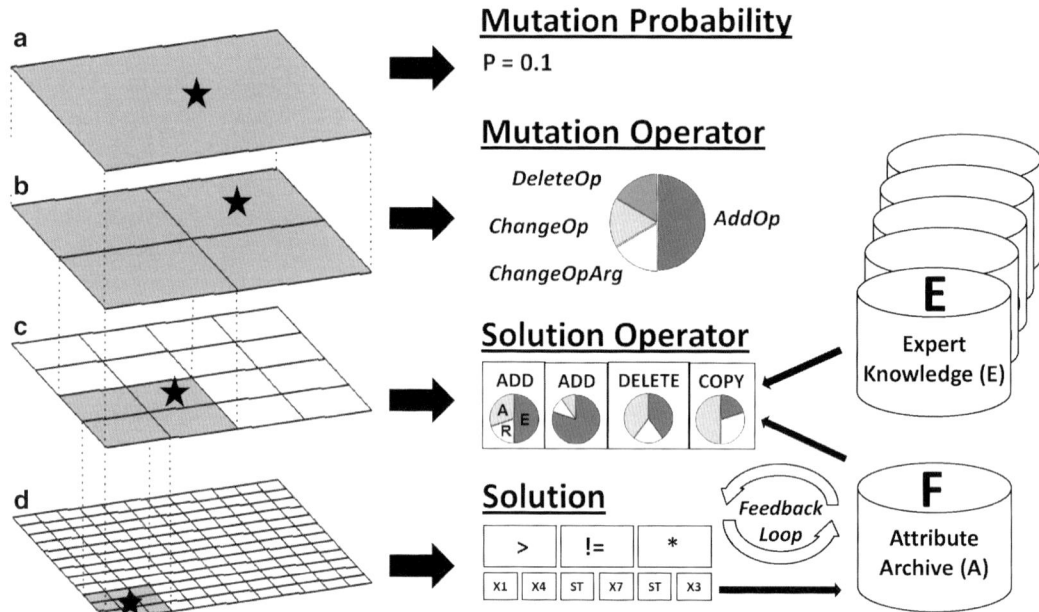

Fig. 1 Visual overview of our computational evolution system for discovering symbolic discriminant functions that differentiate disease subjects from healthy subjects using information about single-nucleotide polymorphisms (SNPs). The hierarchical structure is shown on the *left* while some specific examples at each level are shown in the *middle*. At the lowest level (**d**) is a grid of solutions. Each solution consists of a list of functions and their arguments (e.g., X1 is an attribute or SNP) that are evaluated using a stack (denoted by ST in the solution). The next level up (**c**) is a grid of solution operators that each consists of some combination of the ADD, DELETE, and COPY functions each with their respective set of probabilities that define whether attributes are added, deleted, or copied randomly, using an attribute archive (**f**) or using pre-processed expert knowledge (**e**). The *top two levels* of the hierarchy (**a** and **b**) exist to generate variability in the operators that modify the solutions. This system allows operators of arbitrary complexity to modify solutions. A 12×12 grid is shown here as an example

a lattice of arbitrarily complex solution operators (Fig. 1c), which operate on the solutions in the lower layer. The third layer of the hierarchy contains a lattice of mutation operators (Fig. 1b), which modify the solution operators in the second layer, and the highest layer of the hierarchy governs the rate at which the mutation operators are modified (Fig. 1a). Solution operators can modify solutions randomly, according to an archive or memory (Fig. 1f) or by exploiting one or more sources of expert knowledge that help define building blocks as discussed above (Fig. 1e). We provide details of this base system in the next several subsections.

3.1.1 Genetic Model Representation

Each solution represents a classifier, which takes a set of polymorphisms or other attributes as input and produces an output that can be used to assign case–control status, for example. These solutions are represented as stacks, where each element in the stack consists of a function and two operands (Fig. 1). The function set contains any possible mathematical function. We typically use +, –, *, %, <, <=, >, >=, ==, !=, where % denotes a protected modulus. Operands are either biological measures such as polymorphisms, constants, or the output of another element in the stack. These stack-based models can be easily simplified to trees or mathematical equations that are easier to interpret. We provide some examples later. Defining the function set is important because this defines the set of possible models that can be generated. The broader the function set the more flexible the modeling process will be. This is important for developing genetic models when we don't fully understand the complexity of the genotype–phenotype relationship. Reducing the function set to simple additive functions (e.g., +, –) inherently makes assumptions about the complexity of the models that are likely to explain disease susceptibility. Our general approach is to make as few assumptions as possible. With that said, it is possible to compare and contrast different function sets [26]. This could be useful if the size of the search space is a concern.

3.1.2 Genetic Model Fitness Determination

Each genetic model or CES solution produces an output S_i when applied to attribute values (e.g., genotype codes) for an individual i. Symbolic discriminant analysis [26, 42] is then used to map this output to a classification rule, as follows. The solution is independently applied to the set of cases and controls to obtain two separate distributions of outputs, S^{case} and $S^{control}$, respectively. A classification threshold S_0 is then calculated as the arithmetic mean of the medians of these two distributions. The corresponding solution classifies individual i as case if $S_i > S_0$ and control otherwise. Solution accuracy is assessed through a comparison of predicted and actual clinical endpoints. Specifically, the number of true positives (TP), false positives (FP), true negatives (TN), and false negatives (FN) are used to calculate accuracy as $A = (1/2)(TP/(TP+FN)+TN/(TN+FP))$. This can be a balanced accuracy when the number of cases and controls are not equal [43]. Other solution metrics such as size or complexity can also be saved. Balancing accuracy with measures such as model complexity will be discussed further below in the section on Pareto optimization.

3.1.3 Genetic Model Selection

The population of solutions is organized on a two-dimensional lattice with periodic boundary conditions (*see* **Note 5**). As such, it resembles cellular GP [44]. Each solution occupies a single lattice site, and competes with the solutions occupying the eight spatially adjacent sites (i.e., the Moore neighborhood). Selection is both synchronous and elitist, such that the solution of highest fitness within a given neighborhood is always selected to repopulate the focal site of that neighborhood. Reproduction is either sexual or asexual, as dictated by the evolvable solution operators that reside in the next layer of the hierarchy. The population is typically initialized by randomly generating solutions with 1–15 elements subject to the constraint that they produce a valid output that is not constant for all input. Sensible initialization using expert knowledge has been previously explored and is recommended when the list of attributes is large as in genome-wide studies [35]. This is discussed in more detail below.

3.1.4 Genetic Model Operators

Traditional artificial evolution approaches such as GP use a fixed set of operators that include mutation and recombination, for example. The goal of developing CES was to provide building blocks (i.e., simple functions) for operators that could be combined to create new operators of any arbitrary complexity. We started with the following three basic operator building blocks. The first operator building block, ADD, adds a new function to the list of functions and their arguments that comprise a solution. The second operator building block, DELETE, deletes a function from the list of functions. The third operator, COPY, copies a function from the list of functions from the solution in focus to a solution located within the Moore neighborhood. We have also added other operators such as ALTER that modify a function or argument.

Each of the operator building blocks has a vector of three probabilities associated with it. The first number specifies the probability that the function or argument that is added, deleted, copied, or altered is determined stochastically. The second number specifies the probability that the function or argument change is determined according to an archive of attributes or model pieces that is ranked according to the frequency that they occur in the population of solutions (see below). The third number specifies the probability that the function or argument change is determined according to a particular source of expert knowledge. The ADD, DELETE, COPY, and ALTER operators can combine in any number and order to generate solution operators of any arbitrary complexity.

The higher-level mutation operators described below increase or decrease the size and content of the solution operators.

As with the solutions, each operator is organized on a toroidal grid with a specific X and Y coordinate. Rather than generate one operator for each solution we assigned each operator to a set of solutions. This makes evaluation of the fitness of an operator easier since its positive or negative effect on the solutions can be averaged over multiple solutions. For example, a population of solution operators can be organized in a 6×6 grid (36 total operators) when an 18×18 grid is used for the solutions, and a 12×12 grid (144 total operators) of solution operators can be used when a 36×36 grid is used for the solutions. The fitness of a solution operator is measured by the average change in the fitness of the solutions that the operator modifies from one generation to the next.

3.1.5 Mutation of Genetic Model Operators

An important goal for CES is the ability to generate variability in the operators that modify the genetic models or solutions. To accomplish this goal we developed an additional level in the hierarchy (Fig. 1b) with mutation operators that specifically alter the operators described above. We defined four different fixed mutation operators that are each assigned to a 2×2 grid of solution operators. Solution operators can be modified in the following four ways. First, an operator can have a specific operator building block deleted (DeleteOperator). Second, an operator can have a specific operator building block added (AddOperator). Third, an operator can have a specific operator building block changed (ChangeOperator). Finally, an operator can have its arguments changed (ChangeOperatorArguments). This latter function allows, for example, the range that a DeleteRangeOperation would use. We typically initialize the probabilities with which each of these mutation operators are used to 0.25. These are randomly regenerated at a frequency equal to the overall mutation probability (see below), and their fitness is determined by the change in fitness of the solution operators that they act on.

3.1.6 Mutation Frequency

The top level of the computational evolution system hierarchy (Fig. 1a) is the mutation frequency that controls the probability that one of the four mutation sets in the next level down will mutate a given solution operator two levels down. We typically fix this to 0.1. Note that this frequency does not control the frequency with which an operator modifies a solution in the lowest level. The operator itself controls this when it specifies which solution(s) it will modify.

3.1.7 The CES Archive

Previous studies have demonstrated the utility of archiving GP results for reuse [45]. We have implemented an archive in CES that ranks the attributes by the frequency with which they appear in solutions from the population. These are ranked by their frequencies and then used by the solution operators to decide what gets added, deleted, copied, or altered. Presumably those attributes that show up more frequently are more likely to be important for solving the problem. At the very least, the archive serves as an algorithmic memory of what things have been repeatedly discovered. We used a cumulative archive that updates the previous results each generation. The archive is an important part of the complexity of CES, because it provides a feedback loop between the solutions and the solution operators. An option to archive models has also been implemented but not extensively evaluated.

3.2 Pareto Optimization

Cross-validation methods are difficult within the context of GP and CES due to their stochastic nature [46]. This is because the algorithm can discover completely different models in each division of the data. We have addressed this in several different ways. First, we added small amounts of noise to the data throughout the CES run to keep it from fixating on random patterns in the data that lead to overfitting [36]. This is effective as long as the noise added is less than the signal to be modeled. Second, we explored the use of Pareto optimization to balance the accuracy of the classifier with other measures such as model complexity [34]. We briefly review this approach here and provide an example from real data later.

Pareto optimization [47] offers a viable alternative to cross-validation that is quite effective in the context of GP [48]. Pareto optimization, previously reviewed by Coello et al. [49], balances several different model objectives that are each treated equally. Our first implementation balanced classification accuracy and model size. For a given CES population, the best models, measured by both accuracy and model size, are selected. This subset of Pareto-optimal models is referred to as the Pareto front. Models along the Pareto front are the ones selected for the next generation. This is advantageous because it allows the search algorithm to explore models that have a high accuracy (good) and high complexity (bad) while at the same time exploring those models that have low accuracy (bad) and low complexity (good). This helps promote diversity in the sets of models that are explored, so the algorithm doesn't stall on local minima. Once a final Pareto front is selected, the user can decide on a best model or best set of models that balance these two competing objectives. Users concerned about overfitting might focus on those models that have a lower complexity, for example.

3.3 Sensible Initialization

Most genetic programming methods initialize the starting population of solutions randomly. Previously, we explored the use of sensible initialization within the context of GP [50] and CES [35]. Sensible initialization builds on the idea of identifying and exploiting good building blocks for generating solutions [38]. In other words, starting the initial population with model building blocks that are already believed to be useful might generate better models. Payne et al. [35] showed, using simulated gene–gene interaction data, that initialization of the population using pre-processed ReliefF scores greatly improved the ability of CES to find the complex interaction model above and beyond random initialization or even initializing solutions to ensure all attributes are represented in the population. This kind of expert knowledge-driven initialization could be implemented using any source of pre-processed statistical results or biological knowledge, such as that from prior experimental work or annotations such as Gene Ontology.

3.4 Measures of Interestingness

Central to all GP and CES methods is the function or functions that are used to determine the fitness or quality of the solutions. Fitness for classification problems is usually determined by an error or accuracy measure that reflects the proportion of instances (e.g., subjects) in the data that are correctly labeled (e.g., case–control) by the model. As described above, Pareto optimization provides an intuitive way to incorporate additional fitness measures that can have a big impact on the diversity of solutions generated by CES by rewarding the algorithm for other criteria that might be important to the user. These other criteria can be classified as objective or subjective and collectively grouped into what are called interestingness measures (*see* **Note 6**).

Moore et al. [51] suggested using measures of epistasis as "interesting" qualities of a genetic model. Here, the authors used measures of interaction information as their measures of interestingness [15, 52–54]. Interestingness was introduced into CES in two different ways. First, pre-computed pairwise interaction information scores were used as expert knowledge during the model building process. Second, we used interaction information as an additional axis in a three-way Pareto optimization that also included model error and size. This allowed CES to explore models with high interestingness but low accuracy thus promoting diversity. Interestingness has also been explored for use with data mining [55]. Geng and Hamilton [55] review nine specific criteria for determining whether a model or result is interesting. The first is conciseness or parsimony. The second is coverage (i.e., applies to a large subset of the data). The third is reliability, a measure of the accuracy or error of a classifier. The fourth is peculiarity, a measure of how far away a finding is from others. The fifth is diversity, a measure of how different the elements of a model are. The sixth is novelty (i.e., the result is new). The seventh is surprisingness or how unexpected the result is based on prior knowledge. The eighth is utility or how useful the result is. The final criterion is actionability, which measures how applicable a result is to a particular domain. Each of these criteria can be grouped into objective and subjective categories. For example, conciseness, coverage, reliability, peculiarity, and diversity are all objective measures, because they can be computed using an algorithm or mathematical function. On the other hand, novelty, surprisingness, utility, and actionability are subjective and depend on the experience and knowledge of the particular domain expert. The study by Moore et al. [51] used interaction information as a pre-processed measure of interestingness that could be used to guide the CES. This is an objective measure because a specific measure from the data was computed. However, it can also be seen as a subjective measure because an emphasis is placed on gene–gene interactions. Incorporating measures of interestingness into GP and CES methods moves these algorithms a step closer to a true artificial intelligence, because it allows these algorithms to explore the model space in a much more human-like manner. We provide a summary of the results from this study below.

3.5 Visualization

An important component of any machine learning analysis is the ability to visualize the data and the results. This is essential for moving from machine learning to artificial intelligence, because humans are, by nature, visual creatures that integrate visual cues into almost everything they do. Moore et al. [33] proposed that visualization of CES results will provide important information to the user that can then in turn be used as expert knowledge to help guide the CES toward a useful solution. If true, this would suggest that the ability of the users to interact with the CES in an intuitive visual manner will facilitate the discovery process. The close integration of computational analysis with visualization and human–computer interaction is an emerging discipline called *visual analytics* [56]. This is distinguished from *scientific visualization*, the mathematics and physics of visualizing 3D objects, and *information visualization*, which uses methods such as heat maps to show high-dimensional research results. Heer et al. [57] provide a thorough review of information visualization methods. What makes visual analytics different is the integration of the visualization methods with data analysis. That is, the CES analysis can be launched directly from the visualization and the visualization, in turn, can be changed in a manner that is dependent on the data analysis results. This iterative and synergistic process of visualization and analyzing is facilitated by computer hardware technology that makes it easy for the user to interact with the software. For example, new touch-based computer interfaces such as the Microsoft Surface Computer or the Apple iPad could replace the keyboard and mouse as the preferred interface for visual analytics. All of this combined with a 3D visualization screen or wall provides a modern visual analytics discovery environment that immerses the user in their data and research results. Our ultimate goal is to engineer a CES-based discovery environment for the genetic analysis of complex human disease that combines all of these features.

The goal of the study by Moore et al. [33] was to develop a 3D visualization method that could be used to interactively extract expert knowledge from CES results that could, in turn, improve CES modeling through cascading. Here, the 3D heat map application of Moore et al. [58] that harnesses the power of the Unity 3D video game engine for information visualization and visual analytics was used. The 3D heat map extends the traditional 2D heat map into a third dimension (z-axis) allowing the tops and sides of the resulting 3D bars to be colored, thus permitting additional dimensions of information to be visualized. The video game platform allows interactive exploration of the 3D heat map. Here, the authors used the x-axis for SNPs, the y-axis for CES generations, the z-axis for SNP frequency across best models, the tops of the bars for whether that SNP appeared in the overall best model at a given generation, and the sides of the bars for the accuracy of the best model that SNP appeared in. The visualization was used to select interesting SNPs that were then heavily biased in a consequent CES run (i.e., cascading). The advantage of this approach is that it inserts the human into the CES learning process. We provide an example application of this approach to real data below.

3.6 Interpretation Using Function Mapping

An important criticism of many machine learning methods is that the models represent a black box that is not immediately interpretable. Symbolic functions using GP or CES have the same criticism. The study by Moore et al. [26] introduced a simple method to provide statistical interpretation of symbolic models. The first objective was to visualize the mapping of each set of inputs given a function at a particular node in a GP/CES tree. The novel "function mapping" approach starts by showing the levels (e.g., genotype coding) of each attribute (e.g., SNP) in the leaves of the tree-based model. Also shown is the accuracy associated with that attribute alone as a classifier and its associated odds ratio and 95 % confidence interval. This facilitates a quick assessment of the magnitude of the univariate effects of each attribute. Next, each combination of inputs is shown with the corresponding function at each node in the expression tree. The mapping of the inputs to the range of output values produced by the mathematical function is illustrated. The accuracy and odds ratio for the output of each node is shown. As a whole, the function mapping tree summarizes the mapping of inputs and outputs along with their corresponding effects on the endpoint, so the tree can be decomposed and interpreted. The root node provides the final output values that are then used as discriminant scores to classify cases and controls, for example. We show the distribution of cases and controls for each discriminant score and the corresponding classification label. An application of function mapping for the interpretation of CES models of prostate cancer is provided by Moore et al. [33].

Moore et al. [33] applied CES with the 3D visualization strategy described above to the genetic analysis of prostate cancer. The study population consisted of nationally available genetic data from 2,286 men of European descent (488 nonaggressive, 687 aggressive cases, and 1,111 controls) collected through the Prostate, Lung, Colon, and Ovarian (PLCO) Cancer Screening Trial, a randomized, well-designed, multicenter investigation sponsored and coordinated by the National Cancer Institute (NCI) and their Cancer Genetic Markers of Susceptibility (CGEMS) program. The authors focused on prostate cancer aggressiveness as the endpoint. We used a biological filter to reduce the set of genes to just those involved in apoptosis (programmed cell death), DNA repair, and antioxidation/carcinogen metabolism. These biological processes are hypothesized to play important roles in prostate cancer. A total of 219 SNPs in these genes were studied here.

CES was run 1,000 times with 1,000 generations each on this data set. A penalty of 0.001 to the fitness for each node over 12 and each attribute or SNP over six was applied. This was implemented to put a selective pressure on smaller models. This part of the analysis was called Phase I. The authors then selected "interesting" SNPs from the visualization and provided those to new CES runs as expert knowledge. SNPs were considered interesting if they appeared with high frequency and with some consistency across the 1,000 generations. This new CES was also run 1,000 times for 1,000 generations each. A new penalty of 0.002 was applied to the fitness for each node over 10 and each attribute or SNP over 5. This was called Phase II. The expert knowledge generation from the visualization was repeated, and another 1,000 runs of CES each with 1,000 generations were performed. Finally, a new penalty of 0.003 was applied to the fitness for each node over eight and each attribute or SNP over four. An overall best model at the end of this Phase III run was selected.

Figure 2 illustrates the overall best model selected at the end of Phase III. This model consisted of seven total SNPs in seven genes, ten total nodes, and four constants. This model has a classification accuracy of 0.628. We have highlighted in Fig. 2 the classification accuracy of each node in the tree, thus making it possible to see where the information is coming from. First, it is interesting to note that the very top node is a <= function with a constant that does not provide any additional information. This particular node, or others like it, appear often in the best model across the phases and generations. This is because these nodes create a binary output that can be used as a convenient classifier. Thus, CES has learned to generate models with binary outputs.

An entropy-based analysis revealed evidence of statistical synergy among the branches of the tree indicative of epistasis. Of the seven genes identified in our overall best model, six (AKT3, BCL2,

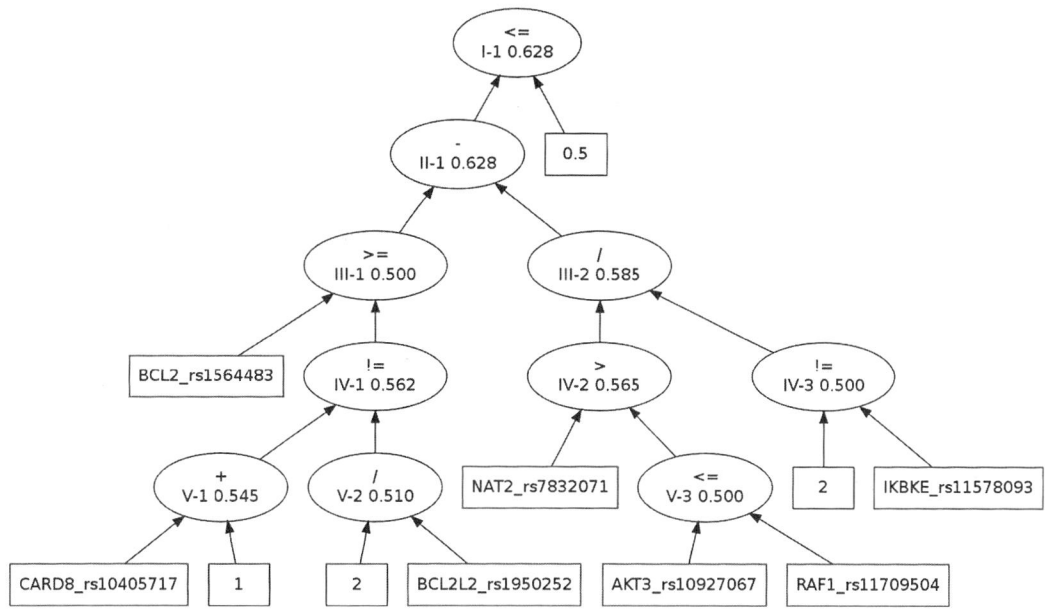

Fig. 2 The overall best model of prostate cancer discovered by CES. The model includes attributes (*rectangles*), constants (*squares*), and mathematical functions or nodes (*ovals*). Each model outputs a discriminant score for each subject in the data set. These scores are then used for classification in a discriminant analysis. The *numbers* shown within each *oval* are the classification accuracies at each level in the tree

BCL2L2, CARD8, IKBKE, NAT2, RAF1) are involved in programmed cell death or apoptosis and the seventh (NAT2) is involved in drug and carcinogen metabolism. All have been associated with prostate cancer or the prostate more generally to varying degrees. For example, there are more than 100 published papers reporting a direct or indirect role for BCL2 in prostate cancer. The IKBKE and NAT2 genes had 18 and 38 publications, respectively, while the others each had two to five publications. Most of these published studies report the results of molecular, cellular, or pharmacologic research. The fact that each of these genes has been previously implicated in prostate pathobiology adds some credibility to the findings. Further, a pathway analysis using the commercial Pathway Studio software reveals predicted or experimentally confirmed interactions among all the proteins suggesting biomolecular interactions between all seven genes in the model. Whether this computationally derived model has actual biological underpinnings remains to be determined. However, CES moves modeling from a one SNP at a time approach to an approach that embraces the complexity of the problem with few assumptions about the underlying model. This is an important step toward the use of computational intelligence for sifting through large volumes of human genetics data.

3.8 Application of CES to Alzheimer's Disease

Moore et al. [51] applied CES with the Pareto strategy described above to the genome-wide genetic analysis of Alzheimer's disease. This study came from the Alzheimer's Disease Neuroimaging Initiative (ADNI) that began on October 1, 2004. The study produced functional MRIs of patients in three categories: those who were neuro-typical, those with mild cognitive impairment, and those with Alzheimer's disease, every 6–12 months. A total of 521,028 single-nucleotide polymorphisms (SNPs) were measured across the human genome in a total of 740 subjects. Here, the authors used neuro-typical patients as the control subjects and those with mild cognitive impairment or Alzheimer's disease as the case subjects creating a binary class or outcome. The goal of the CES modeling exercise was to identify the optimal subset of SNPs and the optimal mathematical model that were predictive of the binary class.

The data were first pre-processed by estimating the interaction information for all pairs of SNPs as described by Moore et al. [15]. The authors considered pairs of SNPs that have higher interaction information more interesting. This pre-processed interestingness measure was used as expert knowledge in the CES solution modifiers and as an additional axis in a three-way Pareto optimization that included complexity and accuracy. Each CES run was conducted with a 36×26 grid of solutions for 2,000 generations. CES was implemented in a hierarchical framework inspired by the age-layered population structure algorithm or ALPS [59]. Here, the authors implemented a depth six binary tree where each node represented a CES run with the leaves of the tree representing the initial runs. Each higher node run was initialized with Pareto optimal solutions from the lower nodes. This was performed ten times with different sets of random seeds.

The overall best model discovered consisted of seven attributes or SNPs achieving an accuracy of 0.738 (Fig. 3). An entropy-based analysis revealed evidence of statistical synergy among the branches of the tree indicative of epistasis. The SNP with the large independent effect (rs429358) was located in the ApoE gene that is a known strong risk factor for Alzheimer's disease. CES was able to improve on this local minimum by adding six additional SNPs to the model that each belong to a relatively strong gene–gene interaction pair. Interestingly, each pair of strong gene–gene interactions was co-located in the model as children of the same function. SNP rs1931073 is in an intergenic region near the PPAP2B gene that is involved with cell adhesion and cell–cell interactions. SNP rs2414325 is in a gene called UNC13C, which is poorly defined functionally. SNP rs7782571 is near the ISPD gene that is mutated in rare diseases such as Walker–Warburg syndrome that is known to have brain anomalies. SNP rs4955208 is in the OSBPL10 gene that codes for an intracellular lipid receptor. SNP rs12209418 is in the PKIB gene, which is a protein kinase inhibitor associated with neuronitis or inflammation of the neurons. SNP rs12785149 is in the FAM107B gene whose function is not well known.

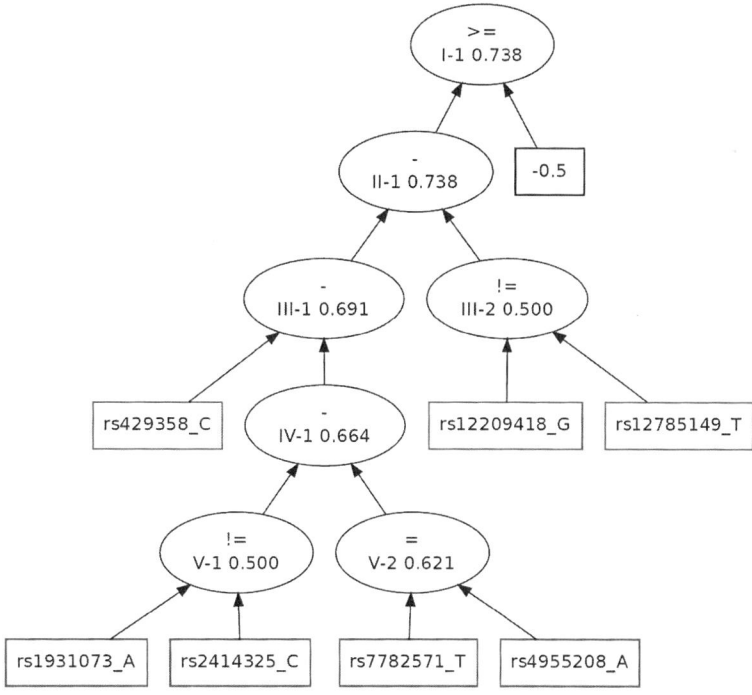

Fig. 3 The overall best model of Alzheimer's disease discovered by CES. The model includes attributes (*rectangles*), constants (*squares*), and mathematical functions or nodes (*ovals*). Each model outputs a discriminant score for each subject in the data set. These scores are then used for classification in a discriminant analysis. The *numbers* shown within each *oval* are the classification accuracies at each level in the tree

Of these seven genes, only ApoE is present in the AlzGene database [60] that provides an unbiased catalog of known genetic risk factors for Alzheimer's disease. The other six may represent new and novel discoveries. Based on their biology, it is plausible that all of them are related to the disease process. For example, OSBPL10 is involved with lipid metabolism, which is a known component of Alzheimer's disease pathobiology. At this point they remain novel hypotheses that will need to be tested with other data.

3.9 Application of CES to Glaucoma

Moore et al. [63] applied CES to the genome-wide genetic analysis of primary open-angle glaucoma (POAG). Data analyzed included more than 650,000 SNPs measured in 1,272 subjects with POAG and 1,057 controls from the Glaucoma Gene Environment Initiative (GLAUGEN). We first pre-processed the data to remove SNPs with more than 10 % missing genotypes and those with a minor allele frequency of less than 0.05. This left 486,726 SNPs for CES analysis. The data were also pre-processed using the ReliefF and interaction information measure of interaction as described above and elsewhere in this volume. As above, the interaction information was used as the third axis in Pareto optimization. CES was run as above with a 36×36 grid of solutions and for 2,000 generations. Here, we implemented a depth 11 binary tree where each node represents an EMERGENT run with the leaves of the tree representing the initial runs. Each higher node run is initialized with Pareto optimal solutions from the lower nodes. We reported the best models discovered at depths 9, 10, and 11.

The overall best model had an accuracy of 0.615 and consisted of six total SNPs. Interestingly, this model simplified to three modules each with two different SNPs. Each pair of SNPs was connected by the functions !=, > and = thus creating binary outputs. These were all added together and the result compared to the constant 1.5 using the > function. This can be interpreted as any two SNPs being true predicts POAG. Two of the SNPs pairs had strong synergistic interactions in the absence of moderate or strong marginal effects. The other pair consisted of SNPs with moderate marginal effects and a moderate interaction. One of these marginal effect SNPs is a previously published GWAS hit for POAG in the gene CDKN2B-AS1. The others are near genes that were not discovered using GWAS (AJAP1, PKHD1, HTR1B, MARCKS, CTIF). Bioinformatics validation of this result using gene–gene relationships inferred across thousands of gene expression experiments and other public knowledge bases (*see* **Note 7**) revealed that six of these genes have inferred biological interactions with the VEGF gene that is a known drug target for eye diseases including POAG. Thus, CES was able to identify sets of genes with epistatic interactions that were not previously described for POAG but that have important biological connections with a gene that is an active drug target.

3.10 Conclusions

We have provided a brief overview of the CES method and some of its extensions. CES is an important step toward using artificial intelligence to dive deeply into large-scale genetic data to identify novel and interesting discoveries that aren't predicted by univariate tests of association. We provided several real data examples that show how CES can reveal patterns of epistasis for complex human diseases. Advantages of this approach include the power to model complex interactions with few assumptions and the ability to exploit expert knowledge. The primary disadvantage is the computational complexity. We think this approach and others like it will complement the toolbox of parametric statistical approaches that are widely used for these kinds of analyses.

4   Notes

The following are some important tips for using CES:

1. Some forms of expert knowledge place a weight on each attribute or SNP that is then used by CES to probabilistically build or modify models. Protein–protein interactions provide

weights on pairs of SNPs or genes. This is handled by forming a lookup table. Pattin et al. [31] used this lookup table to pick new SNPs to insert into a model based on which SNPs were already present. For example, SNP A would have a much higher chance of being inserted into a model if SNP B was already present and if there were high confidence interactions between the protein products of their respective genes. Such lookup tables are very useful for other pairwise measures such as pre-processed measures of interaction information [34].

2. To our knowledge no one has used prior linkage peaks to weight a stochastic search algorithm for machine learning analysis. This would make an interesting approach in the context of genome-wide association studies.

3. ReliefF and its derivatives are powerful approaches for detecting nonadditive interactions among multiple attributes. Although the algorithm itself is relatively simple, it can be difficult to gain an intuition for why it works. Briefly, what makes ReliefF tick is the nearest neighbor algorithm that it uses to assign weights. The distance between any two instances (e.g., cases and/or controls) is computed from the entire vector of genotypes. This by nature takes into account multilocus effects regardless of whether they are additive or nonadditive. A comprehensive review of the baseline ReliefF algorithm is available [61]. See also the ReliefF chapter in this volume.

4. A simple frequency-based imputation approach may work well when there are very few missing genotypes. Of course, genetic data is ideally suited to multivariate imputation methods that use linkage disequilibrium to infer missing genotypes. The freely available PLINK software is commonly used to impute genotype data [62].

5. An advantage of organizing solutions on a grid is that niches can form promoting diversity across the population. This can prevent local minima from rapidly taking over the whole population leading to stalled searches. This is analogous to evolution in spatially organized biological populations.

6. The idea of acknowledging and rewarding models that are "interesting" by some objective or subjective measure is contrary to the agnostic or unbiased approach dictated by genome-wide association studies. The unbiased approach is favored by some because of concern that our existing knowledge base is faulty and might yield models that are false positives. There is the additional concern of type II errors due to the focus on specific areas at the exclusion of others. This is a philosophical debate for which there is likely no right answer. Rather, we recommend using both approaches. First, carry out an unbiased scan of the genome for models not influenced by prior knowledge.

Then carry out a biased search using measures of interestingness to see what new and unexpected results might be found. Both approaches are likely to yield meaningful results.

7. There are two publically available web tools for bioinformatics validation of gene lists derived from machine learning or AI methods. The first is imp.princeton.edu. The second, giant. princeton.edu, is based on the first but can be limited to specific tissue types. In the glaucoma example we limited the analysis to eye tissue which revealed the VEGF connection.

Acknowledgements

This work was supported by National Institutes of Health (NIH) grants AI59694, EY022300, GM103534, GM103506, LM009012, LM010098, and LM011360.

References

1. Moore JH (2003) The ubiquitous nature of epistasis in determining susceptibility to common human diseases. Hum Hered 56:73–82

2. Moore JH (2005) A global view of epistasis. Nat Genet 37:13–14

3. Moore JH, Williams SM (2005) Traversing the conceptual divide between biological and statistical epistasis: systems biology and a more modern synthesis. BioEssays 27:637–646

4. Moore JH, Williams SM (2009) Epistasis and its implications for personal genetics. Am J Hum Genet 85:309–320

5. Moore JH, Asselbergs FW, Williams SM (2010) Bioinformatics challenges for genome-wide association studies. Bioinformatics (Oxford, England) 26:445–455

6. Cordell HJ (2002) Epistasis: what it means, what it doesn't mean, and statistical methods to detect it in humans. Hum Mol Genet 11:2463–2468

7. Cowper-Sal lari R, Cole MD, Karagas MR et al (2011) Layers of epistasis: genome-wide regulatory networks and network approaches to genome-wide association studies. Wiley Interdiscip Rev Syst Biol Med 3:513 526

8. Tyler AL, Asselbergs FW, Williams SM et al (2009) Shadows of complexity: what biological networks reveal about epistasis and pleiotropy. BioEssays 31:220–227

9. Phillips PC (2008) Epistasis—the essential role of gene interactions in the structure and evolution of genetic systems. Nat Rev Genet 9: 855–867

10. Phillips PC (1998) The language of gene interaction. Genetics 149:1167–1171

11. Ritchie MD, Hahn LW, Roodi N et al (2001) Multifactor-dimensionality reduction reveals high-order interactions among estrogen-metabolism genes in sporadic breast cancer. Am J Hum Genet 69:138–147

12. Ritchie MD, Hahn LW, Moore JH (2003) Power of multifactor dimensionality reduction for detecting gene-gene interactions in the presence of genotyping error, missing data, phenocopy, and genetic heterogeneity. Genet Epidemiol 24:150–157

13. Hahn LW, Ritchie MD, Moore JH (2003) Multifactor dimensionality reduction software for detecting gene-gene and gene-environment interactions. Bioinformatics (Oxford, England) 19:376–382

14. Hahn LW, Moore JH (2004) Ideal discrimination of discrete clinical endpoints using multilocus genotypes. In Silico Biol 4:183–194

15. Moore JH, Gilbert JC, Tsai C-T et al (2006) A flexible computational framework for detecting, characterizing, and interpreting statistical patterns of epistasis in genetic studies of human disease susceptibility. J Theor Biol 241: 252–261

16. McKinney BA, Reif DM, Ritchie MD et al (2006) Machine learning for detecting gene-gene interactions: a review. Appl Bioinformatics 5:77–88

17. Eichler EE, Flint J, Gibson G et al (2010) Missing heritability and strategies for finding

the underlying causes of complex disease. Nat Rev Genet 11:446–450

18. Manolio TA, Collins FS, Cox NJ et al (2009) Finding the missing heritability of complex diseases. Nature 461:747–753

19. Koza JR (1992) Genetic programming: on the programming of computers by means of natural selection. MIT Press, Cambridge

20. Banzhaf W, Francone FD, Keller RE et al (1998) Genetic programming: an introduction: on the automatic evolution of computer programs and its applications. Morgan Kaufmann Publishers Inc., San Francisco

21. A Field Guide to Genetic Programming by Riccardo Poli (Paperback)—Lulu. http://www.lulu.com/us/en/shop/riccardo-poli-and-william-b-langdon-and-nicholas-freitag-mcphee/a-field-guide-to-genetic-programming/paperback/product-2502912.html

22. Fogel G, Corne D (2003) Evolutionary computation in bioinformatics. Morgan Kaufmann Publishers, Boston

23. Ritchie MD, White BC, Parker JS et al (2003) Optimization of neural network architecture using genetic programming improves detection and modeling of gene-gene interactions in studies of human diseases. BMC Bioinformatics 4:28

24. Ritchie MD, Motsinger AA, Bush WS et al (2007) Genetic programming neural networks: a powerful bioinformatics tool for human genetics. Appl Soft Comput 7:471–479

25. Jason Moore BW (2007) Genome-wide genetic analysis using genetic programming: the critical need for expert knowledge. Springer, Heidelberg, pp 11–28

26. Moore JH, Barney N, Tsai C-T et al (2007) Symbolic modeling of epistasis. Hum Hered 63:120–133

27. Turner SD, Dudek SM, Ritchie MD (2010) ATHENA: a knowledge-based hybrid backpropagation-grammatical evolution neural network algorithm for discovering epistasis among quantitative trait Loci. BioData Min 3:5

28. Banzhaf W, Beslon G, Christensen S et al (2006) Guidelines: from artificial evolution to computational evolution: a research agenda. Nat Rev Genet 7:729–735

29. Moore JH, Greene CS, Andrews PC et al (2009) Does complexity matter? Artificial evolution, computational evolution and the genetic analysis of epistasis in common human diseases, genetic programming theory and practice VI. Springer, USA, pp 1–19

30. Moore JH, Andrews PC, Barney N et al (2008) Development and evaluation of an open-ended computational evolution system for the genetic analysis of susceptibility to common human diseases. In: Marchiori E, Moore JH (eds) Evolutionary computation, machine learning and data mining in bioinformatics. Springer, Berlin, pp 129–140

31. Pattin KA, Payne JL, Hill DP et al (2011) Exploiting expert knowledge of protein-protein interactions in a computational evolution system for detecting epistasis. In: Riolo R, McConaghy T, Vladislavleva E (eds) Genetic programming theory and practice VIII. Springer, New York, pp 195–210

32. Greene CS, Hill DP, Moore JH (2010) Environmental sensing of expert knowledge in a computational evolution system for complex problem solving in human genetics. In: Riolo R, O'Reilly U-M, McConaghy T (eds) Genetic programming theory and practice VII. Springer, USA, pp 19–36

33. Moore JH, Hill DP, Fisher JM et al (2011) Human-computer interaction in a computational evolution system for the genetic analysis of cancer. In: Riolo R, Vladislavleva E, Moore JH (eds) Genetic programming theory and practice IX. Springer, New York, pp 153–171

34. Moore JH, Hill DP, Sulovari A et al (2013) Genetic analysis of prostate cancer using computational evolution, pareto-optimization and post-processing. In: Riolo R, Vladislavleva E, Ritchie MD et al (eds) Genetic programming theory and practice X. Springer, New York, pp 87–101

35. Payne JL, Greene CS, Hill DP et al (2010) Sensible initialization of a computational evolution system using expert knowledge for epistasis analysis in human genetics. In: Chen Y (ed) Exploitation of linkage learning in evolutionary algorithms. Springer, Berlin, pp 215–226

36. Greene CS, Hill DP, Moore JH (2011) An open-ended computational evolution strategy for evolving parsimonious solutions to human genetics problems. In: Kampis G, Karsai I, Szathmáry E (eds) Advances in artificial life. Darwin meets von Neumann. Springer, Berlin, pp 313–320

37. Goldberg DE (2002) Building blocks, the design of innovation. Springer, USA, pp 59–69

38. Goldberg DE (2002) Design of competent genetic algorithms, the design of innovation. Springer, USA, pp 187–216

39. Pattin KA, Moore JH (2010) Genome-wide association studies for the identification of biomarkers in metabolic diseases. Expert Opin Med Diagn 4:39–51

40. Pattin KA, Moore JH (2008) Exploiting the proteome to improve the genome-wide genetic analysis of epistasis in common human diseases. Hum Genet 124:19–29

41. Pattin KA, Moore JH (2009) Role for protein-protein interaction databases in human genetics. Expert Rev Proteomics 6:647–659

42. Moore JH, Parker JS, Olsen NJ et al (2002) Symbolic discriminant analysis of microarray data in autoimmune disease. Genet Epidemiol 23:57–69

43. Velez DR, White BC, Motsinger AA et al (2007) A balanced accuracy function for epistasis modeling in imbalanced datasets using multifactor dimensionality reduction. Genet Epidemiol 31:306–315

44. Folino G, Pizzuti C, Spezzano G (1999) A cellular genetic programming approach to classification. Proc. of the genetic and evolutionary computation conference GECCO99. Morgan Kaufmann. pp. 1015–1020

45. Vladislavleva E, Smits G, Kotanchek M (2008) Better solutions faster: soft evolution of robust regression models in pareto genetic programming. In: Riolo R, Soule T, Worzel B (eds) Genetic programming theory and practice V. Springer, USA, pp 13–32

46. Moore JH (2003) Cross validation consistency for the assessment of genetic programming results in microarray studies. In: Cagnoni S, Johnson CG, Cardalda JJR et al (eds) Applications of evolutionary computing. Springer, Berlin, pp 99–106

47. Horn J, Nafpliotis N, Goldberg DE (1994) A niched Pareto genetic algorithm for multiobjective optimization. Proceedings of the first IEEE conference on evolutionary computation, 1994. IEEE World Congress on Computational Intelligence, vol. 1. pp. 82–87

48. Smits GF, Kotanchek M (2005) Pareto-front exploitation in symbolic regression. In: O'Reilly U-M, Yu T, Riolo R et al (eds) Genetic programming theory and practice II. Springer, New York, pp 283–299

49. Coello CAC, Lamont GB, Veldhuisen DAV (2007) Evolutionary algorithms for solving multi-objective problems. Springer, New York

50. Greene CS, White BC, Moore JH (2009) sensible initialization using expert knowledge for genome-wide analysis of epistasis using genetic programming, genetic and evolutionary computation conference: [proceedings]/sponsored by ACM SIGEVO. Genet Evol Comput Conf 2009:1289–1296

51. Moore JH, Hill DP, Saykin AJ et al (2014) Exploiting interestingness in a computational evolution system for the genome-wide genetic analysis of Alzheimer's disease. In: Kotanchek M, Riolo R, Moore J (eds) Genetic programming theory and practice XI. Springer, USA, pp 31–45

52. Fan R, Zhong M, Wang S et al (2011) Entropy-based information gain approaches to detect and to characterize gene-gene and gene-environment interactions/correlations of complex diseases. Genet Epidemiol 35:706–721

53. Hu T, Sinnott-Armstrong NA, Kiralis JW et al (2011) Characterizing genetic interactions in human disease association studies using statistical epistasis networks. BMC Bioinformatics 12:364

54. Hu T, Chen Y, Kiralis JW et al (2013) An information-gain approach to detecting three-way epistatic interactions in genetic association studies. J Am Med Inform Assoc 20(4): 630–636

55. Geng L, Hamilton HJ (2006) Interestingness measures for data mining: a survey. ACM Comput Surv 38(3), 5

56. Thomas JJ, Cook KA, National Visualization and Analytics Center (2005) Illuminating the path. IEEE Computer Society, Los Alamitos

57. Heer J, Bostock M, Ogievetsky V (2010) A tour through the visualization zoo. Commun ACM 53:59–67

58. Moore JH, Lari RCS, Hill D, et al (2011) Human microbiome visualization using 3d technology. Pac Symp Biocomput: 154–164

59. G.S. Hornby (2006) ALPS: The age-layered population structure for reducing the problem of premature convergence. Proceedings of the 8th annual conference on genetic and evolutionary computation, ACM, New York, NY, USA. pp. 815–822

60. Bertram L, McQueen MB, Mullin K et al (2007) Systematic meta-analyses of Alzheimer disease genetic association studies: the AlzGene database. Nat Genet 39:17–23

61. Robnik-Šikonja M, Kononenko I (2003) Theoretical and empirical analysis of ReliefF and RReliefF. Mach Learn 53:23–69

62. Purcell S, Neale B, Todd-Brown K et al (2007) PLINK: a tool set for whole-genome association and population-based linkage analyses. Am J Hum Genet 81:559–575

63. Moore JH, Greene CS, Hill DP, Saykin AJ et al (2014) Identification of novel genetic models of glaucoma using the EMERGENT genetic programming-based artificial intelligence system. In: Riolo R, Kotanchek M, Worzel W (eds) Genetic programming theory and practice XII. Springer, New York

INDEX

Jason H. Moore and Scott M. Williams (eds.), *Epistasis: Methods and Protocols*, Methods in Molecular Biology,
vol. 1253, DOI 10.1007/978-1-4939-2155-3, © Springer Science+Business Media New York 2015

Printed by Printforce, the Netherlands